T0237691

INTRODUCTION TO
MATHEMATICAL
LOGIC

Extended Edition

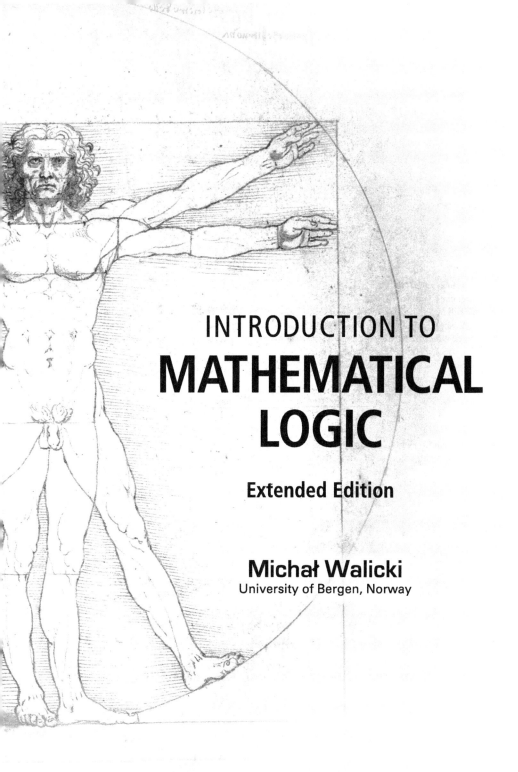

INTRODUCTION TO
MATHEMATICAL
LOGIC

Extended Edition

Michał Walicki
University of Bergen, Norway

World Scientific

NEW JERSEY · LONDON · SINGAPORE · BEIJING · SHANGHAI · HONG KONG · TAIPEI · CHENNAI · TOKYO

Published by

World Scientific Publishing Co. Pte. Ltd.

5 Toh Tuck Link, Singapore 596224

USA office: 27 Warren Street, Suite 401-402, Hackensack, NJ 07601

UK office: 57 Shelton Street, Covent Garden, London WC2H 9HE

Library of Congress Cataloging-in-Publication Data
Names: Walicki, Michał.
Title: Introduction to mathematical logic / by Michał Walicki (University of Bergen, Norway).
Description: Extended edition. | New Jersey : World Scientific, 2016. | Includes index.
Identifiers: LCCN 2016015933| ISBN 9789814719957 (hardcover : alk. paper) |
      ISBN 9789814719964 (pbk. : alk. paper)
Subjects: LCSH: Logic, Symbolic and mathematical.
Classification: LCC QA9 .W334 2016 | DDC 511.3--dc23
LC record available at https://lccn.loc.gov/2016015933

British Library Cataloguing-in-Publication Data
A catalogue record for this book is available from the British Library.

Printed in Singapore

dla Kasi

Acknowledgments

Several lecturers have been using earlier editions of this text and contributed valuable comments and corrections to its present form. I want to specifically thank Tore Langholm, Valentinas Kriaučiukas and Uwe Wolter. Eva Burrows and Sjur Dyrkolbotn deserve thanks for suggesting minor improvements, while Erik Parmann and Roger Antonsen for formulating some exercises.

The first part, on the history of logic, is to a high degree a compilation of various internet resources. I gratefully acknowledge the use of detailed information from Encyclopedia Britannica (`www.britannica.com`), Wikipedia (`en.wikipedia.org`), Leibnitiana (`www.gwleibniz.com`), Stanford Encyclopedia (`plato.stanford.edu`), and apologize the owners of other sources which I might have inadvertently left out. Development and conclusions of this part reflect, however, only the author's views and should not be attributed to other sources.

Michał Walicki

Contents

A HISTORY OF LOGIC

Once upon a time, sitting on a rock in Egypt, Parmenides invented logic. Such a legend might have appealed to people believing in a (rather small) set of well-defined rules constituting *the* logic. This belief had permeated the main-stream thinking at least until the beginning of the 20th century. But even if this medieval story appears now implausible, it reflects the fact that Parmenides was the first philosopher who did not merely propose a vision of reality but who also supported it by an extended argument. He is reported to have had a Pythagorean teacher and, perhaps, his use of argument was inspired by the importance of mathematics to the Pythagorean tradition. Still, he never systematically formulated principles of argumentation and using arguments is not the same as studying them.

"Logical thinking" may be associated roughly with something like correct arguing or reasoning, and the study of logic begins with the attempts to formulate the principles ensuring such correctness. Now, correctness amounts to conformance to some prescribed rules. Identification of such rules, and the ways of verifying conformance to them, begins with Aristotle in the 5th century BC. He defined his logical discourse – a syllogism – as one "in which, certain things being stated something other than what is stated follows of necessity from their being so." This intuition of necessary – unavoidable or mechanical – consequences, embodying the ideal of correctness, both lies at the origin of the discipline of logic and has been the main force driving its development until the 20th century. However, in a quite interesting turn, its concluding chapter (or rather: the chapter at which we will conclude its description) did not establish any consensus about the mechanisms of the human thinking and the necessities founding its correctness. Instead, it provided a precise counterpart of the Aristotelian definition of a process in which, certain things being given, some other follow as their unavoidable, mechanical consequences. This is known as Turing machine and its physical realization is computer.

We will sketch logic's development along the three, intimately connected axes which reflect its three main domains.

(1) The foremost concerns the correctness of arguments and this seems relative to their meaning. Meaning, however, is a vague concept. In order to formulate the rules for construction of correct arguments, one tries to capture it more precisely, and such attempts lead to another, more formal investigation of patterns of arguments.

(2) In order to construct precise and valid *patterns* of arguments one has to determine their "building blocks". One has to identify the basic terms, their kinds and means of combination.

(3) Finally, there is the question of how to *represent* these patterns. Although apparently of secondary importance, it is the answer to this question which puts purely symbolic manipulation in the focus. It can be considered the beginning of modern mathematical logic, which led to the development of the devices for symbolic manipulation – computers.

The first three sections sketch the development along the respective lines until Renaissance beginning, however, with the second point, Section A, following with the first, Section B, and concluding with the third, Section C. Then, Section D indicates the development in the modern era, with particular emphasis on the last two centuries. Section E sketches the basic aspects of modern mathematical logic and its relations to computers.

A. Patterns of reasoning

A.1. Reductio ad absurdum

If Parmenides was only implicitly aware of the general rules underlying his arguments, the same perhaps is not true for his disciple Zeno of Elea (5th century BC). Parmenides taught that there is no real change in the world and that all things remain, eventually, the same one being. In the defense of this heavily criticized thesis, Zeno designed a series of ingenious arguments, known as "Zeno's paradoxes", demonstrating that the contrary assumption leads to absurdity. One of the most known is the story of

Achilles and tortoise competing in a race
Tortoise, being a slower runner, starts some time t before Achilles.

In this time t, it will go some way w_1 towards the goal. Now Achilles starts running but in order to catch up with the tortoise he has to first run the way w_1 which will take him some time t_1 (less than t). In this time, tortoise will again walk some distance w_2 away from the point w_1 and closer to the goal. Then again, Achilles must first run the way w_2 in order to catch the tortoise which, in the same time t_2, will walk some distance w_3 away. Hence, Achilles will never catch the tortoise. But this obviously absurd, so the thesis that the two are really changing their positions cannot be true.

It was only in the 19th century that mathematicians captured and expressed precisely what was wrong with this way of thinking. This, however, does not concern us as much as the fact that *the same form of reasoning* was applied by Zeno in many other stories: assuming a thesis T, he analyzed it arriving at a conclusion C; but C turns out to be absurd – therefore T cannot be true. This pattern has been given the name "reductio ad absurdum" and is still frequently used in both informal and formal arguments.

A.2. ARISTOTLE

Various ways of arguing in political and philosophical debates were advanced by various thinkers. Sophists, often discredited by the "serious" philosophers, certainly deserve the credit for promoting the idea of a correct argument, irrespectively of its subject matter and goal. Horrified by the immorality of sophists' arguing, Plato attempted to combat them by plunging into ethical and metaphysical discussions and claiming that these indeed had a strong methodological logic – the logic of discourse, "dialectic". In terms of development of modern logic there is, however, close to nothing one can learn from that. The formulation of the principles for correct reasoning culminated in ancient Greece with Plato's pupil Aristotle's (384-322 BC) teaching of categorical forms and syllogisms.

A.2.1. CATEGORICAL FORMS

Most of Aristotle's logic was concerned with specific kinds of judgments, later called "categorical propositions", consisting of at most five building blocks: (1) a quantifier ("every", "some", or "no"), (2) a subject, (3) a copula ("is"), (4) perhaps a negation ("not"), and (5) a predicate. Subject, copula and predicate were mandatory, the remaining two elements were optional. Such propositions fall into one of the following forms:

quantifier	subject	copula	(4)	predicate	
Every	A	is		B	: Universal affirmative
Every	A	is	not	B	: Universal negative
Some	A	is		B	: Particular affirmative
Some	A	is	not	B	: Particular negative
	A	is		B	: Singular affirmative
	A	is	not	B	: Singular negative

In the singular judgments A stands for an individual, e.g. "Socrates is a man." These forms gained much less importance than the rest since in most contexts they can be seen as special cases of particular judgments.

A.2.2. Conversions

Sometimes Aristotle adopted alternative but equivalent formulations. The universal negative judgment could also be formulated as "No A is B", while the universal affirmative as "B belongs to every A" or "B is predicated of every A".

Aristotle formulated several such rules, later known as the theory of conversion. To convert a proposition in this sense is to interchange its subject and predicate. Aristotle observed that universal negative and particular affirmative propositions can be validly converted in this way: if "some A is B", then also "some B is A", and if "no A is B", then also "no B is A". In later terminology, such propositions were said to be converted simply (simpliciter). But universal affirmative propositions cannot be converted in this way: if "every A is B", it does not follow that "every B is A". It does follow, however, that "some B is A". Such propositions, which can be converted by interchanging their subjects and predicates and, *in addition*, also replacing the universal quantifier "all" by the existential quantifier "some", were later said to be converted accidentally (per accidens). Particular negative propositions cannot be converted at all: from the fact that some animal is not a dog, it does not follow that some dog is not an animal.

Below, the four figures of syllogism are presented. Aristotle used the laws of conversion to reduce other syllogisms to syllogisms in the first figure. Conversions represent thus the first form of essentially formal manipulation. They provide the rules for:

> *replacing occurrence of one (categorical) form of a statement by another – without affecting the proposition!*

What does "affecting the proposition" mean is another subtle matter. The whole point of such a manipulation is that one changes the concrete appearance of a sentence, without changing its value. The intuition might have been that they essentially mean the same and are interchangeable. In a more abstract, and later formulation, one would say that "not to affect a proposition" is "not to change its truth value" – either both are false or both are true.

Two statements are equivalent if they have the same truth value.

This wasn't exactly the point of Aristotle's but we may ascribe him a lot of intuition in this direction. From now on, this will be a constantly recurring theme in logic. Looking at propositions as thus determining a truth value gives rise to some questions (and severe problems, as we will see.) Since we allow using some "place-holders" – variables – a proposition need not have a unique truth value. "All A are B" depends on what we substitute for A and B. In general, a proposition P may be:

(1) a tautology – P is always true, no matter what we choose to substitute for the "place-holders"; (e.g., "All A are A". In particular, a proposition without any "place-holders", e.g., "all animals are animals", may be a tautology.)
(2) a contradiction – P is never true (e.g., "no A is A");
(3) contingent – P is sometimes true and sometimes false; ("all A are B" is true, for instance, if we substitute "animals" for both A and B, while it is false if we substitute "birds" for A and "pigeons" for B).

A.2.3. SYLLOGISMS

Aristotelian logic is best known for the theory of syllogisms which had remained practically unchanged and unchallenged for approximately 2000 years. In *Prior Analytics*, Aristotle defined a syllogism as a

discourse in which, certain things being stated something other than what is stated follows of necessity from their being so.

In spite of this very general definition, in practice he confined the term to arguments with only two premises and a single conclusion, each of which is a categorical proposition. The subject and predicate of the conclusion each occur in one of the premises, together with a third term (the middle) that is found in both premises but not in the conclusion. A syllogism thus argues that because S(ubject) and P(redicate) are related in certain ways

to some M(iddle) term in the premises, they are related in a certain way
to one another in the conclusion.

The predicate of the conclusion is called the major term, and the premise
in which it occurs is called the major premise. The subject of the conclusion
is called the minor term and the premise in which it occurs is called the
minor premise. This way of describing major and minor terms conforms to
Aristotle's actual practice but was proposed as a definition only by the 6th
century Greek commentator John Philoponus.

Aristotle distinguished three different "figures" of syllogisms, according
to how the middle is related to the other two terms in the premises. He only
mentioned the fourth possibility which was counted as a separate figure
by later logicians. If one wants to prove syllogistically that S(ubject) is
P(redicate), one finds a term M(iddle) such that the argument can fit into
one of the following figures:

    (I)   "M is P" and "S is M" – hence "S is P", or

  (II)   "P is M" and "S is M" – hence "S is P", or

 (III) "M is P" and "M is S" – hence "S is P", or

 (IV) "P is M" and "M is S" – hence "S is P".

Each of these figures can come in various "moods", i.e., each categorical form can come with various quantifiers, yielding a large taxonomy of
possible syllogisms. Since the Middle Ages, one has used the following
abbreviations for the concerned quantifiers:

    A : universal affirmative : all, every

    E : universal negative   : no

    I : particular affirmative : some

    O : particular negative   : some is not, not every.

The following is an example of a syllogism of figure I with the mood A-I-I.
"**Marshal**" is here the middle term and "politician" the major term.

A:	Every	**marshal**	is	a politician.	
I:	Some	soldiers	are	**marshals**.	(A.1)
I:	Some	soldiers	are	politicians.	

Figure A.2 gives examples of syllogisms of all four figures with different
moods. M is the middle term, P the major one and S the minor one. Four
quantifiers, distributed arbitrarily among the three statements of a syllogism, give 64 different syllogisms of each figure and the total of 256 distinct

fig. I:	[M is P]	[S is M]	[S is P]	
A-I-I	Every [M is P]	Some [S is M]	Some [S is P]	*Darii*
A-A-A	Every [M is P]	Every [S is M]	Every [S is P]	*Barbara*
fig. II:	[P is M]	[S is M]	[S is P]	
E-A-E	No [P is M]	Every [S is M]	No [S is P]	*Cesare*
fig. III:	[M is P]	[M is S]	[S is P]	
A-A-I	Every [M is P]	Every [M is S]	Some [S is P]	*Darapti*
A-A-A	Every [M is P]	Every [M is S]	Every [S is P]	–
fig. IV:	[P is M]	[M is S]	[S is P]	
E-A-O	No [P is M]	Every [M is S]	Some [S is not P]	*Fesapo*

Fig. A.2 Examples of syllogisms of each figure and various moods.

syllogisms. Aristotle identified 19 among them which are universally correct or, as we would say today, valid. Validity means here that

no matter what concrete terms are substituted for the variables (P, M, S), if the premises are true then also the conclusion is guaranteed to be true.

For instance, 5 examples in Figure A.2 with the special names in the last column, are valid. The names, given by the medieval scholars to the valid syllogisms, contained exactly three vowels identifying the mood. (The mnemonic aid did not extend further: Celarent and Cesare identify the same mood, so one had to simply remember that the former refers to figure I and the latter to figure II.)

Mood A-A-A in figure III does not yield a valid syllogism. To see this, we find a counterexample. Substituting women for M, female for P and human for S, the premises hold while the conclusion states that every human is female. A counterexample can be found to every invalid syllogism.

Note that a correct application of a valid syllogism does not guarantee truth of the conclusion. (A.1) is such an application, but the conclusion need not be true. It may namely happen that a correct application uses a false assumption, for instance, in a country where the marshal title is not used in the military. In such cases the conclusion may accidentally happen to be true but no guarantees about that can be given. We see again that the main idea is truth preservation in the reasoning process. An obvious, yet nonetheless crucially important, assumption is:

The contradiction principle

Both P and not-P are never true for any proposition P.

This principle seemed (and to many still seems) intuitively obvious and irrefutable – if it were violated, there would be little point in constructing any "truth preserving" arguments. Although most logicians accept it, its status has been questioned and various logics, which do not obey this principle, have been proposed.

A.3. OTHER PATTERNS AND LATER DEVELOPMENTS

Aristotle's syllogisms dominated logic until late Middle Ages. A lot of variations were invented, as well as ways of reducing some valid patterns to others (as in A.2.2). The claim that

all valid arguments can be obtained by conversion and, possibly, reductio ad absurdum from the three (four?) figures

has been challenged and discussed *ad nauseam*.

Early developments (already in Aristotle) attempted to extend the syllogisms to modalities by considering instead of the categorical forms, the propositions of the form "it is possible/necessary that some A are B". Early followers of Aristotle in the 4th/3th BC (Theophrastus of Eresus, Diodorus Cronus, the school of Megarians with Euclid) elaborated on the modal syllogisms and introduced the conditional form of a proposition

if (A is B) then (C is D).

These were further developed by Stoics who also made another significant step. One of great inventions of Aristotle were variables – the use of letters for arbitrary objects or terms. Now, instead of considering only patterns of terms where such variables are place-holders for objects, Stoics started to investigate logic with patterns of propositions. In such patterns, variables would stand for propositions instead of terms. For instance,

*from two propositions: "the first" and "the second", new propositions can be formed, e.g., "the first **or** the second", "**if** the first **then** the second", etc.*

The terms "the first", "the second" were used by Stoics as variables instead of single letters. The truth of such compound propositions may be deter-

mined from the truth of their constituents. We thus get new patterns of
arguments. The Stoics gave the following list of five patterns

(i)		If 1 then 2;	but 1;	therefore 2.	
(ii)		If 1 then 2;	but not 2;	therefore not 1.	
(iii)	Not both 1 and 2;		but 1;	therefore not 2.	(A.3)
(iv)		Either 1 or 2;	but 1;	therefore not 2.	
(v)		Either 1 or 2;	but not 2;	therefore 1.	

Chrysippus, 3th BC, derived many other schemata and Stoics claimed that
all valid arguments could be derived from these patterns. At the time, this
approach seemed quite different from the Aristotelian and a lot of time
went on discussions which is the right one. Stoics' propositional patterns
had fallen into oblivion for a long time, but they re-emerged as the basic
tools of modern propositional logic.

B. A LANGUAGE AND ITS MEANING

Medieval logic had been dominated by Aristotelian syllogisms, but its elab-
orations did not contribute significantly to the theory of formal reasoning.
However, Scholastics developed very sophisticated semantic theories, as we
will see below.

The pattern, or form, of a valid argument is the first and, through
centuries, the fundamental issue in the study of logic. Formation of patterns
depends on the identification of their building blocks, whose character, in
turn, is intimately related to their meaning. The two statements

(1) "all horses are animals", and
(2) "all birds can fly"

are not exactly of the same form. More precisely, this depends on what
a form is. The first says that one class (horses) is included in another
(animals), while the second that all members of a class (birds) have some
property (can fly). Is this grammatical difference essential or not? Or else,
can it be covered by one and the same pattern or not? Can we replace
a noun by an adjective in a valid pattern and still obtain a valid pattern
or not? In fact, the first categorical form subsumes both above sentences,
i.e., from the point of view of the logic of syllogisms, they are considered
as having *the same form*.

Such questions indicate that forms of statements and patterns of reasoning require further analysis of "what can be plugged where" which, in

turn, depends on which words or phrases can be considered as "having similar function", perhaps even as "having the same meaning". What are the objects referred to by various kinds of words? What are the objects referred to by the propositions?

B.1. EARLY SEMANTIC OBSERVATIONS AND PROBLEMS

Certain teachings of the sophists and rhetoricians are significant for the early history of (this aspect of) logic. For example, Prodicus (5th BC) appears to have maintained that no two words can mean exactly the same thing. Accordingly, he devoted much attention to carefully distinguishing and defining the meanings of apparent synonyms, including many ethical terms. On the other hand, Protagoras (5th BC) is reported to have been the first to distinguish different kinds of sentences – questions, answers, prayers, and injunctions. Further logical development addressed primarily propositions, "answers", of which categorical propositions of Aristotle's are the outstanding example. The categorical forms gave a highly sophisticated and very general schema for classifying various terms (possibly, with different grammatical status) as basic building blocks of arguments, i.e., as potential subjects or predicates.

Since logic studies statements, their form as well as patterns in which they enter valid arguments, one of the basic questions concerns the meaning of a proposition. As we indicated earlier, two propositions can be considered equivalent if they have the same truth value. This suggests another law, beside the contradiction principle, namely

The law of excluded middle (tertium non datur)

Each proposition P is either true or false.

There is surprisingly much to say *against* this apparently simple claim. There are modal statements (see B.4) which do not seem to have any definite truth value. Among many early counterexamples, there is the most famous one, which appeared in its usual version in the 4th century BC, and which is still disturbing and discussed by modern logicians:

The liar paradox

The sentence "This sentence is false" does not seem to have any content – it is false if and only if it is true!

Such paradoxes challenge the principle of contradiction or of excluded middle, and call for a closer analysis of the fundamental logical notions.

B.2. THE SCHOLASTIC THEORY OF SUPPOSITION

The character and meaning of various building blocks of a logical language were thoroughly investigated by the Scholastics. Their theory of supposition was meant to answer the question:

> *To what does a given occurrence of a term refer in a given proposition?*

One distinguished three main modes of supposition/reference:

(1) **personal**: In the sentence "Every horse is an animal", the term "horse" refers to individual horses.
(2) **simple**: In the sentence "Horse is a species", the term "horse" refers to a universal (the concept 'horse').
(3) **material**: In the sentence "Horse is a monosyllable", the term "horse" refers to the spoken or written word.

The distinction between (1) and (2) reflects the fundamental duality of individuals and universals which had been one of the most debated issues in medieval philosophy. Point (3), apparently of little significance, marks the increasing attention paid to the *language* and its mere syntax, which slowly becomes the object of study. Today, one easily blurs the distinction between the first two suppositions, for instance, subsuming them under the category of 'use' and opposing to 'mention' which corresponds exactly to (3). Lacking the quotation marks, medieval writers could write, for instance, the example sentence (3) as " Horse taken in the material supposition is a monosyllable." Cumbersome as this may appear to an untrained eye, it disambiguated precisely references to language.

B.3. INTENSION VS. EXTENSION

Besides the supposition theory and its relatives, the logicians of the 14th century developed a sophisticated theory of connotation. The term "black" does not merely denote all black things – it also connotes the quality, blackness, which all such things possess. Connotation is also called "intension" – saying "black" I *intend* blackness. Denotation is closer to "extension" – the collection of all the objects referred to by the term "black". This has become one of the central distinctions in the later development of logic and in the discussions about the entities referred to by words. Its variants recur in most later theories, sometimes as if they were innovations. For instance,

Frege opposes Sinn (sense, concept) to Bedeutung (reference), viewing both as constituting the meaning of a term. De Saussure distinguishes the signified (concept) from the referent (thing), and contrasts both with the signifier (sign). These later variants repeat the medieval understanding of a term which can be represented as follows:

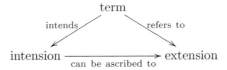

The crux of many problems is that different intensions may refer to (denote) the same extension. The "Morning Star" and the "Evening Star" have different intensions and for centuries were considered to refer to two different stars. As it turned out, these are actually two appearances of one and the same planet Venus. The two terms have the same extension and the insight into this identity is a true discovery, completely different from the empty tautology that "Venus is Venus".

Logic, trying to capture correctness of reasoning and conceptual constructions, might be expected to address the conceptual corner of the above triangle, the connotations or intensions. Indeed, this has been the predominant attitude and many attempts have been made to design a "universal language of thought" in which one could speak directly about the concepts and their interrelations. Unfortunately, the concept of concept is not obvious at all and such attempts never reached any universal consensus. One had to wait until a more tractable ways of speaking about concepts become available. The emergence of modern mathematical logic coincides with the successful coupling of logical language with the precise statement of its meaning in terms of extension. Modern logic still has branches of intensional logic, but its main tools are of extensional nature.

B.4. MODALITIES

In Chapter 9 of *De Interpretatione*, Aristotle discusses the assertion

> *There will be a sea battle tomorrow.*

The problem is that, at the moment when it is made, it does not seem to have any definite truth value – whether it is true or false will become clear tomorrow but until then it is as possible that it will be the one as the other.

This is another example (besides the liar paradox) indicating that adopting the principle of excluded middle, i.e., considering every proposition as having always only one of two possible truth values, may be insufficient.

Besides studying the syllogisms, medieval logicians, having developed the theory of supposition, incorporated into it modal factors. As necessity and possibility are the two basic modalities, their logical investigations reflected and augmented the underlying theological and ethical disputes about God's omnipotence and human freedom. The most important developments in modal logic occurred in the face of such questions as:

(1) whether statements about future contingent events are *now* true or false (the question originating from Aristotle),
(2) whether humans can know in advance *future* contingent events, and
(3) whether God can know such events.

One might distinguish the more ontological character of the first problem from the more epistemic flavour of the latter two, but in all three cases logical modality is linked with time. Thus, for instance, Peter Aureoli (12th/13th century) held that if something is B (for some predicate B) but could be not-B, i.e., is not *necessarily* B, then it might change, in the course of time, from being B to being not-B.

As in the case of categorical propositions, important issues here could hardly be settled before one had a clearer idea concerning the kinds of objects or states of affairs modalities are supposed to describe. In the late 13th century, the link between time and modality was severed by Duns Scotus who proposed a notion of possibility based purely on the notion of semantic consistency. "Possible" means for him logically possible, that is, not involving contradiction. This conception was radically new and had a tremendous influence all the way down to the 20th century. Shortly afterward, Ockham developed an influential theory of modality and time which reconciled the claim that every proposition is either true or false with the claim that certain propositions about the future are genuinely contingent.

Duns Scotus' ideas were revived in the 20th century, starting with the work of Jan Łukasiewicz who, pondering over Aristotle's assertion about tomorrow's sea battle, introduced 3-valued logic – a proposition may be true, or false, or else it may have a third, "undetermined" truth value. Also the "possible worlds" semantics of modalities, introduced by 19 years old Saul Kripke in 1959 (reflecting some ideas of Leibniz and reformulating some insights of Tarski and Jónsson), was based on Scotus' combination of modality with consistency. Today, modal and many-valued logics form a dynamic and prolific field, applied and developed equally by philosophers, mathematicians and computer scientists.

C. A SYMBOLIC LANGUAGE

Logic's preoccupation with concepts and reasoning begun gradually to put more and more severe demands on the appropriate and precise representation of the used terms. We saw that syllogisms used fixed forms of categorical statements with variables – A, B, etc. – representing arbitrary terms (or objects). Use of variables was indisputable contribution of Aristotle to the logical, and more generally mathematical notation. We also saw that Stoics introduced analogous variables standing for propositions. Such notational tricks facilitated more concise, more general and more precise statement of various logical facts.

Following the Scholastic discussions of connotation vs. denotation, logicians of the 16th century felt the increased need for a more general logical language. One of the goals was the development of an ideal logical language that would easily express ideal thought and be more precise than natural language. An important motivation underlying such attempts was the idea of manipulation, in fact, symbolic or even mechanical manipulation of arguments represented in such a language. Aristotelian logic had seen itself as a tool for training "natural" abilities at reasoning. Now one would like to develop methods of thinking that would accelerate or improve human thought or even allow its replacement by mechanical devices.

Among the initial attempts was the work of Spanish soldier, priest and mystic Ramon Lull (1235-1315) who tried to symbolize concepts and derive propositions from various combinations of possibilities. He designed sophisticated mechanisms, known as "Lullian circles", where simple facts, noted on the circumferences of various discs, could be combined by appropriately rotating the discs, providing answers to theological questions. The work of some of his followers, Juan Vives (1492-1540) and Johann Alsted (1588-1683) represents perhaps the first systematic effort at a logical symbolism.

Some philosophical ideas in this direction occurred in 17th century within Port-Royal – a group of anticlerical Jansenists located in Port-Royal outside Paris, whose most prominent member was Blaise Pascal (1623-1662). Elaborating on the Scholastical distinction between intension, or comprehension, and extension, Pascal introduced the distinction between real and nominal definitions. Real definitions aim at capturing the actual concept; they are descriptive and state the essential properties. "Man is a rational animal" attempts to give a real definition of the concept 'man', capturing man's essence. Nominal definitions merely stipulate the conventions by which a linguistic term is to be used. "By monoid we understand a set with a unary operation" is a nominal definition introducing the convention of using a particular word, "monoid", for a given concept. The distinction goes back to the discussions of the 14th century between the nominalism and realism with respect to the nature of universals. But Port-Royal's distinction, accompanied by the emphasis put on usefulness of nominal definitions (in particular, in mathematics), resonated widely signaling a new step on the line marked earlier by the material supposition of the Scholastic theory – the use of language becomes more and more conscious and explicit. The Port-Royal logic itself contained no symbolism, but the philosophical foundation for using symbols by nominal definitions was nevertheless laid.

C.1. The "universally characteristic language"

The goal of a universal language had already been cleary stated by Descartes (1596-1650) who expected it to provide a uniform and general method for any scientific inquiry and then, specifically for mathematics, to form a "universal mathematics". It had also been discussed extensively by the English philologist George Dalgarno (c. 1626-87) and, for mathematical language and communication, by the French algebraist François Viète (1540-1603). But it was Gottfried Leibniz (1646-1716), who gave this idea the most precise and systematic expression. His "lingua characteristica universalis" was an ideal that would, first, notationally represent concepts by displaying the more basic concepts of which they were composed, and then, represent ("iconically") the concept in a way that could be easily grasped by readers, no matter what their native tongue. Leibniz studied and was impressed by the Egyptian and Chinese picture-like symbols for concepts. Although we no longer use his notation, many concepts captured by it re-appear two centuries later in logical texts.

C.2. Calculus of reason

Universal language seems a necessary precondition for another goal which Leibniz proposed for logic. A "calculus of reason" (calculus ratiocinator), based on appropriate symbolism, would

> *involve explicit manipulations of the symbols according to estab-*
> *lished rules by which either new truths could be discovered or pro-*
> *posed conclusions could be checked to see if they could indeed be*
> *derived from the premises.*

Reasoning could then take place in the way large sums are done – mechanically or algorithmically – and thus not be subject to individual mistakes.
Such derivations could be checked by others or performed by machines, a
possibility that Leibniz seriously contemplated. His suggestion that machines could be constructed to draw valid inferences or to check deductions
was followed up in the 19th century by Charles Babbage, William Stanley
Jevons, Charles Sanders Peirce and his student Allan Marquand.

The symbolic calculus that Leibniz devised was motivated by his view
that most concepts were composite: they were collections or conjunctions
of other more basic concepts. Symbols (letters, lines, or circles) were then
used to stand for concepts and their relationships. This resulted in what
is intensional rather than an extensional logic – one whose terms stand for
properties or concepts rather than for the things having these properties.
Leibniz' basic notion of the truth of a judgment was that

> *the concepts making up the predicate are "included in" the concept*
> *of the subject.*

For instance, the judgment 'A zebra is striped and a mammal.' is true
because the concepts forming the predicate 'striped-and-mammal' are "included in" the concept (all possible predicates) of the subject 'zebra'.

What Leibniz symbolized as $A\infty B$, or what we would write today as
$A = B$, was that all the concepts making up concept A also are contained
in concept B, and vice versa.

Leibniz used two further notions to expand the basic logical calculus. In
his notation, $A \oplus B \infty C$ indicates that the concepts in A together with those
in B wholly constitute those in C. Today, we might write this as $A + B = C$
or $A \vee B = C$ – if we keep in mind that A, B, and C stood for concepts or
properties, not for individual things nor sets thereof. Leibniz also used the
juxtaposition of terms, $AB\infty C$ (which we might write as $A \wedge B = C$) to
signify that all the concepts in both A and B constitute the concept C.

A universal affirmative judgment, such as "Every A is B," becomes in
Leibniz' notation $A\infty AB$. This equation states that the concepts included
in the concepts of both A and B are the same as those in A.

The syllogism Barbara:

Every A is B; every B is C; so every A is C,

becomes the sequence of equations: $A\infty AB$; $B\infty BC$; *so* $A\infty AC$.

Notice that this conclusion can be derived from the premises by two simple algebraic substitutions and the associativity of logical multiplication.

$$
\begin{array}{r|l}
1.\ A \infty AB & \text{Every } A \text{ is } B \\
2.\ B \infty BC & \text{Every } B \text{ is } C \\
(1+2)\quad A \infty ABC & \\
\hline
(1)\quad A \infty AC & \text{therefore}:\quad \text{Every } A \text{ is } C
\end{array}
\qquad \text{(C.1)}
$$

As many early symbolic logics, including many developed in the 19th century, Leibniz' system had difficulties with negative and particular statements (Section A.2.1). The treatment of logic was limited and did not include any formalisation of relations nor of quantified statements. Only later Leibniz became keenly aware of the importance of relations and relational inferences. Although Leibniz might seem to deserve the credit for great originality in his symbolic logic – especially in his equational, algebraic logic – such insights were relatively common to mathematicians of the 17th and 18th centuries familiar with the traditional syllogistic logic. For instance, in 1685 Jakob Bernoulli published a work on the parallels of logic and algebra, giving some algebraic renderings of categorical statements. Later symbolic works of Lambert, Ploucquet, Euler, and even Boole – all apparently uninfluenced by Leibniz and Bernoulli – suggest the extent to which these ideas were apparent to the best mathematical minds of the day.

D. 1850-1950 – MATHEMATICAL LOGIC

Leibniz' system and calculus mark the appearance of a formalized, symbolic language which is prone to mathematical manipulation. A bit ironically, the emergence of mathematical logic marks also this logic's divorce, or at least separation, from philosophy. Of course, the discussions of logic have continued both among logicians and philosophers but from now on these groups form two increasingly distinct camps. Not all questions of philosophical logic are important for mathematicians and most of results of mathematical logic have rather technical character which is not always of interest for philosophers.

In this short presentation we have to ignore some developments which did take place between the 17th and 19th centuries. It was only in the

19th century that the substantial contributions were made which created modern logic. Perhaps the most important among those in the first half of the 19th century, was the work of George Boole (1815-1864), based on purely extensional interpretation. It was a real break-through in the old dispute intensional vs. extensional. It did not settle the issue once and for all – for instance Frege, "the father of first order logic" was still in favor of concepts and intensions; and in modern logic there is still a branch of intensional logic. However, Boole's approach was so convincingly precise and intuitive that it was later taken up and become the basis of modern – extensional or set theoretical – semantics.

D.1. GEORGE BOOLE

Although various symbolic or extensional systems appeared earlier, Boole formulated the first logic which was both *symbolic and extensional*. Most significantly, it survived the test of time and is today known to every student of mathematics as well as of computer science or of analytical philosophy as the propositional logic (earlier also as logic or algebra of classes). Boole published two major works, *The Mathematical Analysis of Logic* in 1847 and *An Investigation of the Laws of Thought* in 1854. It was the first of these two works that had the deeper impact. It arose from two streams of influence: the English logic-textbook tradition and the rapid growth of sophisticated algebraic arguments in the early 19th century. German Carl Friedrich Gauss, Norwegian Niels Henrik Abel, French Évariste Galois were major figures in this theoretical appreciation of algebra at that time, represented also in Britain by Duncan Gregory and George Peacock. Such conceptions gradually evolved into abstract algebras of quaternions and vectors, into linear algebra, Galois theory and Boolean algebra itself.

Boole used variables – capital letters – for the *extensions* of terms, to which he referred as classes of "things". This extensional perspective made the Boolean algebra a very intuitive and simple structure which, at the same time, captured many essential intuitions. The universal class – called "the Universe" – was represented by the numeral "1", and the empty class by "0". The juxtaposition of terms (for example, "AB") created a term referring to the intersection of two classes. The addition sign signified the non-overlapping union; that is, "$A + B$" referred to the entities in A or in B; in cases where the extensions of terms A and B overlapped, the expression was "undefined." For designating a proper subclass of a class A, Boole used the notation "vA". Finally, he used subtraction to indicate the removing of terms from classes. For example, "$1 - A$" indicates what one would obtain by removing the elements of A from the universal class – that is, the complement of A (relative to the universe, 1).

Boole offered a very systematic, and almost rigorously axiomatic, presentation. His basic equations included:

$$1A = A \qquad\qquad 0A = 0$$
$$0 + 1 = 1 \qquad\qquad A + 0 = A$$
$$AA = A \qquad\qquad \text{(idempotency)}$$
$$A(BC) = (AB)C \qquad \text{(associativity)}$$
$$AB = BA \qquad A + B = B + A \qquad \text{(commutativity)}$$
$$A(B + C) = AB + AC \qquad A + (BC) = (A + B)(A + C) \quad \text{(distributivity)}$$

A universal affirmative judgment, such as "All A's are B's," can be written using the proper subclass notation as $A = vB$. But Boole could write it also in two other ways: $A = AB$ (as did Leibniz) or $A(1 - B) = 0$. These two interpretations greatly facilitate derivation of syllogisms, as well as other propositional laws, by algebraic substitution. Assuming the distributivity $A(B - C) = AB - AC$, they are in fact equivalent:

$$AB = A \qquad\qquad \text{assumption}$$
$$0 = A - AB \qquad -AB$$
$$0 = A(1 - B) \qquad \text{distributivity}$$

The derivation in the opposite direction (from $0 = A(1 - B)$ to $A = AB$) follows by repeating the steps in the opposite order with adding, instead of subtracting, AB to both sides in the middle. In words, the fact that all A's are B's and that there are no A's which are not B's are equivalent ways of stating the same, which equivalence could be included among Aristotle's conversions, A.2.2. Derivations become now explicitly controlled by the applied axioms. For instance, derivation (C.1) becomes

$$A = AB \qquad\qquad \text{assumption}$$
$$B = BC \qquad\qquad \text{assumption}$$
$$A = A(BC) \qquad \text{substitution } BC \text{ for } B \qquad\qquad \text{(D.1)}$$
$$ = (AB)C \qquad \text{associativity}$$
$$ = AC \qquad\qquad \text{substitution } A \text{ for } AB$$

In contrast to earlier symbolisms, Boole's was extensively developed, exploring a large number of equations and techniques. It was convincingly applied to the interpretation of propositional logic – with terms standing for occasions or times rather than for concrete individual things. Seen in historical perspective, it was a remarkably smooth introduction of the new "algebraic" perspective which dominated most of the subsequent development. *The Mathematical Analysis of Logic* begins with a slogan that could serve as the motto of abstract algebra, as well as of much of formal logic:

> *the validity of the processes of analysis does not depend upon the interpretation of the symbols which are employed, but solely upon the laws of combination.*

D.1.1. FURTHER DEVELOPMENTS; DE MORGAN

Boole's approach was very appealing and quickly taken up by others. In the 1860s Peirce and Jevons proposed to replace Boole's "+" with a simple inclusive union: the expression "$A + B$" was to be interpreted as the class of things in A, in B, or in both. This results in accepting the equation "$1 + 1 = 1$", which is not true of the natural numbers. Although Boole accepted other laws which do not hold in the algebra of numbers (e.g., the idempotency of multiplication $A^2 = A$), one might conjecture that his interpretation of + as *disjoint* union tried to avoid also $1 + 1 = 1$.

At least equally important figure of British logic in the 19th century as Boole was Augustus de Morgan (1806-1871). Unlike most logicians in the United Kingdom, including Boole, de Morgan knew the medieval logic and semantics, as well as the Leibnizian symbolic tradition. His erudition and work left several lasting traces in the development of logic.

In the paper published in 1846 in the *Cambridge Philosophical Transactions*, De Morgan introduced the enormously influential notion of *a possibly arbitrary and stipulated "universe of discourse"*. It replaced Boole's original – and metaphysically a bit suspect – universe of "all things", and has become an integral part of the logical semantics. The notion of a stipulated "universe of discourse" means that, instead of talking about "*The* Universe", one can choose this universe depending on the context. "1" may sometimes stand for "the universe of all animals", and in other contexts for a two-element set, say "the true" and "the false". In the former case, the derivation (D.1) of $A = AC$ from $A = AB; B = BC$ represents an instance of the Barbara syllogism "All A's are B's; all B's are C's; therefore all A's are C's". In the latter case, the equations of Boolean algebra yield the laws of propositional logic where "$A + B$" corresponds to disjunction "A or B", and juxtaposition "AB" to conjunction "A and B". With this reading, the derivation (D.1) represents another reading of Barbara, namely: "If A implies B and B implies C, then A implies C".

Negation of A is simply its complement $1 - A$, and is obviously relative to the actual universe. (It is often written as $\overline{A}$.) De Morgan is known to all students of elementary logic primarily through the de Morgan laws:

$$\overline{AB} = \overline{A} + \overline{B} \quad \text{and dually} \quad \overline{A}\,\overline{B} = \overline{A + B}.$$

Using these laws and some additional, easy facts, like $\overline{B}B = 0$, $\overline{\overline{B}} = B$, we can derive the following reformulation of the reductio ad absurdum "If every A is B then every not-B is not-A":

$$
\begin{array}{r|l}
A = AB & \\
A - AB = 0 & -AB \\
A(1 - B) = 0 & \text{distributivity over } - \\
A\overline{B} = 0 & \overline{B} = 1 - B \\
\overline{A} + B = 1 & \text{deMorgan} \\
\overline{B}(A + B) = \overline{B} & \overline{B}\cdot \\
(\overline{B})(\overline{A}) + \overline{B}B = \overline{B} & \text{distributivity} \\
(\overline{B})(\overline{A}) + 0 = \overline{B} & \overline{B}B = 0 \\
(\overline{B})(\overline{A}) = \overline{B} & X + 0 = X
\end{array}
$$

I.e., if "Every A is B", $A = AB$, than "every not-B is not-A", $\overline{B} = (\overline{B})(\overline{A})$. Or: if "$A$ implies B" then "if B is false (absurd) then so is A".

A series of essays and papers on logic by de Morgan had been published from 1846 to 1862 under the title *On the Syllogism*. (The title indicates his devotion to the philosophical tradition of logic and reluctance to turn it into a mere branch of mathematics). The papers from 1850s are of considerable significance, containing the first extensive discussion of quantified relations since late medieval logic and Jung's massive *Logica hamburgensis* of 1638.

Boole's elegant theory had one serious defect, namely, its inability to deal with relational inferences. De Morgan's first significant contribution to this field was made independently and almost simultaneously with the publication of Boole's first major work. In 1847 de Morgan published his *Formal Logic; or, the Calculus of Inference, Necessary and Probable*. Although his symbolic system was clumsy and did not show the appreciation of abstract algebra that Boole's did, it gave a treatment of relational arguments which was later refined by himself and others. His paper from 1860, *On Syllogism IV and on the logic of relations*, started the sustained interest in the study of relations and their properties. De Morgan observed here that all valid syllogisms could be justified by the copula 'is' being a transitive and convertible (as he calls what today would be named "symmetric") relation, i.e., one for which $A \sim B$ and $B \sim C$ implies $A \sim C$

and, whenever $A \sim B$ then also $B \sim A$. Sometimes the mere transitivity suffices. The syllogism Barbara is valid for every transitive relation, e.g., if A is greater than B and B is greater than C then A is greater than C. In some other cases, also symmetry is needed as, for instance, to verify Cesare of figure II. It says that: if $P \not\sim M$ and $S \sim M$ then $S \not\sim P$. For assuming contrapositively $S \sim P$, then also $P \sim S$ by symmetry which, together with $S \sim M$, implies by transitivity that $P \sim M$.

De Morgan made the point, taken up later by Peirce and implicitly endorsed by Frege, that relational inferences are not just one type reasoning among others but are the core of mathematical and deductive inference and of all scientific reasoning. Consequently (though not correctly, but in the right spirit) one often attributes to de Morgan the observation that all of Aristotelian logic was helpless to show the validity of the inference:

$$\text{\textit{All horses are animals; therefore,}} \qquad \text{(D.2)}$$
$$\textit{every head of a horse is the head of an animal.}$$

This limitation concerns likewise propositional logic of Boole and his followers. From today's perspective, this can be seen more as the limitation of language, which does not provide means for expressing predication. Its appropriate (and significant) extension allows to incorporate analysis of relational arguments. Such an extension, which initially seemed to be a distinct, if not directly opposite approach, was proposed by the German Gottlob Frege, and is today known as first order predicate logic.

D.2. GOTTLOB FREGE

In 1879 the young Gottlob Frege (1848-1925) published perhaps the most influential book on symbolic logic in the 19th century, *Begriffsschrift* ("Conceptual Notation") – the title taken from Trendelenburg's translation of Leibniz' notion of a characteristic language. Frege gives here a rigorous presentation of the role and use of quantifiers and predicates. Frege was apparently familiar with Trendelenburg's discussion of Leibniz but was otherwise ignorant of the history of logic. His book shows no trace of the influence of Boole and little trace of the older German tradition of symbolic logic. Being a mathematician whose speciality, like Boole's, had been calculus, he was well aware of the importance of functions. These form the basis of his notation for predicates and he does not seem to have been aware of the work of de Morgan and Peirce on relations or of older medieval treatments. Contemporary mathematical reviews of his work criticized him for his failure to acknowledge these earlier developments, while reviews written

by philosophers chided him for various sins against reigning idealist conceptions. Also Frege's logical notation was idiosyncratic and problematically
two-dimensional, making his work hardly accessible and little read. Frege
ignored the critiques of his notation and continued to publish all his later
works using it, including his – also little-read – magnum opus, *Grundgesetze
der Arithmetik* (1893-1903; "The Basic Laws of Arithmetic").

   Although notationally cumbersome, Frege's system treated precisely
several basic notions, in the way to be adopted by later logicians. "All
A's are *B*'s" meant for Frege that the concept *A* *implies* the concept *B*, or
that to be *A* implies also to be *B*. Moreover, this applies to arbitrary *x*
which happens to be *A*. Thus the statement becomes: "$\forall x : A(x) \to B(x)$",
where the quantifier $\forall x$ means "for all *x*" and "$\to$" denotes implication. The
analysis of this, and one other statement, can be represented as follows:

Every	horse	is	an animal =
Every x	which is a horse	is	an animal
Every x	if it is a horse	then	it is an animal
$\forall x :$	$H(x)$	$\to$	$A(x)$

Some	animals	are	horses =
Some x's	which are animals	are	horses
Some x's	are animals	and	are horses
$\exists x :$	$A(x)$	$\wedge$	$H(x)$

This was not the way Frege would *write* it but this was the way he would
put it and *think* of it. The Barbara syllogism will be written today in first
order logic following exactly Frege's analysis, though not his notation, as:

$$\Big((\forall x : A(x) \to B(x)) \ \wedge \ (\forall x : B(x) \to C(x))\Big) \ \to \ (\forall x : A(x) \to C(x)).$$

It can be read as: "If every *x* which is *A* is also *B*, and every *x* which is
B is also *C*; then every *x* which is *A* is also *C*." Judgments concerning
individuals can be obtained from the universal ones by substitution. For
instance:

Hugo is				Hugo is	
a horse;	and	Every horse is an animal;	So:	an animal.	
$H(Hugo)$	$\wedge$	$(\forall v : H(v) \to A(v))$			(D.3)
		$H(Hugo) \to A(Hugo)$	$\to$	$A(Hugo)$	

The relational arguments, like (D.2) about horse-heads and animal-heads, can be derived after we have represented the involved statements as follows:

y is a head of some horse $=$			
there is	a horse	and	y is its head
there is an x	which is a horse	and	y is the head of x
$\exists x :$	$H(x)$	$\wedge$	$Hd(y,x)$

y is a head of some animal $=$			
$\exists x :$	$A(x)$	$\wedge$	$Hd(y,x)$

Now, the argument (D.2) will be given the form as in the first line and (very informal) treatment as in the following ones:

$$\forall v(H(v) \rightarrow A(v)) \quad \rightarrow \quad \forall y\Big(\exists x(H(x) \wedge Hd(y,x)) \rightarrow \exists z(A(z) \wedge Hd(y,z))\Big)$$

assume horses are animals   and take an arbitrary horse-head y, e.g., a :

$$\forall v(H(v) \rightarrow A(v)) \quad \rightarrow \quad \exists x\Big(H(x) \wedge Hd(a,x)\Big) \rightarrow \exists z\Big(A(z) \wedge Hd(a,z)\Big)$$

assume horses are animals   and that there is a horse h whose head is a :

$$\forall v(H(v) \rightarrow A(v)) \quad \rightarrow \quad H(h) \wedge Hd(a,h) \rightarrow \exists z\Big(A(z) \wedge Hd(a,z)\Big)$$

but as horses are animals then h is an animal by (D.3),

so $A(h) \wedge Hd(a,h)$

According to the last line, a is an animal-head and since a was an arbitrary horse-head, the claim follows.

In his first writings after the *Begriffsschrift*, Frege defended his own system and attacked bitterly Boolean methods, remaining apparently ignorant of the improvements by Peirce, Jevons, Schröder, and others. His main complaint against Booleans was the artificiality of their notation based on numerals and the failure to develop a genuinely logical notation.

In 1884 Frege published *Die Grundlagen der Arithmetik* ("The Foundations of Arithmetic") and then several important papers on a series of mathematical and logical topics. After 1879 he developed his position that

all of mathematics could be derived from basic logical laws – a
position later known as logicism in the philosophy of mathematics.   (D.4)

This view paralleled similar ideas about the reducibility of mathematics to set theory from roughly the same time. But Frege insisted on keeping them distinct and always stressed that his was an intensional logic of concepts, not of extensions and classes. His views are often marked by hostility to British extensional logic, like that of Boole, and to the general English-speaking tendencies toward nominalism and empiricism. In Britain, however, Frege's work was much admired by Bertrand Russell who promoted Frege's logicist research program – first in the *Introduction to Mathematical Logic* (1903), and then with Alfred North Whitehead, in *Principia Mathematica* (1910-13). Still, Russell did not use Frege's notation and his development of relations and functions was much closer to Schröder's and Peirce's than to Frege's. Frege's hostility to British tradition did not prevent him from acknowledging the fundamental importance of Russell's paradox, which Russell communicated to him in a letter in 1902. The paradox seemed to Frege a shattering blow to his goal of founding mathematics and science in an intensional logic and he expressed his worries in an appendix, hastily added to the second volume of *Die Grundgesetze der Arithmetik*, 1903, which was in press as Russell's letter arrived.

It did not take long before also other mathematicians and logicians started to admire Frege's care and rigour. His derivations were so scrupulous and precise that, although he did not formulate his theories axiomatically, he is sometimes regarded as a founder of the modern, axiomatic tradition in logic. His works had an enormous impact on the mathematical and philosophical logicians of the 20th century, especially, after their translation into English in the 1960s.

D.3. Set theory

As we have seen, the extensional view of concepts began gradually winning the stage with the advances of Boolean algebra. Set theory, founded by Georg Cantor (1845-1918), addresses collections – of numbers, points and, in general, of arbitrary elements, also of other collections – and is thus genuinely extensional. Besides this difference from the traditional logic, oriented more towards the intensional pole of the opposition, the initial development of set theory was completely separate from logic. But already in the first half of the 20th century, symbolic logic developed primarily in interaction with the extensional principles of set theory. Eventually, even Frege's analyses merged with the set theoretical approach to the semantics of logical formalism.

Booleans had used the notion of a set or a class, but did not develop tools for dealing with actually infinite classes. The conception of actual infinities, as opposed to merely potential, unlimited possibilities, was according to Aristotle a contradiction and most medieval philosophers shared this view. It was challenged in Renaissance, e.g., by Galileo, and then also by Leibniz. The problem had troubled 19th century mathematicians, like

Carl Friedrich Gauss and the Bohemian priest Bernhard Bolzano, who devoted his *Paradoxien des Unendlichen* (1851; "Paradoxes of the Infinite") to the difficulties posed by infinities. De Morgan and Peirce had given technically correct characterizations of infinite domains but these were not especially useful and went unnoticed in the German mathematical world. And the decisive development found place in this world.

Infinity – as the "infinitely small", infinitesimal (coming from the *infinitesimus* which, in the Modern Latin of the 17th century, referred to the "infinite-th" element in a series) – entered the mathematical landscape with the integral and derivative calculus, introduced independently by Leibniz and Newton in the 1660s. Infinitesimals have been often severely criticized (e.g., by bishop Berkeley, as the "ghosts of departed quantities") and only in the late 19th century obtained solid mathematical foundations in the work of the French baron Augustin-Louis Cauchy and German Karl Weierstraß. Building now on their discussions of the foundations of the infinitesimals, Germans Georg Cantor and Richard Dedekind developed methods for dealing with the infinite sets of the integers and points on the real number line. First Dedekind and then Cantor used Bolzano's technique of measuring sets by one-to-one mappings. Defining two sets to be "equinumerous" if they are in one-to-one correspondence, Dedekind gave in *Was sind und was sollen die Zahlen?* (1888; "What Are and Should Be the Numbers?") a precise definition of an infinite set:

> *A set is infinite if and only if the whole set can be put into one-to-one correspondence with its proper subset.*

This looks like a contradiction because, as long as we think of finite sets, it indeed is. But take the set of all natural numbers, $\mathbb{N} = \{0, 1, 2, 3, 4, ...\}$ and remove from it 0 getting $\mathbb{N}_1 = \{1, 2, 3, 4, ...\}$. The functions $f : \mathbb{N}_1 \to \mathbb{N}$, given by $f(x) = x - 1$, and $f_1 : \mathbb{N} \to \mathbb{N}_1$, given by $f_1(x) = x + 1$, are mutually inverse and establish a one-to-one correspondence between $\mathbb{N}$ and its proper subset $\mathbb{N}_1$.

A set A is said to be "countable" iff it is equinumerous with $\mathbb{N}$. One of the main results of Cantor was demonstration that there are uncountable infinite sets, in fact, sets "arbitrarily infinite". (For instance, the set $\mathbb{R}$ of real numbers was shown by Cantor to be "genuinely larger" than $\mathbb{N}$.)

Cantor developed the basic outlines of a set theory, especially in his treatment of infinite sets and the real number line. But he did not worry much about rigorous foundations for such a theory nor about the precise conditions governing the concept of a set and the formation of sets. In particular, he did not give any axioms for his theory. The initial attempts to formulate explicitly precise principles, not to mention rigorous axiomatizations, of set theory faced serious difficulties posed by the paradoxes of Russell and the Italian mathematician Cesare Burali-Forti (1897). Some passages in Cantor's writings suggest that he was aware of the potential problems, but he did not address them in a mathematical manner and, consequently, did not propose any technically satisfactory solution. They were first overcome in the rigorous, axiomatic set theory – initially, by Ernst Zermelo in 1908, and in its final version of Zermelo and Fraenkel in 1922.

D.4. 20TH CENTURY LOGIC

The first half of the 20th century was the most active period in the history of logic. The late 19th century work of Frege, Peano and Cantor, as well as Peirce's and Schröder's extensions of Boole's insights, had broken new ground and established new international communication channels. A new alliance – between logic and mathematics – emerged, gathering various lines of the late 19th century's development. Common to them was the effort to use symbolic techniques, sometimes called "mathematical" and sometimes "formal". Logic became increasingly mathematical in two senses. On the one hand, it attempted to use symbolic methods that had come to dominate mathematics, addressing the questions about

(1) the applications of the axiomatic method,
(2) a consistent theory of properties/relations (or sets),
(3) a logic of quantification.

On the other hand, it served the analysis of mathematics, as a tool in

(4) defining mathematical concepts,
(5) precisely characterizing mathematical systems, and
(6) describing the nature of mathematical proof.

This later role of logic – as a meta-mathematical and eventually foundational tool – followed Frege's logicism and dictated much of the development in the first decades of the 20th century.

D.4.1. LOGICISM

An outgrowth of the theory of Russell and Whitehead, and of most modern set theories, was a stronger articulation of Frege's logicism, claim (D.4) p.24, according to which mathematical operations and objects are really purely logical constructions. Consequently, the question what exactly pure logic is and whether, for example, set theory is really logic in a narrow sense has received increased attention. There seems little doubt that set theory is not *only* logic in the way in which, for example, Frege viewed it, i.e., as a formal theory of properties. Cantorian set theory engenders a large number of transfinite sets, i.e., nonphysical, nonperceived abstract objects. For this reason it has been regarded – by some as suspiciously, by others as endearingly – Platonistic. Still others, such as Quine, have only pragmatically endorsed set theory as a convenient way of organizing our world, especially if this world contains some elements of transfinite mathematics. The controversies about the status of infinite sets notwithstanding, it is thanks to them that, today, set theory as a foundation for most (even all) mathematical disciplines is hardly questioned. Mathematical theorems – whether in finitary mathematics, or else in topology or analysis – can, at least in principle, be formulated and proven in the language of set theory.

But the first decades of the 20th century displayed a strong finitist *Zeitgeist*, comparable to the traditional scepticism against actual infinities, and embodied now in various criticisms of transfinite set theory. Already Kronecker in 19th century, opposing Weierstraß and Cantor, declared that God made only integers, while everything else – in particular, of infinitary character – is the work of man. The same spirit, if not technical development, was represented by the constructivism (known as intuitionism) of Dutch Brouwer and Heyting, or by formalism searching for a finitary representation of mathematics in Hilbert's program, named so after the German mathematician David Hilbert (1862-1943). This program asked for an axiomatization of the whole of mathematics as a logical theory in order to prove formally that it is consistent. Even for those who did not endorse this logicist program, logic's goal was closely allied with techniques and goals in mathematics, such as giving an account of formal systems or of the nature of nonempirical proof. The program stimulated much activity in the first decades of the 20th century. It waned, however, after Austrian Kurt Gödel demonstrated in 1931 that logic could not provide a foundation for mathematics nor even a complete account of its formal systems. He proved namely a theorem which, interpreted informally, says something like:

Gödel's (second) incompleteness theorem

Any logical theory, satisfying reasonable and rather weak conditions, cannot be consistent and prove all its logical consequences.

Gödel's first incompleteness theorem shows that no effective formal system can prove all true arithmetical facts. Thus mathematics cannot be reduced to a provably complete and consistent logical theory. Interestingly, the proof shows that in any formal theory satisfying its conditions, one can write an analogue of the liar paradox, namely, the sentence "I am not provable in this theory", which cannot be provable unless the theory is inconsistent.

In spite of this negative result, logic has remained closely allied with mathematical foundations and principles. In particular, it has become a mathematical discipline. Traditionally, its task has been understanding of valid arguments, in particular, those formulated in natural language. It had developed tools needed for describing concepts, propositions, and arguments and – especially, as the "logical patterns" – for assessing argument's quality. During the first decades of the 20th century, logic become gradually more and more occupied with the historically somewhat foreign role of analyzing arguments in only one field, mathematics. The philosophical and linguistic task of developing tools for analyzing arguments in some natural language, or else for analyzing propositions as they are actually (and perhaps necessarily) conceived by humans, was almost completely lost. This task was, to some extent, taken over by analytical philosophy and by scattered efforts attempting to reduce basic principles of other disciplines – such as physics, biology, and even music – to axioms, usually, in set theory or first order logic. But even if they might have shown that it could be done, at least in principle, they were not very enlightening: one does not better or more usefully understand a bacteria, an atom or an animal by being told that it is a certain set or a (model of) certain axiomatic theory. Such efforts, at their zenith in the 1950s and '60s, had virtually disappeared in the '70s. Logic has become a formal discipline with its relations to natural reasoning seriously severed. Instead, it found multiple applications in the field which originated from the same motivations and had been germinating underneath the developments of logic – the field of purely formal manipulations and mechanical reasoning, arising from the same finitist *Zeitgeist* of the first half of the 20th century: computer science. Its emergence from and dependence on logic will become even clearer after we have described the basic elements of modern, formal logical systems.

E. MODERN SYMBOLIC LOGIC

Already Aristotle and Euclid were aware of the notion of a rigorous logical theory, in the sense of a – possibly axiomatic – specification of its theorems. Then, in the 19th century, the crises in geometry could be credited with renewing the attention for very careful presentations of these theories and other aspects of formal systems.

Euclid designed his *Elements* around 10 axioms and postulates which one could not resist accepting as obvious (e.g., "an interval can be prolonged indefinitely", "all right angles are equal"). Assuming their truth, he deduced some 465 theorems. The famous postulate of the parallels was

> **The fifth postulate**
> *If a straight line falling on two straight lines makes the interior angles on the same side less than the two right angles, the two straight lines, if produced indefinitely, meet on that side on which the angles are less than the two right angles.*

This postulate, even if reformulated, was somehow less intuitive and more complicated than others. Through hundreds of years mathematicians had unsuccessfully tried to derive it from the others until, in the 19th century, they started to reach the conclusion that it must be independent from the rest. This meant that one might as well drop it! That was done independently by the Russian Nicolai Lobachevsky in 1829 and the Hungarian János Bolayi in 1832. (Gauss, too, considered this move, but he never published his ideas on this subject.) What was left was a *new axiomatic system*. The big question about what this subset of axioms possibly described was answered by Lobachevsky and Bolayi who created its models, which satisfied all the axioms except the fifth – the first non-Euclidean geometries. This first exercise in what in the 20th century became "model theory", can be considered the beginning of modern axiomatic approach. For the discovery of non-Euclidean geometries unveiled the importance of admitting the possibility of manipulating the axioms which, perhaps, are not given by God and intuition but may be chosen with some freedom.

E.1. FORMAL LOGICAL SYSTEMS: SYNTAX

Although set theory and the type theory of Russell and Whitehead were considered to be logic for the purposes of the logicist program, a narrower sense of logic re-emerged in the mid-20th century as what is usually called the "underlying logic" of these systems. It does not make any existential assumptions (as to what kinds of mathematical objects do or do not exist) and concerns only rules for propositional connectives, quantifiers, and nonspecific terms for individuals and predicates. (An interesting issue is whether the privileged relation of identity, denoted "=", is a part of logic: most researchers have assumed that it is.) In the early 20th century and especially after Alfred Tarski's (1901-1983) work in the 1920s and '30s, a formal logical system was regarded as being composed of three parts, all of which could be rigorously described:

(1) the syntax (or notation);
(2) the rules of inference (or the patterns of reasoning);
(3) the semantics (or the meaning of the syntactic symbols).

One of the fundamental contributions of Tarski was his analysis of the concept of 'truth' which, in the above three-fold setting is given a precise treatment as a particular

relation between syntax (language) and semantics (the world).

The Euclidean, and then non-Euclidean geometry were built as axiomatic-deductive systems (point 2). The other two aspects of a formal system identified by Tarski were present too, but much less emphasized: notation was very informal, relying often on drawings; the semantics was only intuitive. Tarski's work initiated rigorous study of all three aspects.

E.1.1. THE LANGUAGE

First, there is the notation:

the rules of formation for terms and for well-formed formulas.

A formal language is simply a set of words (well formed formulae, wff), that is, strings over some given alphabet (set of symbols) and is typically specified by the rules of formation. For instance:

- the alphabet $\Sigma = \{\Box, \triangle, \rightarrow, -, (,)\}$
- the rules for forming words of the language L:
  - $\Box, \triangle$ are in L
  - if A, B are in L then so are $-A$ and $(A \rightarrow B)$.

This specification allows us to conclude that, for instance, $\triangle$, $-\Box$, $(\triangle \rightarrow -\Box)$, $-(\Box \rightarrow -\triangle)$ all belong to L, while $\Box\triangle$, $()$ or $\Box \rightarrow$ do not.

Previously, notation was often a haphazard affair in which it was unclear what could be formulated or asserted in a logical theory and whether expressions were finite or were schemata standing for infinitely long wffs. Now, the theory of notation itself became subject to exacting treatment, starting with the theory of strings of Tarski, and the work of the American Alonzo Church. Issues that arose out of notational questions include definability of one wff by another (addressed in Beth's and Craig's theorems, and in other results), creativity, and replaceability, as well as the expressive power and complexity of different logical languages (gathered, e.g., in Chomsky hierarchy).

E.1.2. REASONING – AXIOMATIC SYSTEM

The second part of a logical system consists of

> the axioms and rules of inference, or other ways of identifying what counts as a theorem.

This is what is usually meant by the logical "theory" proper: a (typically recursive) description of the theorems of the theory, including axioms and every wff derivable from axioms by admitted rules. Using the language L, one migh, for instance, define the following theory T:

Axioms	Rules
i) $\square$	R1) $\dfrac{(A \to B) \; ; \; (B \to C)}{(A \to C)}$
ii) $(\triangle \to -\square)$	R2) $\dfrac{(A \to B) \; ; \; A}{B}$
iii) $(A \to --A)$	
iv) $(--A \to A)$	R3) $\dfrac{(A \to B) \; ; \; -B}{-A}$

Upper case letters are schematic variables for which we can substitute arbitrary formulae of our language L. We can now perform symbolic derivations, starting with axioms and applying the rules, so that correctness can be checked mechanically. For instance:

$$
\text{R3}\;\cfrac{\text{R2}\;\cfrac{\cfrac{\text{iii}}{(\square \to --\square)} \quad \cfrac{\text{i}}{\square}}{--\square} \quad \cfrac{\text{ii}}{(\triangle \to -\square)}}{-\triangle} \quad \cfrac{\cfrac{\text{iii}}{(-\triangle \to ---\triangle)}}{---\triangle}\;\text{R2} \tag{E.1}
$$

Thus, $---\triangle$ is a theorem of our theory, and so is $-\triangle$ which is obtained by the (left) subderivation ending with the application of rule R3.

A formal description of a language, together with a specification of a theory's theorems (derivable propositions), are often called the "syntax" of the theory. This may be somewhat misleading when compared to the practice in linguistics, which would limit syntax to the narrower issue of grammaticality. The term "calculus" is sometimes chosen to emphasize the purely syntactic, uninterpreted nature of reasoning system.

E.1.3. SEMANTICS

The last component of a logical system is the semantics for such a theory and language, a specification of

what the terms of a theory refer to, and how the basic operations and connectives are to be interpreted in a domain of discourse, including truth conditions for the formulae in this domain.

Consider, as an example the rule R1 from the theory T above. It is merely a "piece of text" and its symbols allow almost unlimited interpretations. We may, for instance, take $A, B, C, \ldots$ to denote propositions and $\rightarrow$ an implication. (Note how rules R2 and R3 capture then Stoics' patterns (i) and (ii) from (A.3), p. 9.) But we may likewise let $A, B, C, \ldots$ stand for sets and $\rightarrow$ for set-inclusion. The following give then examples of applications of this rule under these two interpretations:

If	it's nice	then	we'll leave	$\{1,2\} \subseteq \{1,2,3\}$
If	we leave	then	we'll see a movie	$\{1,2,3\} \subseteq \{1,2,3,5\}$
If	it's nice	then	we'll see a movie	$\{1,2\} \subseteq \{1,2,3,5\}$

The rule is "sound" with respect to these interpretations – when applied to these domains in the prescribed way, it represents a valid argument. In fact, R1 expresses transitivity of $\rightarrow$ and will be sound for every transitive relation interpreting $\rightarrow$. This is just a more formal way of expressing de Morgan's observation that the syllogism Barbara is valid for all transitive relations.

A specification of a domain of objects (de Morgan's "universe of discourse"), and of the rules for interpreting the symbols of a logical language in this domain such that all the theorems of the logical theory are true is said to be a "model" of the theory. The two suggested interpretations are models of rule R1. (To make them models of the whole theory T would require more work, in particular, finding appropriate interpretation of $\Box$, $\triangle$ and $-$, such that the axioms become true and all rules sound. For the propositional case, one could for instance let $-$ denote negation, $\Box$ 'true' and $\triangle$ 'false'.)

If we chose to interpret the formulae of L as events and $A \to B$ as, say, "A is independent from B", the rule would not be sound. Such an interpretation would not give a model of the theory or, what amounts to the same, if the theory were applied to this part of the world, we could not trust its results. The next subsection describes some further concepts arising with the formal semantics.

E.2. FORMAL SEMANTICS

Formal semantics, or model theory, relates the mere syntax to the whole of mathematics by connecting the syntactic expressions with potentially unlimited number of mathematical entities. It is more complex than the logical syntax alone and has a more complicated history, which often seems insufficiently understood. Certainly, Frege's notion that propositions refer to (bedeuten) "the true" or "the false" – and this for complex propositions as a function of the truth values of simple propositions – counts as semantics. This intuition underlies the ancient law of excluded middle and is likewise reflected in the use of letters for referring to the values 1 and 0, that started with Boole. Although modal propositions and paradoxes pose severe problems for this view, it dominates most of the logic, perhaps, because it provides a relatively simple and satisfactory model for a very significant portion of mathematical and natural discourse. Medieval theories of supposition formulated many useful semantic observations. In the 19th century, both Peirce and Schröder occasionally gave brief demonstrations of the independence of certain postulates using models in which some postulates were true, but not others. This was also the technique used by the inventors of non-Euclidean geometry.

The first significant and general result of a clearly model theoretic character is usually accepted to be a result discovered by Löwenheim in 1915 and strengthened by Skolem in the 1920s.

Löwenheim-Skolem theorem

A theory that has a model at all, has a countable model.

That is to say, if there exists some model of a theory (i.e., an application of it to some domain of objects), then there is sure to be one with a domain no larger than the natural numbers. This theorem is in some ways a shocking result, since it implies that any consistent formal theory of anything – no matter how hard it tries to address the phenomena unique to a field such as biology, physics, or even sets or just real numbers – can just as well be understood as being about natural numbers: it says nothing more about the actually intended field than it says about natural numbers.

E.2.1. Consistency

The second major result in formal semantics, Gödel's completeness theorem of 1930 (see E.2.2 below), required even for its description, let alone its proof, more careful development of precise metalogical concepts about logical systems than existed earlier. One question for all logicians since Boole, and certainly since Frege, had been:

> *Is the theory consistent? In its purely syntactic analysis, this amounts to the question: Is a contradictory sentence (of the form "A and not-A") derivable?*

In most cases, the equivalent semantic counterpart of this is the question:

> *Does the theory have a model at all?*

For a logical theory, consistency means that a contradictory theorem cannot be derived in the theory. But since logic was intended to be a theory of necessarily true statements, the goal was stronger: a theory is Post-consistent (named after Emil Post) if every theorem is valid – that is, if no theorem is a contradictory or a contingent statement. (In nonclassical logical systems, one may define many other interestingly distinct notions of consistency; these notions were not distinguished until the 1930s.) Consistency was quickly acknowledged as a desired feature of formal systems. Earlier assumptions about consistency of various theories of propositional and first order logic turned out to be correct. A proof of the consistency of propositional logic was first given by Post in 1921. Although the problem itself is rather simple, the original difficulties concerned the lack of precise syntactic and semantic means to characterize consistency. The first clear proof of the consistency of the first order predicate logic is found in the book of David Hilbert and Wilhelm Ackermann, *Gründzuge der theoretische Logik* ("Principles of theoretical logic") from 1928. Here, in addition to a precise formulation of consistency, the main problem was also a rigorous statement of first order predicate logic as a formal theory.

Consistency of more complex systems proved elusive. Hilbert had observed that there was no proof that even Peano arithmetics was consistent, while Zermelo was concerned with demonstrating that set theory was consistent. These questions received an answer that was not what was hoped for. Although Gerhard Gentzen (1909-1945) showed that Peano arithmetics is consistent, he needed for this purpose stronger assumptions than those of Peano arithmetics. Thus "true" consistency of arithmetics still depends on the consistency of the extended system used in the proof. But this system cannot prove its own consistency and this is true about any system, satisfying some reasonably weak assumptions. This is the content of Gödel's second incompleteness theorem, which put a definite end to Hilbert's program of proving formally the consistency of mathematics.

E.2.2. COMPLETENESS

In their book from 1928 Hilbert and Ackermann also posed the question of whether a logical system and, in particular, first order predicate logic, was "complete", i.e.,

> *whether every valid proposition – that is, every proposition that is true in all intended models – is provable in the theory.*

In other words, does the formal theory describe all the noncontingent truths of its subject matter? Some idea of completeness had clearly accompanied Aristotle's attempts to collect *all* human knowledge and, in particular, *all* valid arguments or, in geometry, Euclid's attempts to derive *all* true theorems from a minimal set of axioms. Completeness of a kind had also been a guiding principle of logicians since Boole – otherwise they would not have sought numerous axioms, risking their mutual dependence and even inconsistency. But all these earlier writers have lacked the semantic terminology to specify what their theory was about and wherein "aboutness" consists. In particular, they lacked the precise grasp of the *"all* truths" which they tried to capture. Even the language of Hilbert and Ackermann from 1928 is not perfectly clear by modern standards.

Post had shown the completeness of propositional logic in 1921 and Gödel proved the completeness of first order predicate logic in his doctoral dissertation of 1930. In many ways, however, explicit consideration of issues in semantics, along with the development of many of the concepts now widely used in formal semantics and model theory, first appeared in a paper by Alfred Tarski, *The Concept of Truth in Formalized Languages*, which was published in Polish in 1933 and became widely known through its German translation of 1936. Introducing the idea of a sentence being "true in" a model, the paper marked the beginning of modern model theory. Even if the outlines of how to model propositional logic had been clear to the Booleans and to Frege, one of Tarski's crucial contributions was an application of his general theory to the semantics of the first order logic (now termed the set-theoretic, or Tarskian, interpretation). Relativity of truth to a model suggests choosing the models with some freedom (recall de Morgan's stipulated universe of discourse). Specifying precisely the class of intended models for a theory allows then to ask about proposition's "validity", i.e., its truth in all intended models. Completeness amounts to the syntactic derivability of every valid proposition, and this definition applies unchanged to propositional and first order logic, as well as to any other logical system.

Although the specific theory of truth Tarski advocated has had a complex and debated legacy, his techniques and precise language for discussing semantic concepts – such as consistency, completeness, independence – having rapidly entered the literature in the late 1930s, remained in the center of the subsequent development of logic and analytic philosophy. This influence accelerated with the publication of his works in German and then in English, and with his move to the United States in 1939.

E.3. COMPUTABILITY AND DECIDABILITY

The underlying theme of the whole development we have sketched is the attempt to *formalize* reasoning, hopefully, to the level at which it can be performed mechanically. The idea of mechanical reasoning has always been present, if not always explicitly, in the logical investigations and could be almost taken as their primary, if only ideal, goal. Intuitively, "mechanical" involves some blind following of the rules and such a blind rule following is the essence of a symbolic system as described in E.1.2. This "mechanical blindness" follows from the fact the language and the rules are unambiguously defined. Consequently, correctness of the application of a rule to an actual formula can be verified mechanically. One can easily see that all rule applications in the derivation (E.1) are correct and equally easily that, for instance, $\frac{(\square \to \triangle) \; ; \; \triangle}{\square}$ is not a correct application of any rule from T.

Logic was supposed to capture correct reasoning and correctness amounts to conformance to some accepted rules. A symbolic reasoning system is an ultimately precise expression of this view of correctness which also makes its verification a purely mechanical procedure. Such a mechanism is possible because all legal moves and restrictions are expressed in the syntax: the language, axioms and rules. In other words, it is exactly the uninterpreted nature of symbolic systems which leads to mechanization of reasoning. Naturally enough, once the symbolic systems were defined and one became familiar with them, i.e., in the beginning of the 20th century, the questions about mechanical computability were raised by the logicians. The answers led to the design and use of computers – devices for symbolic, that is, uninterpreted manipulation.

E.3.1. WHAT DOES IT MEAN TO "COMPUTE MECHANICALLY"?

In the 1930s this question acquired the ultimately precise, mathematical meaning. Developing the concepts from Hilbert's school, in his Princeton lectures 1933-34 Gödel introduced the schemata for so-called "recursive functions" working on natural numbers. Some time later Alonzo Church proposed the famous thesis

> **Church's thesis**
>
> *A function is (mechanically) computable if and only if it can be defined using only recursive functions.*

This may sound astonishing – why just recursive function are to have such a special significance? The answer comes from the work of Alan Turing who introduced "devices" which came to be known as Turing machines. Although defined as conceptual entities, one could easily imagine that such devices could be actually built as physical machines performing exactly the operations suggested by Turing. The machines could, for instance, recognize whether a string had some specific form and, generally, compute functions. The functions which could be computed on Turing machines were shown to be exactly the recursive functions! Even more significant for us may be the fact that there is a well-defined sublogic of first order logic in which proving a theorem amounts to computing a recursive function, that is, which can code all possible computer programs. This subset comprises Horn formulae, namely, the conditional formulae of the form

$$\text{If } A_1 \text{ and } A_2 \text{ and ... and } A_n \text{ then } C. \qquad (\text{E.2})$$

Such rules might be claimed to have more "psychological plausibility" than recursive functions. But they are computationally equivalent. With a few variations and additions, the formulae (E.2) give the syntax of an elegant programming language PROLOG. Thus, in the wide field of logic, there is a small subdomain providing sufficient means to study the issues of computability. (Such connections are much deeper and more intricate but we cannot address them all here.)

Church's thesis remains only a *thesis*, claiming that the informal and intuitive notion of mechanical computability is formalized exactly by the notion of recursive functions (or their equivalents, like Horn formulae or Turing machines). The fact that they are exactly the functions computable on the physical computer lends this thesis a lot of plausibility. Moreover, so far nobody has managed to introduce a notion of computability which would be intuitively acceptable, physically realizable and, at the same time, would exceed the capacities of Turing machines. A modern computer program, with all its tricks and sophistication is, as far as its power and possibilities are concerned, *nothing more* than a Turing machine, a set of Horn formulae. Thus, logical results, in particular the negative theorems stating the limitations of logical formalisms, determine also the ultimate limits of computers' capabilities as exemplified below.

E.3.2. Decidability

By the 1930s almost all work in the foundations of mathematics and in symbolic logic was being done in a standard first order predicate logic, often extended with axioms or axiom schemata of set-theory. This underlying logic consisted of a theory of classical truth functional connectives, such as "and", "not" and "if...then..." (propositional logic, as with Stoics or Boole) and first order quantification permitting propositions that "all" and "at least one" individual satisfy a certain formula (Frege). Only gradually in the 1920s and '30s did a conception of a "first order" logic, and of more expressive alternatives, arise.

Formal theories can be classified according to their expressive or representational power, depending on their language (notation) and reasoning system (inference rules). Propositional logic allows merely manipulation of simple, propositional patterns, combined with operators like "or", "and", (A.3), p.9. First order logic allows explicit reference to, and quantification over, individuals, such as numbers or sets, but not quantification over properties of these individuals. For instance, the statement "for all x: if x is man then x is human" is first order. But the following one is second order, involving quantification over properties P, R: "for every x and any properties P, R: if P implies R and x is P then x is R."[1] (Likewise, the fifth

---

[1] Note a vague analogy of the distinction between first order quantification over individu-

postulate of Euclid is not finitely axiomatisable in the first order language but is rather a schema or second order formulation.)

The question "why should one bother with less expressive formalisms, when more expressive ones are available?" should appear quite natural. The answer lies in the fact that increasing expressive power of a formalism clashes with another desired feature, namely:

decidability
there exists a finite mechanical procedure for determining whether a proposition is, or is not, a theorem of the theory.

The germ of this idea is present in the law of excluded middle claiming that every proposition is either true or false. But decidability adds to it the requirement which can be expressed only with the precise definition of a finite mechanical procedure, of computability. This is the requirement that not only the proposition must be true/provable or not: there must be a terminating algorithm which can be run (on a computer) to decide which is the case. (In E.1.2 we have shown that, for instance, $-\triangle$ is a theorem of the theory T defined there. But if you were now to tell whether $(- - \triangle \rightarrow (-\square \rightarrow \square))$ is a theorem, you might have hard time trying to find a derivation and even harder trying to prove that no derivation of this formula exists. Decidability of a theory means that there is a computer program capable to answer every such question.)

The decidability of propositional logic, through the use of truth tables, was known to Frege and Peirce; its proof is attributable to Jan Łukasiewicz and Emil Post independently in 1921. Löwenheim showed in 1915 that first order predicate logic with only single-place predicates was decidable and that the full theory was decidable if the first order predicate calculus with only two-place predicates was decidable. Further developments were made by Thoralf Skolem, Heinrich Behmann, Jacques Herbrand, and Willard Quine. Herbrand showed the existence of an algorithm which, if a theorem of the first order predicate logic is valid, will determine it to be so; the difficulty, then, was in designing an algorithm that in a finite amount of time would determine that propositions were invalid. (We can easily imagine a machine which, starting with the specified axioms, generates all

---

als and second order quantification over properties to the distinction between extensional and intensional aspects from B.3. Since in the extensional context, a property P is just a set of individuals (possessing P), the intensional or property-oriented language becomes higher-order, having to address not only individuals but also sets thereof. Third-order language allows then to quantify over sets of sets of individuals, etc.

possible theorems by simply generating all possible derivations – sequences of correct rule applications. If the formula is provable, the machine will, sooner or later, find a proof. But if the formula is not provable, the machine will keep for ever since the number of proofs is, typically, infinite.) As early as the 1880s, Peirce seemed to be aware that the propositional logic was decidable but that the full first order predicate logic with relations was undecidable. The fact that first order predicate logic (in any general formulation) was *undecidable* was first shown definitively by Alan Turing and Alonzo Church independently in 1936. Together with Gödel's (second) incompleteness theorem and the earlier Löwenheim-Skolem theorem, the Church-Turing theorem of the undecidability of the first order predicate logic is one of the most important, even if "negative", results of 20th century logic.

Many facts about the limits of computers arise as consequences of these negative results. For instance, it is not (and never will be!) possible to write a computer program which, given an arbitrary first order theory T and some formula f, is guaranteed to terminate giving the answer "Yes" if f is a theorem of T and "No" if it is not. A more mundane example is the following. One can easily write a computer program which for some inputs does not terminate. It might be therefore desirable to have a program U which could take as input another program P (a piece of text just like "usual" input to any program) and description of its input d and *decide* whether P run on d would terminate or not. Such a program U, however, will never be written as the problem is undecidable.

F. SUMMARY

The idea of correct thinking is probably as old as thinking itself. With Aristotle there begins the process of explicit formulation of the rules, patterns of reasoning, conformance to which would guarantee correctness. This idea of correctness has been gradually made precise and unambiguous leading to the formulation of (the general schema for defining) symbolic languages, the rules of their manipulation and hence criteria of correct "reasoning". It is, however, far from obvious that the result indeed captures the natural reasoning as performed by humans. The need for precision led to the complete separation of the reasoning aspect (syntactic manipulation) from its possible meaning. The completely uninterpreted nature of symbolic systems makes their relation to the real world highly problematic. Moreover, as one has arrived at the general schema of defining formal systems, no unique system has arisen as *the right* one and their variety seems surpassed only by the range of possible application domains. The discussions about which rules actually represent human thinking can probably continue indefinitely. Most significantly, this purely syntactic character of formal reasoning systems provided the basis for a precise definition of the old theme of logical investigations: the unavoidable consequence, which now appears co-extensional, if not synonymous, with the mechanical computability.

The question whether human mind can be reduced to such a mechanical computation and simulated by a computer is still discussed by the philosophers and cognitive scientists. Also, much successful research is driven by the idea, if not the explicit goal, of such a reduction. The "negative" results as those quoted at the end of the last section, established by human mind and demonstrating limitations of logic and computers, suggest that human cognition may not be reducible to mechanical computation. In particular, such a reduction would imply that all human thinking could be expressed as applications of simple rules like (E.2) on p. 38. Its possibility has not been disproved but it does not appear plausible. Yet, as computable functions correspond only to a small part of logic, even if *this* reduction turns out impossible, the question of reduction of thinking to logic at large would still remain open. Few researchers seem to believe in such reductions and, indeed, one need not believe in them to study logic. In spite of its philosophical roots and apparently abstract character, it turned out to be the fundamental tool in the development, and later in the use and management, of the most useful appliance of the 20th century – the computer.

Part I
ELEMENTS OF SET THEORY

Chapter 1

SETS, FUNCTIONS, RELATIONS

1.1 SETS AND FUNCTIONS

———————————— a background story ————————————

A *set* is an arbitrary collection of arbitrary objects, called its *members*. One should take these two occurrences of "arbitrary" seriously. Firstly, sets may be finite, e.g., the set C of cars on the parking lot outside the building, or infinite, e.g. the set N of numbers greater than 5.

Secondly, any objects can be members of sets. We can talk about sets of cars, blood-cells, numbers, Roman emperors, etc. We can also talk about the set X whose elements are: my car, your mother and number 6. (Not that such a set is necessarily useful for any purpose, but it is possible to collect these various elements into one set.) In particular sets themselves can be members of other sets. We can, for instance, form the set whose elements are: your favorite pen, your two best friends and *the set N*. This set will have *4* elements, even though the set N itself is infinite.

A set with only one element is called a *singleton*, e.g., the set containing only planet Earth. There is one special and very important set – the *empty* set – which has no members. If it seems startling, you may think of the set of all square circles or all numbers x such that $x < x$. This set is mainly a mathematical convenience – defining a set by describing the properties of its members in an involved way, we may not know from the very beginning what its members are. Eventually, we may find that no such objects exist, that is, that we defined an empty set. It also makes many formulations simpler since, without the assumption of its existence, one would often had to take special precautions for the case a set happened to contain no elements.

It may be legitimate to speak about a definite set even if we do not know exactly its members. The set of people born in 1964 may be hard to determine exactly but it is a well defined object because, at least in principle, we can determine membership of any object in this set. Similarly, we will say that the set R of red objects is well defined even if we certainly do not know all its members. But confronted with a new object, we can determine if it belongs to R or not (assuming, that we do not dispute the meaning of the word "red").

There are four basic means of specifying a set.

(1) If a set is finite and small, we may list all its elements, e.g., $S = \{1, 2, 3, 4\}$ is a set with four elements.

(2) A set can be specified by determining a property which makes objects qualify as its elements. The set R of red objects is specified in this way. The set S can be described as 'the set of natural numbers greater than 0 and less than 5'.

(3) A set may be obtained from other sets. For instance, given the set S and the set $S' = \{3, 4, 5, 6\}$ we can form a new set $S'' = \{3, 4\}$ which is the *intersection* of S and S'. Given the sets of odd $\{1, 3, 5, 7, 9, ...\}$ and even numbers $\{0, 2, 4, 6, 8, ...\}$ we can form a new set $\mathbb{N}$ by taking their *union*.

(4) Finally, a set can be also given by means of the rules for generating its elements. For instance, the set $\mathbb{N}$ of natural numbers $\{0, 1, 2, 3, 4, ...\}$ can be described as follows: 0 belongs to $\mathbb{N}$, if n belongs to $\mathbb{N}$ then also $n + 1$ belongs to $\mathbb{N}$ and, finally, nothing else belongs to $\mathbb{N}$.

In this chapter we will use mainly the first three ways of describing sets. In particular, we will use various set building operations as in point 3. In the later chapters, we will constantly encounter sets described by the last method. One important point is that the properties of a set are entirely independent from the way the set is described. Whether we just say 'the set of natural numbers' or the set $\mathbb{N}$ as defined in point 2. or 4., we get the same set. Another thing is that *studying and proving* properties of a set may be easier when the set is described in one way rather than another. ⎯⎯⎯⎯⎯⎯⎯⎯⎯⎯

We start by introducing some notational conventions for sets and some basic operations allowing to obtain new sets from other ones.

Definition 1.1 Given some sets S and T we write:

$x \in S$ - x is a member (element) of S

$S \subseteq T$ - S is a subset of T .... for all x : if $x \in S$ then $x \in T$

$S \subset T$ - $S \subseteq T$ and $S \neq T$ .... for all x : if $x \in S$ then $x \in T$

and for some $x : x \in T$ and $x \notin S$

Set building operations:

$\varnothing$ - empty set .... for any $x : x \notin \varnothing$

$S \cup T$ - union of S and T .... $x \in S \cup T$ iff $x \in S$ or $x \in T$ [1]

$S \cap T$ - intersection of S and T .... $x \in S \cap T$ iff $x \in S$ and $x \in T$

$S \setminus T$ - difference of S and T .... $x \in S \setminus T$ iff $x \in S$ and $x \notin T$

$\overline{S}$ - complement of S; given a universe U of all elements $\overline{S} = U \setminus S$

$S \times T$ - Cartesian product of S and T .... $x \in S \times T$ iff $x = \langle s, t \rangle$ and
$s \in S$ and $t \in T$

$\mathcal{P}(S)$ - the power set of S .... $x \in \mathcal{P}(S)$ iff $x \subseteq S$

Also, $\{x \in S : \mathsf{Prop}(x)\}$ denotes the set of those $x \in S$ which have the specified property Prop.

Sets may be members of other sets. For instance $\{\varnothing\}$ is the set with one element – which is the empty set $\varnothing$. In fact, $\{\varnothing\} = \mathcal{P}(\varnothing)$. It is different from the set $\varnothing$ which has no elements. $\{\{a, b\}, a\}$ is a set with two elements: a and the set $\{a, b\}$. Also $\{a, \{a\}\}$ has two different elements: a and $\{a\}$. In particular, the power set contains only sets as elements: $\mathcal{P}(\{a, \{a, b\}\}) = \{\varnothing, \{a\}, \{\{a, b\}\}, \{a, \{a, b\}\}\}$.

In the definition of Cartesian product, we used the notation $\langle s, t \rangle$ to denote an *ordered* pair whose first element is s and second t. In set theory, all possible objects are modeled as sets. An ordered pair $\langle s, t \rangle$ is then represented as the set with two elements – both being sets – $\{\{s\}, \{s, t\}\}$. Why not $\{\{s\}, \{t\}\}$ or, even simpler, $\{s, t\}$? Because elements of a set are not ordered. Thus $\{s, t\}$ and $\{t, s\}$ denote the same set. Also, $\{\{s\}, \{t\}\}$ and $\{\{t\}, \{s\}\}$ denote the same set (but different from the set $\{s, t\}$). In ordered pairs, on the other hand, the order does matter – $\langle s, t \rangle$ and $\langle t, s \rangle$ are different pairs. This ordering is captured by the representation $\{\{s\}, \{s, t\}\}$. We have here a set with two elements $\{A, B\}$ where $A = \{s\}$ and $B = \{s, t\}$. The relationship between these two elements tells us which is the first and which the second: $A \subseteq B$ identifies the member of A as the first element of the pair, and then the element of $B \setminus A$ as the second one (provided that $A \neq B$). Thus $\langle s, t \rangle = \{\{s\}, \{s, t\}\} \neq \{\{t\}, \{s, t\}\} = \langle t, s \rangle$.

---

[1] The abbreviation "iff" stands for two-ways implication "if and only if".

The set operations $\cup$, $\cap$, and $\setminus$ obey some well known laws:

(1) Empty set, the universe

$$A \cup \varnothing = A$$
$$A \cap \overline{\varnothing} = A$$

(2) Associativity

$$(A \cup B) \cup C = A \cup (B \cup C)$$
$$(A \cap B) \cap C = A \cap (B \cap C)$$

(3) Commutativity

$$A \cup B = B \cup A$$
$$A \cap B = B \cap A$$

(4) Distributivity

$$A \cup (B \cap C) = (A \cup B) \cap (A \cup C)$$
$$A \cap (B \cup C) = (A \cap B) \cup (A \cap C)$$

(5) Complement

$$A \cap \overline{A} = \varnothing$$
$$A \cup \overline{A} = \overline{\varnothing}$$

(6) de Morgan

$$\overline{(A \cup B)} = \overline{A} \cap \overline{B}$$
$$\overline{(A \cap B)} = \overline{A} \cup \overline{B}$$

(7)   a) $A \subseteq B$ iff $A \cup B = B$ iff $A \cap B = A$ b) $A \setminus B = A \cap \overline{B}$

Note that the "universe" U can be any set, chosen depending on the application context. It is therefore convenient to denote it simply as the complement of the empty set, $\overline{\varnothing}$.

Remark 1.2 [Venn's diagrams]

It is very common to represent sets and set relations by means of Venn's diagrams – overlapping figures, typically, circles or rectangles. On the left in the figure below, we have two sets A and B in some universe $\overline{\varnothing} = U$. Their intersection $A \cap B$ is marked as the area belonging to both by both vertical and horizontal lines. If we take A to represent Armenians and B bachelors, the darkest region in the middle represents Armenian bachelors. The region covered by only vertical, but not horizontal, lines is the set difference $A \setminus B$ – Armenians who are not bachelors. The whole region covered by either vertical or horizontal lines represents all those who are either Armenian or are bachelors.

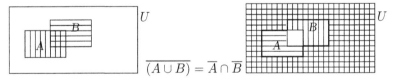

$$\overline{(A \cup B)} = \overline{A} \cap \overline{B}$$

Now, the white region is the complement of the set $A \cup B$ (in the universe U) – all those who are neither Armenians nor bachelors. The diagram to the right is essentially the same but was constructed in a different way. Here, the region covered with vertical lines is the complement of A – all non-Armenians. The region covered with horizontal lines represents all non-bachelors. The region covered with *both* horizontal and vertical lines is the intersection of these two complements – all those who are neither Armenians nor bachelors. The two diagrams illustrate the first de Morgan law since the white area on the left, $\overline{(A \cup B)}$, is exactly the same as the area covered with both horizontal and vertical lines on the right, $\overline{A} \cap \overline{B}$. □

Venn's diagrams may be handy tool to visualize simple set operations. However, the equalities above can be also seen as a (not yet quite, but almost) formal system allowing one to derive various other set equalities. The rule for performing such derivations is 'substitution of equals for equals', known also from elementary arithmetics. For instance, idempotency of $\cap$, i.e., $A = A \cap A$ follows by $A \stackrel{(1)}{=} A \cap \overline{\varnothing} \stackrel{(5)}{=} A \cap (A \cup \overline{A}) \stackrel{(4)}{=} (A \cap A) \cup (A \cap \overline{A}) \stackrel{(5)}{=} (A \cap A) \cup \varnothing \stackrel{(1)}{=} A \cap A$. The fact that, for an arbitrary set $A : A \subseteq A$ amounts then to a single application of (7).a: $A \subseteq A$ iff $A \cap A = A$, while $(A \cup B) \cup C = (C \cup A) \cup B$ follows by $(A \cup B) \cup C \stackrel{(3)}{=} C \cup (A \cup B) \stackrel{(2)}{=} (C \cup A) \cup B$. Exercises contain more elaborate examples.

In addition to the set building operations from the Definition 1.1, one often encounters also *disjoint union* of sets A and B, written $A \uplus B$ and defined as $A \uplus B = (A \times \{0\}) \cup (B \times \{1\})$. The idea is to use 0, resp. 1, as indices to distinguish the elements originating from A and from B. If $A \cap B = \varnothing$, this would not be necessary, but otherwise the "disjointness" of this union requires that the common elements be duplicated. E.g., for $A = \{a, b, c\}$ and $B = \{b, c, d\}$, their union is $A \cup B = \{a, b, c, d\}$ while the disjoint union $A \uplus B = \{\langle a, 0 \rangle, \langle b, 0 \rangle, \langle c, 0 \rangle, \langle b, 1 \rangle, \langle c, 1 \rangle, \langle d, 1 \rangle\}$, which can be thought of as $\{a_0, b_0, c_0, b_1, c_1, d_1\}$.

Definition 1.3 For two sets S and T, a *function* f from S to T, $f : S \to T$, is a subset of $S \times T$ such that

- whenever $\langle s, t \rangle \in f$ and $\langle s, t' \rangle \in f$, then $t = t'$, and
- for each $s \in S$ there is some $t \in T$ such that $\langle s, t \rangle \in f$.

A subset of $S \times T$ that satisfies the first condition above but not necessarily the second, is called a *partial* function from S to T.

For a function $f : S \to T$, the set S is called the *source* or *domain* of the function, and the set T its *target* or *codomain*.

The second point of this definition means that function is *total* – for each argument (element $s \in S$), the function has some value, i.e., an element $t \in T$ such that $\langle s, t \rangle \in f$. Sometimes this requirement is dropped and one speaks about *partial functions* which may have no value for some arguments but we will be for the most concerned with total functions.

Example 1.4
Let $\mathbb{N}$ denote the set of natural numbers $\{0, 1, 2, 3, ...\}$. The mapping $f :$ $\mathbb{N} \to \mathbb{N}$ defined by $f(n) = 2n$ is a function. It is the set of all pairs $f = \{\langle n, 2n \rangle : n \in \mathbb{N}\}$. If we let M denote the set of all people, then the set of all pairs $father = \{\langle m, m's \text{ father} \rangle : m \in M\}$ is a function assigning to each person his/her father. A mapping '*children*', assigning to each person his/her children is not a function $M \to M$ for two reasons. For the first, a person may have no children, while saying "function" we mean a total function. For the second, a person may have more than one child. These problems may be overcome if we considered it instead as a function $M \to \mathcal{P}(M)$ assigning to each person the set (possibly empty) of all his/her children.   □

Notice that although intuitively we think of a function as a mapping assigning to each argument some value, the definition states that it is actually a set (a subset of $S \times T$ is a set). The restrictions put on this set are exactly what makes it possible to think of this set as a mapping. Nevertheless, functions – being sets – can be elements of other sets. We may encounter situations involving sets of functions, e.g. the set T^S of all functions from set S to set T, which is just the set of all subsets of $S \times T$, each satisfying the conditions of Definition 1.3.

Remark 1.5 [Notation]
A function f associates with each element $s \in S$ a *unique* element $t \in T$. We write this t as $f(s)$ – the value of f at point s. When S is finite (and small) we may sometimes write a function as a set $\{\langle s_1, t_1 \rangle, \langle s_2, t_2 \rangle, ..., \langle s_n, t_n \rangle\}$ or else as $\{s_1 \mapsto t_1, s_2 \mapsto t_2, ..., s_n \mapsto t_n\}$.

If f is given then by $f[s \mapsto p]$ we denote the function f' which is the same as f for all arguments $x \neq s : f'(x) = f(x)$, while $f'(s) = p$.   □

Definition 1.6   A function $f : S \to T$ is

injective iff whenever $f(s) = f(s')$ then $s = s'$;
surjective iff for all $t \in T$ there exists an $s \in S$ such that $f(s) = t$;
bijective iff it is both injective and surjective.

Injectivity means that no two distinct elements from the source set are mapped to the same element in the target set; surjectivity that each element in the target is an image of some element from the source.

Example 1.7
The function $father : M \to M$ is neither injective – brothers have the

same father, nor surjective – not everybody is a father of somebody. The following drawing gives some more examples:

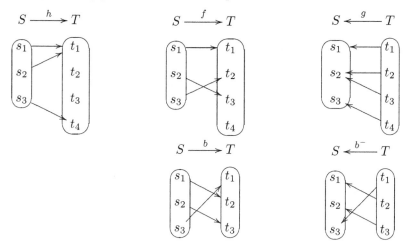

Here h is neither injective nor surjective, f is injective but not surjective, $g : T \to S$ is surjective but not injective. b and b^- are both injective and surjective, i.e, bijective.   □

Soon we will see the particular importance of bijections. There may exist several bijections between two sets S and T. If there is at least one bijection between the two sets, we write $S \rightleftharpoons T$. The following lemma gives another criterion for a function to be a bijection.

Lemma 1.8 $f : S \to T$ is a bijection if and only if there exists an *inverse* function $f^- : T \to S$, such that for all $s \in S : f^-(f(s)) = s$ and for all $t \in T : f(f^-(t)) = t$.

Proof. We have to show two implications:
only if) If f is bijective, we can define f^- simply as the set of pairs $f^- \stackrel{\text{def}}{=} \{\langle t, s \rangle : t = f(s)\}$.
if) If $f(s) = f(s')$ then $s = f^-(f(s)) = f^-(f(s')) = s'$, i.e., f is injective. Also, for any $t \in T$, taking $s = f^-(t)$ we obtain $f(s) = f(f^-(t)) = t$, i.e., f is surjective.   QED (1.8)

In Example 1.7 both b and b^- are bijective. They act as mutual inverses satisfying the conditions of the above lemma: for each $s \in S : b^-(b(s)) = s$ and for each $t \in T : b(b^-(t)) = t$.

Definition 1.9 For any set U and $A \subseteq U$, the *characteristic function* of A (relatively to U), denoted f_A, is the function

$$\{\langle x, 1 \rangle : x \in A\} \ \cup \ \{\langle x, 0 \rangle : x \in \overline{A}\}.$$

Hence $f_A(x) = 1$ iff $x \in A$. Note that f_A is a function from U to $\{1, 0\}$, where $\{1, 0\}$ is a(ny) set with exactly two elements.

Let $f_- : \mathcal{P}(U) \to 2^U$ denote the function sending each subset A of U to its characteristic function f_A (the notation 2^U stands for the set of all functions from U to a two-element set, e.g., to $\{1, 0\}$).

Proposition 1.10 For every set $U : \mathcal{P}(U) \rightleftharpoons 2^U$.

> **Proof.** If $A \neq B$ then some x is such that either $x \in A \setminus B$ or $x \in B \setminus A$. In either case, $f_A(x) \neq f_B(x)$, i.e., $f_A \neq f_B$, so f_- is injective.
> On the other hand, every function $g \in 2^U$ is the characteristic function of the subset $A_g = \{x \in U : g(x) = 1\} \subseteq U$, so f_- is surjective. Thus, we have a bijection $f_- : \mathcal{P}(U) \rightleftharpoons 2^U$.   QED (1.10)

1.2 Relations

Definition 1.11 A *relation R on sets* $S_1 \ldots S_n$ is a subset $R \subseteq S_1 \times \ldots \times S_n$. A binary relation *on a set* S is a subset of $S^2 = S \times S$ and, generally, an n-ary relation on S is a subset of $S^n = \underbrace{S \times S \times \ldots \times S}_{n}$.

This definition makes *any* subset of $S \times T$ a relation. Definition 1.3 of function, on the other hand, required this set to satisfy some additional properties. Hence a function is a special case of relation, namely, a relation which relates each element of S with exactly one element of T.

Binary relations are sets of *ordered* pairs, i.e., $\langle s, t \rangle \neq \langle t, s \rangle$ – if $s \in S$ and $t \in T$, then the former belongs to $S \times T$ and the latter to $T \times S$, which are different sets. It is common to write the fact that s and t stand in a relation R, $\langle s, t \rangle \in R$, as sRt or as $R(s, t)$.

In general, relations may have arbitrary arities, for instance a subset $R \subseteq S \times T \times U$ is a ternary relation, etc. As a particular case, a *unary* relation (also called a predicate) on S is simply a subset $R \subseteq S$. In the following we are speaking only about binary relations (unless explicitly stated otherwise).

Example 1.12

Functions, so to speak, map elements of one set onto the elements of another set (possibly the same). Relations *relate* some elements of one set with some elements of another. Recall the problem from Example 1.4 with treating *children* as a function $M \rightarrow M$. This problem is overcome by treating *children* as a binary relation on the set M of all people. Thus $children(p, c)$ holds if an only if c is a child of p. Explicitly, this relation is $children = \{\langle p, c \rangle : p, c \in M$ and c is a child of $p\}$.   □

Definition 1.13 Given two relations $R \subseteq S \times T$ and $P \subseteq T \times U$, their *composition* is the relation $R ; P \subseteq S \times U$, defined as the set of pairs
$$R ; P = \{\langle s, u \rangle \in S \times U : \text{there is a } t \in T \text{ with } R(s, t) \text{ and } P(t, u)\}.$$

As functions are special cases of relations, the above definition allows us to form the composition $g ; f$ of functions $g : S \rightarrow T$ and $f : T \rightarrow U$, namely, the function from S to U given by the equation $(g ; f)(s) = f(g(s))$.

Definition 1.14 A relation $R \subseteq S \times S$ is:

connected	iff	for all pairs of distinct $s, t \in S : R(s, t)$ or $R(t, s)$
reflexive	iff	for all $s \in S : R(s, s)$
irreflexive	iff	for no $s \in S : R(s, s)$
transitive	iff	when $R(s_1, s_2)$ and $R(s_2, s_3)$ then $R(s_1, s_3)$
symmetric	iff	when $R(s, t)$ then $R(t, s)$
asymmetric	iff	when $R(s, t)$ then not $R(t, s)$
antisymmetric	iff	when $R(s, t)$ and $R(t, s)$ then $s = t$
equivalence	iff	it is reflexive, transitive and symmetric.

Numerous connections hold between these properties: every irreflexive, transitive relation is also asymmetric, and every asymmetric relation is also antisymmetric. Note also that just one relation is both connected, symmetric and reflexive, namely the universal relation $S \times S$ itself.

The most common (and smallest) example of equivalence is the *identity* relation $id_S = \{\langle s, s \rangle : s \in S\}$. The relation $\rightleftharpoons$ – existence of a bijection – is also an equivalence relation on any set (collection) of sets. An equivalence relation $\sim$ on a set S allows us to *partition* S into disjoint *equivalence classes* $[s] = \{s' \in S : s' \sim s\}$.

Given a relation $R \subseteq S \times S$, we can form its *closure* with respect to one or more of these properties. For example, the *reflexive closure* of R, written $\underline{R}$, is the relation $R \cup id_S$. (It may very well happen that R already is reflexive, $R = \underline{R}$, but then we do have that $\underline{R} = \underline{\underline{R}}$.) The *transitive closure* of R, written R^+, can be thought of as the infinite union
$$R \cup (R ; R) \cup (R ; R ; R) \cup (R ; R ; R ; R) \cup (R ; R ; R ; R ; R) \cup \ldots$$
and can be defined as the least relation R^+ such that $R^+ = R \cup (R^+ ; R)$.

1.3 ORDERING RELATIONS

Of particular importance are the *ordering* relations which we will use extensively from Chapter 2 on. Often we assume an implicit set S and talk only about relation R. However, it is important to remember that talking about a relation R, we are actually considering a *set with structure*, namely a pair $\langle S, R \rangle$.

Definition 1.15 $\langle S, R \rangle$ is a *quasiorder* (or *preorder*), QO, iff R is
   transitive : $R(x, y) \wedge R(y, z) \rightarrow R(x, z)$
    reflexive : $R(x, x)$
$\langle S, R \rangle$ is a *weak partial order*, wPO, iff R is
    a quasiorder : $R(x, y) \wedge R(y, z) \rightarrow R(x, z)$
               : $R(x, x)$
  antisymmetric : $R(x, y) \wedge R(y, x) \rightarrow x = y$
$\langle S, R \rangle$ is a *strict partial order*, sPO, iff R is
   transitive : $R(x, y) \wedge R(y, z) \rightarrow R(x, z)$
   irreflexive : $\neg R(x, x)$
A *total order*, TO, is a PO which is connected : $x \neq y \rightarrow R(x, y) \vee R(y, x)$

A QO allows loops, for instance the situations like $R(a_1, a_2)$, $R(a_2, a_3)$, $R(a_3, a_1)$ for distinct a_1, a_2, a_3. PO forbids such situations: by applications of transitivity we have

- $R(a_2, a_1)$ as well as $R(a_1, a_2)$, which for wPO's imply $a_1 = a_2$.
- $R(a_1, a_1)$, which is impossible for sPO's.

Obviously, given a wPO, we can trivially construct its strict version (by making it irreflexive) and vice versa. We will therefore often say "partial order" or PO without specifying which one we have in mind. Instead of writing $R(x, y)$ for an ordering, we often use the infix notation $x \leq y$ for the weak version and $x < y$ for the strict one, i.e., $x \leq y$ means $x < y$ or $x = y$, and $x < y$ means $x \leq y$ and $x \neq y$.

Example 1.16
Consider the set of all people and their ages.
(1) The relation 'x is older than y' is an sPO on the set of all people : it is transitive, irreflexive (nobody is older than himself) and asymmetric: if 'x is older than y' then 'y is not older than x'.
(2) The relation 'x is not younger than y' is a QO. It is not a wPO since it is not antisymmetric. (There are different people of the same age.)

The weak version of the relation in 1. is the relation 'x is older than y or x, y is the same person', which is clearly different from the relation in 2. The relation in 1. is not a TO – of two *different* persons there may be none who is older than the other.   □

Example 1.17
Given a set of symbols, for instance $\Sigma = \{a, b, c\}$, the set Σ^* contains all *finite* strings over Σ, e.g., *a, aa, abba, aaabbababab*, ... There are various natural ways of ordering Σ^*. Let $s, p \in \Sigma^*$

(1) Define $s \prec_Q p$ iff $length(s) < length(p)$. This gives an sPO. The weak relation $s \preceq_Q p$ iff $length(s) \leq length(p)$ will be a QO but not a wPO since now any subset containing strings of equal length will form a loop.

(2) Define $s \prec_P p$ iff s is a prefix of p. (Prefix is an initial segment.) We now obtain a wPO because any string is its own prefix and if both s is a prefix of p and p is a prefix of s the two must be the same string. This is not, however, a TO : neither of the strings a and bc is a prefix of the other.

(3) Suppose that the set Σ is totally ordered, for instance, let Σ be the Latin alphabet with the standard ordering $a \prec b \prec c \prec d$.... We may then define the *lexicographic* TO on Σ^* as follows:
$s \prec_L p$ iff either s is a prefix of p or else the two have a longest common prefix u (possibly empty) such that $s = uv$, $t = uw$ and $head(v) \prec head(w)$, where *head* is the first symbol of the argument string. This defines the usual ordering used, for instance, in dictionaries.   □

HOMOMORPHISMS . . . . [optional]
The following notions will not be used extensively, but it is often important to realize that functions between ordered sets should consider not only the elements of the sets but also the relation between these elements.

Definition 1.18 Given two orderings $P = \langle S_1, R_1 \rangle$ and $Q = \langle S_2, R_2 \rangle$, a *homomorphism* $h : P \to Q$ is an *order preserving* function, i.e., a function $h : S_1 \to S_2$ such that $R_1(x, y)$ implies $R_2(h(x), h(y))$.
An *order-isomorphism* is a bijection $h : S_1 \rightleftharpoons S_2$ such that both h and h^- are homomorphisms.

Thus, an order-isomorphism is a bijection which, in addition, preserves the structure of the isomorphic relations. One often encounters two ordered sets which are bijective but not order-isomorphic.

Example 1.19
Given two 4-element sets A and B, consider two sPO's $\mathbf{A} = \langle A, \prec_A \rangle$ and $\mathbf{B} =$

$\langle B, \prec_B \rangle$ as shown below:

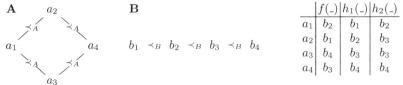

	$f(\_)$	$h_1(\_)$	$h_2(\_)$
a_1	b_2	b_1	b_2
a_2	b_1	b_2	b_3
a_3	b_4	b_3	b_3
a_4	b_3	b_4	b_4

Any injective $f : A \to B$ is also a bijection, i.e., $A \rightleftharpoons B$. However, a homomorphism $h : \mathbf{A} \to \mathbf{B}$ must also satisfy the additional condition, e.g., $a_1 \prec_A a_2$ requires that also $h(a_1) \prec_B h(a_2)$. Thus, for instance, f from the table is not a homomorphism $\mathbf{A} \to \mathbf{B}$ while h_1 and h_2 are.

Now, although the underlying sets A and B are isomorphic, there is no *order-isomorphism* between $\mathbf{A} = \langle A, \prec_A \rangle$ and $\mathbf{B} = \langle B, \prec_B \rangle$ because there is no *homomorphism* from the latter to the former. $\mathbf{B}$ is connected, i.e., a total order while $\mathbf{A}$ is not – any homomorphism would have to preserve the relation $\prec_B$ between arbitrary two elements of B, but in A there is no relation $\prec_A$ between a_2 and a_3.
.... [end optional]

1.4 INFINITIES

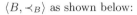 a background story

Imagine a primitive shepherd who possesses no idea of number or counting – when releasing his sheep from the cottage in the morning he wants to find the means of figuring out if, when he collects them in the evening, all are back or, perhaps, some are missing.

He can, and most probably did, proceed as follows. Find a stick and let the sheep leave one by one, marking each on the stick, e.g., by something like /. When all the sheep have left, there will be as many marks on the stick as there were sheep. On their return, do the same: let them go through the gate one by one. For each sheep, erase one mark – e.g., set a \making one / into ×. When all the sheep are inside, check if there are any /-marks left. If no, i.e., there are only × on the stick, then everything is ok – as many sheep returned home as had left in the morning. If yes, then some sheep are missing.

Notice, that the shepherd still does not know "how many" sheep he has – he still does not have the idea of a number. But, perhaps a bit paradoxically, he has the idea of *two equal numbers*! This idea is captured by the correspondence, in fact, several functions: the first is a morning-function m which for each sheep from the set S_M of all sheep, assigns a new / on the stick – all the marks obtained in the morning form a set M. The other, evening-function e, assigns to each returning sheep (from the set of returning sheep S_E) another mark \.

Superimposing the evening marks \ onto the morning marks /, i.e., forming ×-marks, amounts to comparing the number of elements in the sets M and E by means of a function c assigning to each returning sheep marked by \, a morning sheep /.

$$S_M \overset{m}{\leftrightarrow} M \overset{c}{\leftrightarrow} E \overset{e}{\leftrightarrow} S_E$$

In fact, this simple procedure may be considered as a basis of counting – comparing the number of elements in various sets. In order to ensure that the two sets, like S_M and M, have equal number of elements, we have to insist on some properties of the involved function m. Each sheep must be given a mark (m must be total) and for two distinct sheep we have to make two distinct marks (m must be injective). The third required property – that of surjectivity – follows automatically in the above procedure, since the target set M is formed only along as we mark the sheep. In short, the shepherd knows that there are *as many* sheep *as* the morning marks on the stick because he has a bijective function between the respective sets. For the same reason, he knows that the sets of returning sheep and evening marks have the same number of elements and, finally, establishing a bijection between the sets M and E, he rests satisfied. (Violating a bit the profiles of the involved functions, we may say that the composite function $e \, ; c \, ; m : S_E \to E \to M \to S_M$ turns out to be bijective.)

You can now easily imagine what is going on when the shepherd discovers that some sheep are missing in the evening. He is left with some /-marks which cannot be converted into ×-marks. The function $c : E \to M$ is injective but not surjective. This means that the set E has strictly fewer elements than the set M.

These ideas do not express immediately the concept of a number as we are used to it. (They can be used to do that.) But they do express our intuition about one number being equal to, smaller or greater than another number. What is, perhaps, most surprising is that they work equally well when the involved sets are infinite, thus allowing us to compare "the number of elements" in various infinite sets. _____

In this section, we consider only sets and set functions, in particular, bijections.

Definition 1.20   Two sets S and T have *equal cardinality* iff they are bijective $S \rightleftharpoons T$.
The *cardinality* of a set S, $|S|$, is the equivalence class $[S] = \{T : T \rightleftharpoons S\}$.

This is not an entirely precise definition of cardinality which, as a matter of fact, is the *number associated with* such an equivalence class. The point is that it denotes the intuitive idea of *the number of elements* in the set – all bijective sets have the same number of elements.

Definition 1.21 $|S| \leq |T|$ iff there exists an injective function $f : S \to T$.

It can be shown that this definition is consistent, i.e., that if $|S_1| = |S_2|$ and $|T_1| = |T_2|$ then there exists an injective function $f_1 : S_1 \to T_1$ iff there exists injective function $f_2 : S_2 \to T_2$. A set S has cardinality strictly less than a set T, $|S| < |T|$, iff there exists an injective function $f : S \to T$ *but there exists no* such surjective function.

Example 1.22
$|\varnothing| = 0$, $|\{\varnothing\}| = 1$, $|\{\{\varnothing\}\}| = 1$, $|\{\varnothing, \{\varnothing\}\}| = 2$.
$|\{a, b, c\}| = |\{\bullet, \#, +\}| = |\{0, 1, 2\}| = 3$.
For finite sets, all operations from Definition 1.1 yield sets with possibly different cardinality:

1. $|\{a, b\} \cup \{a, c, d\}| = |\{a, b, c, d\}|$ $|S \cup T| \geq |S|$
2. $|\{a, b\} \cap \{a, c, d\}| = |\{a\}|$ $|S \cap T| \leq |S|$
3. $|\{a, b\} \setminus \{a, c, d\}| = |\{b\}|$ $|S \setminus T| \leq |S|$
4. $|\{a, b\} \times \{a, d\}| = |\{\langle a, a\rangle, \langle a, d\rangle, \langle b, a\rangle \langle b, d\rangle\}|$ $|S \times T| = |S| * |T|$
5. $|\mathcal{P}(\{a, b\})| = |\{\varnothing, \{a\}, \{b\}, \{a, b\}\}|$ $|\mathcal{P}(S)| > |S|$

□

From certain assumptions ("axioms") about sets it can be proven that the relation $\leq$ on cardinalities has the properties of a weak TO, i.e., it is reflexive (obvious), transitive (fairly obvious), antisymmetric (not so obvious) and total (less obvious).

SCHRÖDER-BERNSTEIN THEOREM .... [optional]
As an example of how intricate reasoning may be needed to establish such "not quite but almost obvious" facts, we show that $\leq$ is antisymmetric.

Theorem 1.23   For arbitrary sets X, Y, if there are injections $i : X \to Y$ and $j : Y \to X$, then there is a bijection $f : X \to Y$ (i.e., if $|X| \leq |Y|$ and $|Y| \leq |X|$ then $|X| = |Y|$).

Proof. If the injection $i : X \to Y$ is surjective, i.e., $i(X) = Y$, then i is a bijection and we are done. Otherwise, we have $Y_0 = Y \setminus i(X) \neq \varnothing$ and we apply j and i repeatedly as follows

$$Y_0 = Y \setminus i(X) \qquad X_0 = j(Y_0)$$
$$Y_{n+1} = i(X_n) \qquad X_{n+1} = j(Y_{n+1})$$
$$Y^* = \bigcup_{n=0}^{\omega} Y_n \qquad X^* = \bigcup_{n=0}^{\omega} X_n$$

i.e.,

$$
\begin{array}{ccccccccc}
Y_0 & & Y_1 & & Y_2 & & & \cdots & Y^* \\
\uparrow i & \searrow j & \uparrow i & \searrow j & \uparrow i & \searrow j & \uparrow i & \searrow j & \\
X & & X_0 & & X_1 & & X_2 & \cdots & X^*
\end{array}
$$

So we can divide both sets into disjoint components as in the diagram below.

$$
\begin{array}{ccccc}
Y & = & Y^* & \cup & (Y \setminus Y^*) \\
j \downarrow \uparrow i & & \Updownarrow j & & \Updownarrow i \\
X & = & X^* & \cup & (X \setminus X^*)
\end{array}
$$

We show that the respective restrictions of j and i are bijections. First, $j : Y^* \to X^*$ is a bijection (it is injective, and the following equation shows that it is surjective):

$$j(Y^*) = j\left(\bigcup_{n=0}^{\omega} Y_n\right) = \bigcup_{n=0}^{\omega} j(Y_n) = \bigcup_{n=0}^{\omega} X_n = X^*.$$

By Lemma 1.8, $j^- : X^* \to Y^*$, defined by $j^-(x) = y : j(y) = x$ is a bijection too. Furthermore:

$$i(X^*) = i\left(\bigcup_{n=0}^{\omega} X_n\right) = \bigcup_{n=0}^{\omega} i(X_n) = \bigcup_{n=0}^{\omega} Y_{n+1} = \bigcup_{n=1}^{\omega} Y_n = Y^* \setminus Y_0. \qquad (1.24)$$

Below, the first equality holds since i is injective, the second one by (1.24) and since $i(X) = Y \setminus Y_0$ (definition of Y_0), and the last one since $Y_0 \subseteq Y^*$:

$$i(X \setminus X^*) = i(X) \setminus i(X^*) = (Y \setminus Y_0) \setminus (Y^* \setminus Y_0) = Y \setminus Y^*,$$

i.e., $i : (X \setminus X^*) \to (Y \setminus Y^*)$ is a bijection. We obtain a bijection $f : X \to Y$ defined by

$$f(x) = \begin{cases} i(x) & \text{if } x \in X \setminus X^* \\ j^-(x) & \text{if } x \in X^* \end{cases}$$

QED (1.23)

A more abstract proof

The construction of the sets X^* and Y^* in the above proof can be subsumed under a more abstract formulation implied by Claims 1. and 2. below.

Claim 1. For any set X, if $h : \mathcal{P}(X) \to \mathcal{P}(X)$ is monotonic, i.e. such that, whenever $A \subseteq B \subseteq X$ then $h(A) \subseteq h(B)$; then there is a set $T \subseteq X : h(T) = T$.
  We show that $T = \bigcup\{A \subseteq X : A \subseteq h(A)\}$.

a) $T \subseteq h(T)$: for each $t \in T$ there is an $A : t \in A \subseteq T$ and $A \subseteq h(A)$. But then $A \subseteq T$ implies $h(A) \subseteq h(T)$, and so $t \in h(T)$.

b) $h(T) \subseteq T$: from a) $T \subseteq h(T)$, so $h(T) \subseteq h(h(T))$ which means that $h(T) \subseteq T$ by definition of T.

Claim 2. Given injections i, j define $\_^* : \mathcal{P}(X) \to \mathcal{P}(X)$ by $A^* = X \setminus j(Y \setminus i(A))$. If $A \subseteq B \subseteq X$ then $A^* \subseteq B^*$.
Follows trivially from injectivity of i and j. $A \subseteq B$, so $i(A) \subseteq i(B)$, so $Y \setminus i(A) \supseteq Y \setminus i(B)$, so $j(Y \setminus i(A)) \supseteq j(Y \setminus i(B))$, and hence $X \setminus j(Y \setminus i(A)) \subseteq X \setminus j(Y \setminus i(B))$.

3. Claims 1 and 2 imply that there is a $T \subseteq X$ such that $T = T^*$, i.e., $T = X \setminus j(Y \setminus i(T))$. Then $f : X \to Y$ defined by $f(x) = \begin{cases} i(x) & \text{if } x \in T \\ j^-(x) & \text{if } x \notin T \end{cases}$ is a bijection. We have $X = j(Y \setminus i(T)) \cup T$ and $Y = (Y \setminus i(T)) \cup i(T)$, and obviously j^- is a bijection between $j(Y \setminus i(T))$ and $Y \setminus i(T)$, while i is a bijection between T and $i(T)$.. [end optional]

The cardinality of every finite set is a natural number, representing the number of elements in the set. The apparently empty Definition 1.20 becomes more significant when we look at the infinite sets.

Definition 1.25 A set S is *infinite* iff there exists a proper subset $T \subset S$ such that $S \rightleftharpoons T$.

Example 1.26
Denote the cardinality of the set of natural numbers by $|\mathbb{N}| \overset{\text{def}}{=} \aleph_0$. (Sometimes it is also written ω, although axiomatic set theory distinguishes between the *cardinal number* $\aleph_0$ and the *ordinal number* ω. Ordinal number is a more fine-grained notion than cardinal number, but we shall not worry about this.) We have, for instance, that $|\mathbb{N}| = |\mathbb{N} \setminus \{0\}|$, as shown below to the left. In fact, the cardinality of $\mathbb{N}$ is the same as the cardinality of the even natural numbers! It is easy to see that the pair of functions $f(n) = 2n$ and $f^-(2n) = n$, as shown to the right, forms a bijection:

$$
\begin{array}{llllll}
\{ & 0 & 1 & 2 & 3 & \ldots \\
  & \updownarrow & \updownarrow & \updownarrow & \updownarrow & \cdots \\
\{ & 1 & 2 & 3 & 4 & \ldots
\end{array}
\qquad\qquad
\begin{array}{llllllll}
 & \{ & 0 & 1 & 2 & 3 & 4 & 5 & \ldots \\
f\downarrow & & \updownarrow & \updownarrow & \updownarrow & \updownarrow & \updownarrow & \updownarrow & \uparrow f^- \\
 & \{ & 0 & 2 & 4 & 6 & 8 & 10 & \ldots
\end{array}
$$

In general, when $|S| = |T| = \aleph_0$ and $|P| = n < \aleph_0$, we have

$$|S \cup T| = \aleph_0 \qquad |S \setminus P| = \aleph_0 \qquad |S \times T| = \aleph_0.$$

The following drawing illustrates a possible bijection $\mathbb{N} \rightleftharpoons \mathbb{N} \times \mathbb{N}$:

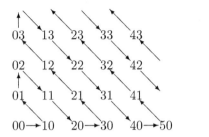

A bijection $S \rightleftharpoons \mathbb{N}$ amounts to an *enumeration* of the elements of S. Thus, if $|S| \leq \aleph_0$ we say that S is *enumerable* or *countable*; in case of equality, we say that it is *countably infinite*. Now, the question "are there any uncountable sets?" was answered by the founder of modern set theory:

Theorem 1.27 [Georg Cantor] For any set $A : |A| < |\mathcal{P}(A)|$.

Proof. The construction applied here shows that the contrary assumption – $A \rightleftharpoons \mathcal{P}(A)$ – leads to a contradiction. Obviously, $|A| \leq |\mathcal{P}(A)|$, since the inclusion defined by $f(a) = \{a\}$ is an injective function $f : A \to \mathcal{P}(A)$. So assume the equality $|A| = |\mathcal{P}(A)|$, i.e., a corresponding $F : A \to \mathcal{P}(A)$ which is both injective and surjective. Define the subset of A by $B \stackrel{\text{def}}{=} \{a \in A : a \notin F(a)\}$. Since $B \subseteq A$, so $B \in \mathcal{P}(A)$ and, since F is surjective, there is a $b \in A$ such that $F(b) = B$. Is b in B or not? Each of the two possible answers yields a contradiction:

(1) $b \in F(b)$ means $b \in \{a \in A : a \notin F(a)\}$, which means $b \notin F(b)$
(2) $b \notin F(b)$ means $b \notin \{a \in A : a \notin F(a)\}$, which means $b \in F(b)$.

QED (1.27)

Corollary 1.28 There is no greatest cardinal number.

In particular, $\aleph_0 = |\mathbb{N}| < |\mathcal{P}(\mathbb{N})| < |\mathcal{P}(\mathcal{P}(\mathbb{N}))| < \ldots$ Theorem 1.27 proves that there exist uncountable sets, but are they of any interest? Another theorem of Cantor shows that such sets have been around in mathematics for quite a while.

Theorem 1.29 The set $\mathbb{R}$ of real numbers is uncountable.

Proof. Since $\mathbb{N} \subset \mathbb{R}$, we know that $|\mathbb{N}| \leq |\mathbb{R}|$. The *diagonalisation* technique introduced here by Cantor, reduces the assumption that $|\mathbb{N}| = |\mathbb{R}|$ ad absurdum. If $\mathbb{R} \rightleftharpoons \mathbb{N}$ then, certainly, we can enumerate any subset of $\mathbb{R}$. Consider only the closed interval $[0,1] \subset \mathbb{R}$. If it is countable, we can list all its members, writing them in decimal expansion (each r_{ij} is a digit):

$$
\begin{aligned}
n_1 &= \quad 0.\ \mathbf{r_{11}}\ r_{12}\ r_{13}\ r_{14}\ r_{15}\ r_{16}\ .... \\
n_2 &= \quad 0.\ r_{21}\ \mathbf{r_{22}}\ r_{23}\ r_{24}\ r_{25}\ r_{26}\ .... \\
n_3 &= \quad 0.\ r_{31}\ r_{32}\ \mathbf{r_{33}}\ r_{34}\ r_{35}\ r_{36}\ .... \\
n_4 &= \quad 0.\ r_{41}\ r_{42}\ r_{43}\ \mathbf{r_{44}}\ r_{45}\ r_{46}\ .... \\
n_5 &= \quad 0.\ r_{51}\ r_{52}\ r_{53}\ r_{54}\ \mathbf{r_{55}}\ r_{56}\ .... \\
n_6 &= \quad 0.\ r_{61}\ r_{62}\ r_{63}\ r_{64}\ r_{65}\ \mathbf{r_{66}}\ .... \\
&\quad\ \ \vdots
\end{aligned}
$$

Form a new real number r by replacing each $\mathbf{r_{ii}}$ with another digit, for instance, let $r = 0.r_1 r_2 r_3 r_4...$, where $r_i = \mathbf{r_{ii}} + 1 \bmod 10$. Then r cannot be any of the listed numbers $n_1, n_2, n_3, ...$ For each such number n_i has a digit $\mathbf{r_{ii}}$ at its i-th position which is different from the digit r_i at the i-th position in r.   QED (1.29)

"SETS" WHICH ARE NOT SETS

Among the set building operations introduced in Definition 1.1, the power set operation $\mathcal{P}(\_)$ has proven particularly potent. However, the most peculiar one is the *comprehension* operation, namely, the one allowing us to form a set of elements satisfying some property $\{x : \mathrm{Prop}(x)\}$. Although apparently very natural, its unrestricted use leads to severe problems.

Russell's Paradox.
Let $U = \{x : x \notin x\}$. Is $U \in U$? Each possible answer leads to absurdity:
(1) $U \in U$ means that U is one of x in U, i.e., $U \in \{x : x \notin x\}$, so $U \notin U$,
(2) $U \notin U$ means that $U \in \{x : x \notin x\}$, so $U \in U$.   □

The paradox arises because in the definition of U we did not specify what kind of x's we are gathering. The most common solution simply excludes such definitions by requiring that x's must already belong to some set. This is the formulation we used in Definition 1.1, where we said that if S is a set then $\{x \in S : \mathrm{Prop}(x)\}$ is a set too. The "definition" of U does not respect this format and hence is not considered a valid description of a set.

EXERCISES 1.

EXERCISE 1.1 Let $A = \{a, \varnothing, \{\varnothing\}\}$, $B = \{\{\varnothing\}, d\}$, $C = \{d, e, f\}$.

(1) Write the sets: $A \cup B$, $A \cap B$, $A \cup (B \cap C)$
(2) Is a a member of $\{A, B\}$, of $A \cap B$, of $A \cup B$?
(3) Write the sets $A \times B$ and $B \times A$.

EXERCISE 1.2 Construct $\mathcal{P}(S_i)$ for $S_1 = \{0, 1, 2\}$ and $S_2 = \{0, 1, \{0, 1\}\}$.

EXERCISE 1.3 Using the set theoretic equalities (page 48), show that:

(1) $A \subseteq B$ iff $\overline{B} \subseteq \overline{A}$
(2) $A \setminus (B \cup C) = A \cap (\overline{B} \setminus C)$
(3) $A \cap (B \setminus A) = \varnothing$ and $A \cap (A \setminus B) = A \setminus B$.
(4) $((A \cup C) \cap (B \cup \overline{C})) \subseteq (A \cup B)$

Here, show first some lemmata:
a) $A \cap B \subseteq A$
b) if $A \subseteq B$ then $A \subseteq B \cup C$
c) if $A_1 \subseteq X$ and $A_2 \subseteq X$ then $A_1 \cup A_2 \subseteq X$.
Expand then the expression $(A \cup C) \cap (B \cup \overline{C})$ to one of the form $X_1 \cup X_2 \cup X_3 \cup X_4$, show that each $X_i \subseteq A \cup B$ and use Lemma c).

EXERCISE 1.4 Specify each of the two infinite sets
$\quad S_1 = \{5, 10, 15, 20, 25, ...\}$ and $S_2 = \{3, 4, 7, 8, 11, 12, 15, 16, 19, 20, ...\}$
by defining two properties P_1, P_2, such that $S_i = \{x \in \mathbb{N} : P_i(x)\}$.

EXERCISE 1.5 Let $f : A \to B, g : B \to C$ be functions. Give proofs for the correct statements below and counter-examples for the incorrect ones:

(1) If both f and g are injective, then so is $f ; g$.
(2) If both f and g are surjective, then so is $f ; g$.
(3) If $f ; g$ is surjective, then so is f.
(4) If $f ; g$ is surjective, then so is g.
(5) If $f ; g$ is injective, then so is f.
(6) If $f ; g$ is injective, then so is g.

EXERCISE 1.6 Consider the following two relations on the set $\{a, b, c\}$:
$\quad R = \{\langle a, a \rangle, \langle a, b \rangle, \langle b, b \rangle, \langle b, a \rangle, \langle c, c \rangle\}$ and $S = \{\langle a, b \rangle, \langle a, c \rangle, \langle a, b \rangle\}$.
Say, for R and S, which properties from Definition 1.14 it has.

EXERCISE 1.7 Two sets A, B are disjoint if $A \cap B = \varnothing$. Is the relation $\sim$, defined by $A \sim B$ iff A and B are disjoint, an equivalence relation?

EXERCISE 1.8 Let P, Q be equivalence relations on some set S.

(1) Show that $P \cap Q$ is an equivalence relation on S.
(2) Show that $P \cup Q$ need not be an equivalence relation on S.

EXERCISE 1.9 Given a QO $\langle S, \leq \rangle$, let $[s] \stackrel{\text{def}}{=} \{t \in S \mid s \leq t \text{ and } t \leq s\}$ for each $s \in S$ and $[S] \stackrel{\text{def}}{=} \{[s] \mid s \in S\}$.

(1) Show that for every $s \in S$, $\langle [s], \leq \rangle$ is an equivalence.
(2) Define $\langle [S], \leq' \rangle$ by $[s] \leq' [t]$ iff $s \leq t$. Show that $\leq'$ is a well-defined (for all $s_1, s_2 \in [s]$ and $t_1, t_2 \in [t] : s_1 \leq t_1$ iff $s_2 \leq t_2$) wPO on $[S]$.

EXERCISE 1.10 Prove the claims cited below Definition 1.14, that

(1) Every sPO (irreflexive, transitive relation) is asymmetric.
(2) Every asymmetric relation is antisymmetric – and irreflexive.
(3) If R is connected, symmetric and reflexive, then $R(s,t)$ for every pair s, t. What about a relation that is connected, symmetric and transitive? In what way does this depend on the cardinality of S?

EXERCISE 1.11 Define the weak version of the sPO $\prec_Q$ from Example 1.17.(1). Make sure that the defined relation is a partial order.

EXERCISE 1.12 Let lec be defined on sets by: $lec(S,T)$ iff $|S| \leq |T|$ (Definition 1.21). Which type of ordering (from Definition 1.15) is it?

EXERCISE 1.13 Assuming $\prec \subseteq \Sigma \times \Sigma$ to be a total ordering of a finite alphabet Σ, show that its lexicographic extension, as defined in Example 1.17.(3), is also a total order on Σ^*.

EXERCISE 1.14 For any (non-empty) collection of sets C, show that

(1) the inclusion relation $\subseteq$ is a wPO on C
(2) $\subset$ is its strict version
(3) $\subseteq$ is not (necessarily) a TO on C.

EXERCISE 1.15 If $|S| = n$ for some $n \in \mathbb{N}$, what will be $|\mathcal{P}(S)|$?

EXERCISE 1.16 Let A be a countable set.

(1) If also B is countable, show that:
    (a) the disjoint union $A \uplus B$ is countable (specify its enumeration, assuming the existence of the enumerations of A and B);
    (b) the union $A \cup B$ is countable (specify an injection into $A \uplus B$).
(2) If B is uncountable, can $A \times B$ ever be countable?

EXERCISE 1.17 Let A be a countable set. Show that A has countably many *finite* subsets, proceeding as follows:

(1) Show first that for every $n \in \mathbb{N}$, the set $\mathcal{P}^n(A)$ of finite subsets – with exactly n elements – of A is countable.
(2) Using a technique similar to the one from Example 1.26, show that the union $\bigcup_{n \in \mathbb{N}} \mathcal{P}^n(A)$ of all these sets is countable.

Chapter 2

INDUCTION

2.1 WELL-FOUNDED RELATIONS

——————————————— *a background story* ———————————————

The ancients had many ideas about the basic structure and limits of the world. According to one of them our world – the earth – rested on a huge tortoise. The tortoise itself couldn't just be suspended in a vacuum – it stood on the backs of several elephants. The elephants stood all on a huge disk which, in turn, was perhaps resting on the backs of some camels. And camels? Well, the story obviously had to stop somewhere because, as we notice, one could produce new sub-levels of animals resting on other objects resting on yet other animals, resting on ... indefinitely. The idea is not well founded because such a hierarchy has no well defined beginning. Any attempt to provide the last, the most fundamental level is immediately met with the question "And what is beyond that?"

The same problem of the lacking foundation can be encountered when one tries to think about the beginning of time. When was it? Physicists may say that it was Big Bang. But then one immediately asks "OK, but what was before?". Some early opponents of the Biblical story of creation of the world – and thus, of time as well – asked "What did God do before He created time?". St. Augustine, realising the need for a definite answer which, however, couldn't be given in the same spirit as the question, answered "He kept preparing the hell for those asking such questions."

One should be wary here of the distinction between the beginning and the end, or else, between moving backward and forward. For sure, we imagine that things, the world may continue to exist indefinitely in the future – this idea does not cause much trouble. But our intuition is uneasy with things which do not have any beginning, with chains of events extending indefinitely backwards, whether it is a backward movement along the dimension of time or causality.

Such *non well-founded* chains are hard to imagine and even harder to do anything with – all our thinking, activity, constructions have to start from some basis. Having an idea of a beginning, one will often be able to develop it into a description of the ensuing process. One will typically say: since the beginning was so-and-so, such-and-such had to follow since it is implied by the properties of the beginning. Then, the properties of this second stage, imply some more, and so on. But having nothing to start with, we are left without foundation to perform any intelligible acts.

Mathematics has no problems with chains extending infinitely in both directions. Yet, it has a particular liking for chains which do have a beginning, for relations which are well-founded. As with our intuition and activity otherwise, the possibility of arranging the items in a way which identifies the least, first, starting elements, gives a mathematician a lot of powerful tools. We will study in this chapter some fundamental tools of this kind. Almost all our presentation in the subsequent chapters will be based on well-founded orderings. ___

Definition 2.1 Given a binary relation $< \subseteq S \times S$ and $T \subseteq S$:
- $x \in T$ is a $<$-*minimal element* of T iff there is no element smaller than x, i.e., for all $y \in T$, if $y < x$ then $y = x$.
- $<$ is *well-founded* on S iff each $\varnothing \neq T \subseteq S$ has a minimal element.

The set of natural numbers with the standard ordering $\langle \mathbb{N}, < \rangle$ is well-founded, but its extension to the set of all integers, $\langle \mathbb{Z}, < \rangle$, is not – the subset of negative integers has no $<$-minimal element. Intuitively, well-foundedness means that the ordering has a "basis", minimal "starting points", and can be traversed by going always to the "next" minimal points, "just after" the current ones. This is captured by the following lemma.

Lemma 2.2 $\langle S, < \rangle$ is well-founded iff there is no infinite decreasing sequence, i.e., no $\{a_n\}_{n \in \mathbb{N}} \subseteq S$ such that for all $n : a_n > a_{n+1}$.

Proof. We have to show two implications.

$\Leftarrow$) If $\langle S, < \rangle$ is not well-founded, then let $T \subseteq S$ be a subset without a minimal element. Let $a_1 \in T$ – since it is not minimal, we can find

$a_2 \in T$ such that $a_1 > a_2$. Again, a_2 is not minimal, so we can find $a_3 \in T$ such that $a_2 > a_3$. Continuing this process we obtain an infinite descending sequence $a_1 > a_2 > a_3 > \ldots$

$\Rightarrow$) If there is such a sequence $a_1 > a_2 > a_3 > \ldots$ then, obviously, the set $\{a_n : n \in \mathbb{N}\} \subseteq S$ has no minimal element.   QED (2.2)

A reflexive relation will be typically called well-founded when its irreflexive version is. For instance, for $\leq = \{\langle a, a \rangle, \langle b, b \rangle, \langle c, c \rangle, \langle a, b \rangle, \langle b, a \rangle\}$, the sequence $a \geq a \geq a \ldots$ does not contradict well-foundedness of $\leq$ but $a > b > a > b > \ldots$ does – the set $\{a, b\}$ has no minimal element.

Arbitrary binary relation may happen to be well-founded but, usually, well-foundedness plays its role along some ordering relation.

Example 2.3

Consider the orderings on finite strings as defined in Example 1.17.

(1) The relation $\prec_Q$ is a well-founded sPO; there is no way to construct an infinite sequence of strings with ever decreasing lengths!

(2) The relation $\prec_P$ is a well-founded PO : any subset of strings will contain element(s) such that none of their prefixes (except the strings themselves) are in the set. For instance, a and bc are $\prec_P$-minimal elements in $S = \{ab, abc, a, bcaa, bca, bc\}$.

(3) The relation $\prec_L$ is not well-founded, since there exist infinite descending sequences like
$$\ldots \prec_L \; aaaab \; \prec_L \; aaab \; \prec_L \; aab \; \prec_L \; ab \; \prec_L \; b.$$

In order to construct any such descending sequence, however, there is a need to introduce ever longer strings as we proceed towards infinity. Hence the alternative ordering below is also of interest.

(4) Define $s \prec_{L'} p$ iff $s \prec_Q p$ or ($length(s) = length(p)$ and $s \prec_L p$), i.e., sequences are ordered primarily by length and secondarily by the previous lexicographic order. The ordering $\prec_{L'}$ is indeed well-founded and, in addition, connected, i.e., a well-founded TO.   □

Definition 2.4 A well-ordering, WO, is a well-founded TO.

Notice that well-founded ordering is *not* the same as well-ordering. The former can be a PO which is not a TO. The requirement that a WO $= \langle S, < \rangle$ is a TO implies that each subset of S has not only a minimal element but a *unique* minimal element (called also minimum, or least element).

Example 2.5

The set of natural numbers with the "less than" relation, $\langle \mathbb{N}, < \rangle$, is an sPO. It is also a TO (one of two distinct natural numbers must be smaller than the other) and well-founded (any non-empty set of natural numbers contains a least element). □

Although sets like $\mathbb{N}$ or $\mathbb{Z}$ have the natural orderings, these are not the only possible orderings of these sets. In particular, for a given set S there may be several different ways of imposing a well-founded ordering on it.

Example 2.6

That $<$ fails to be WO on $\mathbb{Z}$ does not mean that $\mathbb{Z}$ cannot be made into a WO. One has to come up with another ordering. For instance, let $|x|$ for an $x \in \mathbb{Z}$ denote the absolute value of x (i.e., $|x| = x$ if $x \geq 0$ and $|x| = -x$ if $x < 0$.) Say that $x \prec y$ if either $|x| < |y|$ or ($|x| = |y|$ and $x < y$), i.e., we order $\mathbb{Z}$ as $0 \prec -1 \prec 1 \prec -2 \prec 2 \ldots$. This $\prec$ is clearly a WO on $\mathbb{Z}$.

Of course, there may be many different WO's on a given set. Another WO on $\mathbb{Z}$ could be obtained by swapping the positive and negative integers with the same absolute values, i.e., $0 \prec' 1 \prec' -1 \prec' 2 \prec' -2 \ldots$. □

2.1.1 INDUCTIVE PROOFS

Well-founded relations play a central role in many contexts because they lead to a particularly convenient proof technique – proof by induction – which we now proceed to study.

———————————— a background story ————————————

A typical problem is to show that all elements of some given set S satisfy some property, call it P, i.e., that for all $x \in S : P(x)$. How one can try to prove such a fact depends on how the set S is described.

A special case is when S is finite and has only few elements – in this case, we can just start proving $P(x)$ for each x separately.

A more common situation is that S has infinitely many elements. Let $S = \{2i : i \in \mathbb{Z}\}$ and show that each $x \in S$ is an even number. Well, this is trivial by the way we have defined the set. Let x be an arbitrary element of S. Then, by definition of S, there is some $i \in \mathbb{Z}$ such that $x = 2i$. But this means precisely that x is an even number and, since x was assumed arbitrary, the claim holds for all $x \in S$.

Of course, in most situations, the relation between the definition of S and the property we want to prove isn't that simple. Then the question arises: How to ensure that we check the property for *all* elements of S and that we can do it in finite time? The idea of proof by induction answers this question in a particular way. It tells us that we have to find some well-founded ordering of the elements of S and then proceed in a prescribed fashion: one shows the statement for the minimal elements and then proceeds to greater elements in the ordering. The trick is that the strategy ensures that only finitely many steps of the proof are needed in order to conclude that the statement holds for *all* elements of S.

Inductive proof is not guaranteed to work in all cases and, particularly, it depends heavily on the choice of the ordering. It is, nevertheless, a very powerful proof technique which will be of crucial importance for all the rest of the material we will study. _____

The most abstract statement of the inductive proof strategy is as follows.

Theorem 2.7 Let $\langle S, < \rangle$ be well-founded and $T \subseteq S$. Assume that: for all $x \in S$: if (for all $y \in S : y < x \rightarrow y \in T$) then $x \in T$. Then $T = S$.

Proof. Assume that T satisfies the condition but $T \neq S$, that is, $S \setminus T \neq \varnothing$. Since S is well-founded, $S \setminus T$ has a minimal element x. Since x is minimal in $S \setminus T$, any $y < x$ must be in T. But then the condition implies $x \in T$. This is a contradiction – we cannot have both $x \in T$ and $x \in S \setminus T$ – showing that $T = S$.   QED (2.7)

This theorem of *induction* underlies the following proof strategy for showing properties of sets equipped with some well-founded relation.

Idea 2.8 [**Inductive proof**] Let $\langle S, < \rangle$ be well-founded and $P \subseteq S$ be a predicate. Suppose we want to prove that every $x \in S$ has the property P – that $P(x)$ holds for all $x \in S$, i.e., that the sets $T = \{x \in S : P(x)\}$ and S are equal. Proceed as follows:

INDUCTION :: Let x be an arbitrary element of S. Assuming that
        for all $y < x : P(y)$ holds,   (IH)
        prove that this implies that also $P(x)$ holds.
CLOSURE :: If you managed to show this, you may conclude $S = T$,
        i.e., $P(x)$ holds for all $x \in S$.

Observe that the hypothesis in the INDUCTION step, called the *induction hypothesis*, IH, allows us to assume $P(y)$ for all $y < x$. Since $\langle S, < \rangle$ is well-founded, there are some elements x, minimal in the whole S, for which no such y exists. For these minimal x's we have no hypothesis and simply have to show that the claim $P(x)$ holds for them without any assumptions. This part of the proof is called the BASIS of induction.

Example 2.9

Consider the natural numbers above 1, i.e. the set $\mathbb{N}_2 = \{n \in \mathbb{N} : n \geq 2\}$. We prove the existence part of the *fundamental theorem of arithmetic*: for each $n \in \mathbb{N}_2 : P(n)$, where $P(n)$ stands for 'n is a product of prime numbers'. (The full theorem states also uniqueness of such a factorization.) First we have to decide which well-founded ordering on $\mathbb{N}_2$ to use – the most obvious first choice is to try the natural ordering $<$, that is, we prove the statement on the well-founded ordering $\langle \mathbb{N}_2, < \rangle$:

BASIS :: Since 2 – the minimal element in $\mathbb{N}_2$ – is a prime number, we have $P(2)$.

IND. :: So let $n > 2$ and assume IH: that $P(k)$ holds for every $k < n$. If n is prime, $P(n)$ holds trivially. So, finally, assume that n is a non-prime number greater than 2. Then $n = x * y$ for some $2 \leq x, y < n$. By IH, $P(x)$ and $P(y)$, i.e., x and y are products of primes. Hence, n is a product of primes.

CLSR. :: So $P(n)$ for all $n \in \mathbb{N}_2$. $\square$

Example 2.10

For any number $x \neq 1$ and any $n \in \mathbb{N}$, we want to show: $1+x+x^2+\ldots+x^n = \frac{x^{n+1}-1}{x-1}$. There are two different sets involved (of x's and of n's), so we first try the easiest way – we attempt induction on the well-founded ordering $\langle \mathbb{N}, < \rangle$, which is simply called "induction on n":

BASIS :: For $n = 0$, we have $1 = \frac{x-1}{x-1} = 1$.

IND. :: Let $n' > 0$ be arbitrary, i.e., $n' = n + 1$. We expand the left hand side of the equality:

$$1 + x + x^2 + \ldots + x^n + x^{n+1} = (1 + x + x^2 + \ldots + x^n) + x^{n+1}$$

$$\text{(by IH since } n < n' = n + 1) = \frac{x^{n+1} - 1}{x - 1} + x^{n+1}$$

$$= \frac{x^{n+1} - 1 + (x - 1)x^{n+1}}{x - 1}$$

$$= \frac{x^{n+1+1} - 1}{x - 1}.$$

Notice that here we have used much weaker IH – the hypothesis that the claim holds for $n = n' - 1$ (implied by IH) is sufficient to establish the induction step. CLOSURE yields the claim *for all $n \in \mathbb{N}$*. $\square$

The proof rule from Idea 2.8 used in the above examples may be written more succinctly as

$$\frac{\forall x(\forall y(y < x \to P(y)) \to P(x))}{\forall x P(x)} \tag{2.11}$$

where "$\forall x$" is short for "for all $x \in \mathbb{N}_2$" (in 2.9), resp. "for all $x \in \mathbb{N}$" (in 2.10) and the horizontal line indicates that the sentence below can be inferred from the sentence above.

Example 2.12
A *convex n-gon* is a polygon with n sides and each interior angle less than $180°$. A triangle is a convex 3-gon and, as you should know from the basic geometry, the sum of the interior angles of a triangle is $180°$. Now, show by induction that the sum of interior angles of any convex n-gon is $(n-2)180°$.

The first question is: induction *on what?* It seems natural to try induction on the number n of sides, i.e., to consider the well-founded ordering $<$ on n-gons in which $X < Y$ iff X has fewer sides than Y.

The basis case is: let X be an arbitrary triangle, i.e., 3-gon. We use the known result that the sum of interior angles of any triangle is indeed $180°$.

For the induction step: let $n > 3$ be arbitrary number and X an arbitrary convex n-gon. Selecting two vertices with one common neighbour vertex between them, we can always divide X into a triangle X_3 and $(n-1)$-gon X_r, as indicated by the dotted line on the drawing below.

X_r has one side less than X so, by induction hypothesis, we have that the sum of its angles is $(n-3)180°$. Also by IH, the sum of angles in X_3 is $180°$. At the same time, the sum of the angles in the whole X is simply the sum of angles in X_3 and X_r. Thus it equals $(n-3)180° + 180° = (n-2)180°$ and the proof is complete. □

The simplicity of the above examples is not only due to the fact that the problems are easy but also that the ordering is easy to identify. In general, however, there may be different orderings on a given set and then the first challenge of an inductive proof is the choice of an appropriate one.

Example 2.13
We want to prove that for all integers $z \in \mathbb{Z}$:
$$\begin{cases} 1 + 3 + 5 + \ldots + (2z - 1) = z^2 & \text{if } z > 0 \\ (-1) + (-3) + (-5) + \ldots + (2z + 1) = -(z^2) & \text{if } z < 0 \end{cases}.$$

We show examples of two proofs using different orderings.

(1) For the first, this looks like two different statements, so we may try to prove them separately for positive and negative integers. Let's do it:

BASIS :: For $z = 1$, we have $1 = 1^2$.
   IND. :: Let $z' > 1$ be arbitrary, i.e., $z' = z + 1$ for some $z > 0$:
$$1 + 3 + \ldots + (2z' - 1) = 1 + 3 + \ldots + (2z - 1) + (2(z+1) - 1)$$
   (by IH since $z < z' = z + 1$) $= z^2 + 2z + 1 = (z+1)^2 = (z')^2$

The proof for $z < 0$ is entirely analogous, but now we have to reverse the ordering: we start with $z = -1$ and proceed along the negative integers considering $z \prec z'$ iff $|z| < |z'|$, where $|z|$ denotes the absolute value of z (i.e., $|z| = -z$ for $z < 0$). For $z, z' < 0$, we thus have that $z \prec z'$ iff $z > z'$.

BASIS :: For $z = -1$, we have $-1 = -(-1)^2$.
   IND. :: Let $z' < -1$ be arbitrary, i.e., $z' = z - 1$ for some $z < 0$.
$$-1 - 3 + \ldots + (2z' + 1) = -1 - 3 + \ldots + (2z + 1) + (2(z-1) + 1)$$
   (by IH since $z \prec z'$) $= -|z|^2 - 2|z| - 1$
$$= -(|z|^2 + 2|z| + 1) = -(|z| + 1)^2$$
$$= -(z - 1)^2 = -(z')^2$$

The second part of the proof makes it clear that the well-founded ordering on the whole $\mathbb{Z}$ we have been using was not the usual $<$. We have, in a sense, ordered both segments of positive and negative numbers independently in the following way (the arrow $x \to y$ indicates the ordering $x \prec y$):

$$1 \longrightarrow 2 \longrightarrow 3 \longrightarrow 4 \longrightarrow 5 \longrightarrow \ldots$$

$$-1 \longrightarrow -2 \longrightarrow -3 \longrightarrow -4 \longrightarrow -5 \longrightarrow \ldots$$

The upper part coincides with the typically used ordering $<$ but the lower one was the matter of more specific choice.

   Notice that the above ordering is different from another, natural one, which orders two integers $z \prec' z'$ iff $|z| < |z'|$. This ordering (shown below) makes, for instance, $3 \prec' -4$ and $-4 \prec' 5$. The ordering above did not relate any pair of positive and negative integers with each other.

(2) The problem can be stated a bit differently. We have to show that

$$\text{for all } z \in \mathbb{Z} : \begin{cases} 1 + 3 + 5 + \ldots + (2z - 1) = z^2 & \text{if } z > 0 \\ -(1 + 3 + 5 + \ldots + (2|z| - 1)) = -(z^2) & \text{if } z < 0 \end{cases}.$$

Thus formulated, it becomes obvious that we only have one statement to prove: we show the first claim (in the same way as we did it in point (1)) and then apply trivial arithmetics to conclude from $x = y$ that also $-x = -y$.

This, indeed, is a smarter way to prove the claim. Is it induction? Yes it is. We prove it first for all positive integers – by induction on the natural ordering of the positive integers. Then, we take an arbitrary negative integer z and observe (assume induction hypothesis!) that we have already proved $1 + 3 + \ldots + (2|z| - 1) = |z|^2$. The well-founded ordering on $\mathbb{Z}$ we are using in this case orders first all positive integers along $<$ (for proving the first part of the claim) and then, for any $z < 0$, puts z after $|z|$ but unrelated to other $n > |z| > 0$ – the induction hypothesis for proving the claim for such a $z < 0$ is, namely, that the claim holds for $|z|$. The ordering is shown on the left:

As a matter of fact, the structure of this proof allows us to view the used ordering in yet another way. We first prove the claim for *all* positive integers. Thus, when proving it for an arbitrary negative integer $z < 0$, we can assume the stronger statement than the one we are actually using, i.e., that the claim holds for *all* positive integers. This ordering puts all negative integers after all positive ones as shown on the right in the figure above.

Note that none of the orderings encountered in this example is total. □

2.2 INDUCTIVE DEFINITIONS

We have introduced the general idea of inductive proof over an *arbitrary* well-founded ordering defined on an arbitrary set. The idea of induction – a kind of stepwise construction of the whole from a "basis" by repetitive applications of given rules – can be applied not only for constructing proofs but also for constructing, that is defining, sets. We now illustrate this technique of *definition by induction* and then (Subsection 2.2.3) show how it gives rise to the possibility of using a special case of the inductive proof strategy – the *structural induction* – on sets defined in this way.

——————————————— a background story ———————————————

Suppose I make a simple statement, for instance, (1) 'John is a nice person'. Its truth may be debatable – some people may think that, on the contrary, he is not nice at all. Pointing this out, they might say – "No, he is not, you only think that he is". So, to make my statement less definite I might instead say (2) 'I think that 'John is a nice person''. In the philosophical tradition one would say that (2) expresses a reflection over (1). But now, (2) is a new statement, and so I can reflect over it again: (3) 'I think that 'I think that 'John is nice'''. It isn't perhaps obvious why I should make this kind of statement, but I certainly can make it and, with some effort, perhaps even attach some meaning to it. Then, I can just continue: (4) 'I think that (3)', (5) 'I think that (4)', etc. The further we go, the less idea we have what one might possibly intend with such expressions. Philosophers used to spend time analysing their possible meaning – the possible meaning of such repetitive acts of reflection over reflection over reflection ... over something. In general, they agree that such an infinite regress does not yield anything intuitively meaningful and should be avoided.

In the daily discourse, we hardly ever carry such a process beyond the level (2) – higher levels do not make any meaningful contribution to a conversation. Yet they are possible for purely linguistic reasons – each statement obtained in this way is grammatically correct. And what is 'this way'? Simply:

BASIS ::   Start with some statement, e.g., (1) 'John is nice'.
STEP ::   Whenever you have produced some statement (n) – at first, it is just (1), but after a few steps, you have some higher statement (n) – you may produce a new statement by prepending (n) with 'I think that ...'. Thus you obtain a new, (n+1), statement 'I think that (n)'.

Anything generated by this rule happens to be grammatically correct and the infinite chain of such statements is called an infinite regress.

The crucial point here is that we do not start with some set which we analyse. We are *defining a new set* – the set of statements $\{(1), (2), (3), \ldots\}$ – in a peculiar way. The idea of induction – stepwise construction from a "basis" – is not applied here for proving properties of a given set but for defining a new one. _____

One describes often sets using abbreviations like $E = \{0, 2, 4, 6, \ldots\}$ or $T = \{1, 4, 7, 10, 13, \ldots\}$. The dots "..." indicate that you are assumed to have figured out what the subsequent elements will be – and that there will be infinitely many of them. It is assumed that you have figured out *the rule* by which to generate all the elements. The same sets may be defined more precisely using the respective rule explicitly:

$$E = \{2 * n : n \in \mathbb{N}\} \text{ and } T = \{3 * n + 1 : n \in \mathbb{N}\}. \tag{2.14}$$

Another way to describe these rules is as follows. The set E is defined by:

BASIS :: $0 \in E$ and,
  STEP :: whenever an $x \in E$, then also $x + 2 \in E$.
CLSR. :: Nothing else belongs to E.

The other set is defined similarly:

BASIS :: $1 \in T$ and,
  STEP :: whenever an $x \in T$, then also $x + 3 \in T$.
CLSR. :: Nothing else belongs to T.

Here we are not so much defining the whole set by one static formula, as we did in (2.14), but are specifying the rules for *generating new* elements from some elements which we already have included in the set. Not all formulae allow equivalent formulation in terms of such generation rules. Yet, many sets of interest can be given by means of such generation rules, that is, by an *inductive definition*. They will play a central role in the whole book.

Idea 2.15 An *inductive definition* of a set S consists of

BASIS :: a list of some (at least one) elements $B \subseteq S$,
  IND. :: one or more rules to construct new elements of S from already existing elements,
CLSR. :: the statement that S contains only the elements obtained from the basis by the induction steps.

  Saying that something is defined by induction, the closure condition becomes implicitly assumed and is not stated.

Example 2.16
The finite strings Σ^* over Σ, Example 1.17, can be defined inductively, staring with the empty string, ϵ, i.e., the string of length 0, as follows:

BASIS :: $\epsilon \in \Sigma^*$
  IND. :: if $s \in \Sigma^*$ then $xs \in \Sigma^*$ for all $x \in \Sigma$

Constructors are the empty string ϵ and the operations prepending an element in front of a string x_-, for all $x \in \Sigma$. Notice that 1-element strings like x will be here represented as $x\epsilon$. □

Example 2.17

The finite non-empty strings Σ^+ over alphabet Σ are defined by starting with a different basis.

BASIS :: $x \in \Sigma^+$ for all $x \in \Sigma$
  IND. :: if $s \in \Sigma^+$ then $xs \in \Sigma^+$ for all $x \in \Sigma$ □

Often, one is not interested in all strings but only in their subset. Such subsets are called *languages* and, typically, are defined by induction.

Example 2.18

Define the set of strings $\mathbb{N}$ over $\Sigma = \{0, s\}$:

BASIS :: $0 \in \mathbb{N}$
  IND. :: If $n \in \mathbb{N}$ then $sn \in \mathbb{N}$

This language is the basis of the formal definition of natural numbers. The constructors are 0 and the operation of prepending the symbol 's' to the left. (The 's' signifies the "successor" function corresponding to $n + 1$.) Notice that we do not obtain the set $\{0, 1, 2, 3 \ldots\}$ but $\{0, s0, ss0, sss0 \ldots\}$, which is a kind of unary representation of natural numbers. For instance, the strings $00s, s0s0 \notin \mathbb{N}$, i.e., $\mathbb{N} \neq \Sigma^*$.   □

Example 2.19

(1) Let $\Sigma = \{a, b\}$ and let us define the language $L \subseteq \Sigma^*$ containing only the strings starting with a number of a's followed by the equal number of b's, i.e., $\{a^n b^n : n \in \mathbb{N}\}$.

    BASIS :: $\epsilon \in L$
     IND. :: if $s \in L$ then $asb \in L$

    Constructors of L are ϵ and the operation adding an a in the beginning and a b at the end of a string $s \in L$.

(2) Here is a more complicated language over $\Sigma = \{a, b, c, (,), \neg, \rightarrow\}$ with two rules of generation.

    BASIS :: $a, b, c \in L$
     IND. :: if $s \in L$ then $\neg s \in L$
            if $s, r \in L$ then $(s \rightarrow r) \in L$

    By the closure, we have, for instance, '(' $\notin L$ and $(\neg b) \notin L$.   □

In Section 2.1, e.g., Example 2.13, we saw that a given set may be *endowed* with various well-founded orderings. Such an ordering enables us to use the powerful technique of proof by induction according to Theorem 2.7. The usefulness of inductive *definitions* is related to the fact that such an ordering may be obtained for free – the resulting set obtains implicitly a well-founded ordering induced by its very definition as follows.[1]

Idea 2.20 [**Induced wf order**] For an inductively defined set S, define a function $f : S \to \mathbb{N}$ as follows:

BASIS :: Let $S_0 = B$ and for all $b \in S_0 : f(b) \overset{\text{def}}{=} 0$.
  IND. :: Given S_i, let S_{i+1} be the union of S_i and all the elements $x \in S \setminus S_i$ which can be obtained according to one of the rules from some elements $y_1, \ldots, y_n$ of S_i. For each such new $x \in S_{i+1} \setminus S_i$, let $f(x) \overset{\text{def}}{=} i + 1$.
CLSR. :: The actual ordering is then $x \prec y$ iff $f(x) < f(y)$.

The function f is essentially counting the minimal number of steps – consecutive applications of the rules allowed by the induction step of Definition 2.15 – needed to obtain a given element of the inductive set.

Example 2.21
Refer to Example 2.16. Since the induction step amounts there to increasing the length of a string by 1, following the above idea, we would obtain the ordering on strings $s \prec p$ iff $length(s) < length(p)$. □

2.2.1 "1-1" Definitions

A common feature of the above examples is the impossibility of deriving an element in more than one way. In Example 2.17, for instance, the only way to derive abc is to start with c and then add b and a to the left in sequence. One apparently tiny modification changes this state of affairs:

Example 2.22
The finite non-empty strings Σ^+ can also be defined inductively as follows.

BASIS :: $x \in \Sigma^+$ for all $x \in \Sigma$
  IND. :: if $s \in \Sigma^+$ and $p \in \Sigma^+$ then $sp \in \Sigma^+$ □

According to this example, abc can be derived by concatenating either a and bc, or ab and c. We often say that the former definitions are 1-1, while

---

[1] In fact, an inductive definition imposes at least two such orderings of interest, but here we consider just one.

the latter is not. Given a 1-1 inductive definition of a set S, there is an easy way to define new functions on S – again by induction.

Idea 2.23 [**Inductive function definition**] Let S be defined inductively from basis B and some construction rules. To define a function f on elements of S do the following:

BASIS :: Specify the value $f(x)$ for each x in B.
  IND. :: For each way an $x \in S$ can be obtained from some $y_1 \ldots y_n \in S$,
          specify how to obtain $f(x)$ from the values $f(y_1) \ldots f(y_n)$.
CLSR. :: The closure of S ensures that f is defined for all $x \in S$.

When the set's definition is 1-1, there is only one way to obtain each x, and this guarantees consistency of such a function definition. The next few examples illustrate this case.

Example 2.24
We define the *length* and the concatenation functions of finite strings by induction on the definition in Example 2.16 as follows:

BASIS :: $length(\epsilon) = 0$ BASIS :: $\epsilon \cdot t = t$
  IND. :: $length(xs) = length(s) + 1$ IND. :: $xs \cdot t = x(s \cdot t)$ □

Example 2.25
In Example 2.16 strings were defined by a left append (prepend) operation which we wrote as juxtaposition xs. A corresponding right append operation $\vdash$, adding an element at the end of a string, can be defined as shown to the left. The operation reversing a string is then defined to the right:

  BASIS :: $\epsilon \vdash y = y\epsilon$ BASIS :: $\epsilon^R = \epsilon$
    IND. :: $xs \vdash y = x(s \vdash y)$ IND. :: $(xs)^R = s^R \vdash x$ □

The operation $\vdash$ does not quite fit the format of Idea 2.23, taking two arguments – a symbol and a string. It is possible to give a more general version that covers such cases as well, but we shall not do so here. The definition below also apparently goes beyond the Idea 2.23, but in order to make it fit we merely have to think of addition, for instance in $m + n$, as an application of the one-argument function *add* n to the argument m.

Example 2.26
Using the definition of $\mathbb{N}$ from Example 2.18, we can define the plus operation for all $n, m \in \mathbb{N}$:

BASIS :: $0 + n = n$
IND. :: $s(m) + n = s(m + n)$

It is not obvious that this is the usual addition. For instance, does it hold that $n + m = m + n$? We shall verify this in an exercise.

We can use this definition to calculate the sum of two arbitrary natural numbers represented as elements of $\mathbb{N}$. For instance, $2 + 3$ would be processed as follows:

$$ss0 + sss0 \mapsto s(s0 + sss0) \mapsto ss(0 + sss0) \mapsto ss(sss0) = sssss0. \qquad \Box$$

Note that Idea 2.23 of inductive function definition is guaranteed to work only when the set is given by a 1-1 definition. Imagine that we tried to define a version of the length function in Example 2.24 by induction on the definition in Example 2.22 as follows: $len(x) = 1$ for $x \in \Sigma$, while $len(ps) = len(p) + 1$. This would give different (and hence mutually contradictive) values for $len(abc)$, depending on which way we choose to derive abc. Idea 2.23 can be applied also when the definition of the set is not 1-1, but then one has to ensure that the function value for every element is independent from the way it is derived. In the present case, the two equations $len(x) = 1$ and $len(ps) = len(p) + len(s)$ provide a working definition, but it takes some reasoning to check that this is indeed the case.

2.2.2: RECURSIVE PROGRAMMING .... [optional]
If you are not familiar with the basics of programming you may skip this subsection and go directly to Subsection 2.2.3. No new concepts are described here but merely relation between inductive definitions and recursive programming.

All basic data structures in computer science are defined inductively – any instance of a List, Stack, Tree, etc., is generated in finitely many steps by applying some basic constructor operations. These operations themselves may vary from one programming language to another, or from one application to another, but they always capture the inductive structure of these data types. We give here but two simple examples which illustrate the inductive nature of two basic data structures and show how this leads to the elegant technique of recursive programming.

1. Lists
A List (to simplify matters, we assume that we store only integers) is a sequence of 0 or more integers. The idea of a sequence or, more generally, of a linked structure is captured by pointers between objects storing data. Thus, one would define objects of the form shown to the left

```
List
    int x;
    List next;
```

The list $\langle 3, 7, 5 \rangle$ would contain 3 `List` objects, plus the additional `null` object at the end, as shown to the right. The declaration tells us that a `List` is:

(1) either a `null` object (the default value for pointers), or

(2) an integer (stored in the current `List` object) followed by another `List` object.

But this is exactly an inductive definition, namely, the one from Example 2.16, the only difference being that of the language used.

1a. From inductive definition to recursion

The above is also a 1-1 definition and thus gives rise to natural recursive programming over lists. The idea of recursive programming is to traverse a structure in the order opposite to the way we imagine it built along its inductive definition. We start at some point of the structure and proceed to its subparts until we reach the basis case. For instance, the function computing length of a list is programmed recursively to the left:

```
int length(List L)            int sum(List L)
  IF (L is null) return 0;      IF (L is null) return 0;
  ELSE return 1+length(L.next);  ELSE return L.x+sum(L.next);
```

It should be easy to see that the pseudo-code on the left is nothing more than the inductive definition of the function from Example 2.24. Instead of the mathematical formulation used there, it uses the operational language of programming to specify:

(1) the value of the function in the basis case (which also terminates the recursion) and then

(2) the way to compute the value in the non-basis case from the value for some subcase which brings recursion "closer to" the basis.

The same schema is applied in the function to the right which computes the sum of all integers in the list.

Abstractly, `List`s can be viewed as finite strings. You may rewrite the earlier definitions, e.g., concatenation from Example 2.24, for `List`s as represented here.

1b. Equality of lists

Inductive 1-1 definition of a set (here, of `List` objects) gives also rise to the obvious recursive function checking equality. Two lists are equal iff

(1) they have the same structure (i.e., the same number of elements) and

(2) respective elements in both lists are equal.

The corresponding function checks, in lines 1-2, point (1) and ensures termination upon reaching the basis case. Line 3 checks point (2) and, if elements are equal, recursion proceeds to check the rest of the lists in line 4:

```
boolean equal(List L, R)
  IF (L is null AND R is null) return TRUE;
  ELSE IF (L is null OR R is null) return FALSE;
  ELSE IF (L.x ≠ R.x) return FALSE;
  ELSE return equal(L.next, R.next);
```

2. Trees

Another very common data structure is binary tree BT. (Again, we simplify presentation by assuming that we only store integers as data.) Unlike in a list, each node (except the null ones) has two successors (called "children"), left and right:

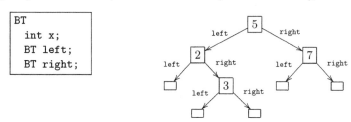

The inductive definition says that a binary tree BT is

(1) either a null object, or

(2) an integer (stored in the current BT object) with two pointers (left and right) to other (always distinct) BT objects.

2a. From inductive definition to recursion

To compute the sum of integers stored in a given tree, we have to compute the sums stored in its left and right subtrees and add to them the value stored at the current node itself. Similarly, the height of the tree (the length of its longest branch) is computed from the heights of its subtrees. The recursive functions reflecting the inductive definition are:

```
int sum(BT T)
  IF (T is null) return 0;
  ELSE return (sum(T.left) + T.x + sum(T.right));

int height(BT T)
  IF (T is null) return 0;
  ELSE return 1 + max(height(T.left),height(T.right));
```

In both functions, the first line detects the basis case, while the second one computes the non-basic case, descending recursively down the tree structure.

2b. Equality of binary trees

Using the *operational* intuition of inductive generation might suggest that the definition of binary tree is not 1-1. The tree from the example drawing could be generated in various orders, first the left subtree and then the right one, or other way around, and so on down the tree.

However, this operational intuition, relaying on the *temporal ordering* of construction steps is not what matters for a definition to be 1-1. There is *only one logical* way to obtain this tree: we must have *both* its left *and* its right subtree, and *only then* we can make this tree. That is, there are *unique* elements (left and right subtree, and the integer to be stored in the root) and the *unique* rule to be applied (put left subtree to the left, right to the right, and the integer in the node). The definition is 1-1.

Equality of binary trees follows naturally: two trees are equal iff
    (1) they have the same structure and
    (2) respective elements stored at respective nodes in both trees are equal.
Having the same structure amounts here to the following condition: trees $T1$ and
$T2$ have the same structure iff:
    (1a) both are `null` or
    (1b) $T1 = (L1, x1, R1)$, $T2 = (L2, x2, R2)$, i.e., neither is `null`, and *both*
            $L1, L2$ have the same structure *and* $R1, R2$ have the same structure.
This is clearly an inductive definition of 'having the same structure', giving us
the recursive pseudo-code for checking equality of two binary trees. In addition
to the equality of structures, the function ensures also, in line 3, that elements x,
stored at respective nodes in both trees, are equal:

```
boolean equal(BT T1,T2)
   IF (T1 is null AND T2 is null) return TRUE;
   ELSE IF (T1 is null OR T2 is null) return FALSE;
   ELSE IF (T1.x ≠ T2.x) return FALSE;
   ELSE return ( equal(T1.left, T2.left) AND
                         equal(T1.right,T2.right) );
```

Ending now this programming excursion, we return to the proof strategy arising
from inductive definitions of sets.................................[end optional]

2.2.3   Proofs by structural induction

Since, according to Idea 2.20, an inductive definition of a set induces a well-
founded ordering, it allows us to perform inductive proofs of the properties
of this set. This is called proof by *structural induction* – the word "structural" indicating that the proof, so to speak, proceeds along the structure
of the set imposed by its inductive definition. In contrast to the definitions
of functions on inductive sets from Idea 2.23, this proof strategy works for
all inductively defined sets, regardless of whether the definitions are 1-1.

    Proof by structural induction is just a special case of the inductive proof
Idea 2.8. Simply, because any inductive definition of a set induces a well-
founded ordering on this set according to the Idea 2.20. Using this ordering
leads to the following proof strategy:

Idea 2.27 [Proof by structural induction] Suppose that, given a set
S defined inductively from basis B, we want to prove that each element $x \in S$
has the property P – that $P(x)$ holds for all $x \in S$. Proceed as follows:

Basis :: Show that $P(x)$ holds for all $x \in B$.
  Ind. :: For each way an $x \in S$ can be constructed from one or
           more $y_1, \ldots, y_n \in S$, show that the induction hypothesis
           $P(y_1), \ldots, P(y_n)$ implies $P(x)$.

CLSR. :: If you managed to show this, then the closure property of S allows you to conclude that $P(x)$ holds for all $x \in S$.

It is straightforward to infer the proof rule in Idea 2.27 from the one in Idea 2.8. We assume that Idea 2.8 holds. Then, assuming BASIS and INDUCTION step of 2.27, we merely have to prove the INDUCTION step in Idea 2.8. So let x be an arbitrary member of S and assume the IH that $P(y)$ holds for all $y \prec x$. There are two possible cases: Either $f(x) = 0$, in which case $x \in S_0 = B$, and $P(x)$ follows from the BASIS part of Idea 2.27. Otherwise $x \in S_{i+1} \setminus S_i$ for some $i \in \mathbb{N}$. Then x can be obtained from some $y_1, \ldots, y_n \in S_i$. Since these are all less than x in the sense of $\prec$, by the IH $P(y_1)$ and $\ldots$ and $P(y_n)$. But then $P(x)$ follows from the INDUCTION part of Idea 2.27, and the argument is complete.

The great advantage of inductively defined sets is that, in order to prove their properties, one need not know the details of the induced ordering but merely has to follow the steps of the definition, as given in Idea 2.27.

Example 2.28
The set of natural numbers was defined inductively in Example 2.18. The induced well-founded ordering will say that $s^n 0 \prec s^m 0$ iff $n < m$. Thus, writing the natural numbers in the usual way $0, 1, 2, 3, \ldots$, and replacing sn by $n + 1$, the induced ordering is the standard ordering $< -$ the structural induction on this set will be the usual mathematical induction. That is:

BASIS :: Show that $P(0)$ holds.

IND. :: Assuming the induction hypothesis $P(n)$, show that it implies also $P(n+1)$.

It is the most common form of induction used in mathematics. The proof in Example 2.10 used this form, observed there as a "weaker induction hypothesis" (than the one allowed by the general formulation from Idea 2.8). Using structural induction, we show that for all $n \in \mathbb{N}$:

$$1 + 2 + \ldots + n = \frac{n(n+1)}{2}.$$

BASIS :: $n = 0 : 0 = \dfrac{0(0+1)}{2} = 0$

IND. :: Assume the IH: $1 + 2 + \ldots + n = \dfrac{n(n+1)}{2}$. Then

$$1 + \ldots + n + (n+1) = \frac{n(n+1)}{2} + (n+1) = \frac{(n+1)(n+2)}{2}. \quad \Box$$

The proof rule for natural numbers used above can be written more succinctly as follows.

$$\frac{P(0) \ \& \ \forall x(P(x) \to P(x+1))}{\forall x P(x)}. \tag{2.29}$$

This rule can be sometimes more cumbersome to use than the rule (2.11). To see this, try to use it to prove the prime number theorem which was shown in Example 2.9 using the general induction schema for natural numbers (i.e., the rule (2.11)). The difference between the two concerns, actually, only the "easiness" of carrying out the proof – as you are asked to show in Exercise 2.23, the two proof rules have the same power.

Example 2.30
We show that concatenation from Example 2.24 is associative, i.e., for all strings $s \cdot (t \cdot p) = (s \cdot t) \cdot p$. We proceed by structural induction on the first argument:

BASIS :: $\epsilon \cdot (t \cdot p) = t \cdot p = (\epsilon \cdot t) \cdot p$
IND. :: $xs \cdot (t \cdot p) = x(s \cdot (t \cdot p)) \overset{\text{IH}}{=} x((s \cdot t) \cdot p) = (x(s \cdot t)) \cdot p = (xs \cdot t) \cdot p.$ □

Example 2.31
Define the set U inductively:

BASIS :: $\varnothing \in U$
IND. :: if $S \in U$ then $\mathcal{P}(S) \in U$

and show that for all $S \in U : S \notin S$. What is the ordering $\prec$ induced on the set U by this definition? Well, $\varnothing$ is the least element and then, for any set S we have that $S < \mathcal{P}(S)$. $\prec$ is the transitive closure of $<$. The nice thing about structural induction is that one actually need not have a full insight into the structure of the induced ordering but merely has to follow the inductive definition of the set.

BASIS :: Since for all $X : X \notin \varnothing$ so, in particular, $\varnothing \notin \varnothing$.
IND. :: Assume IH : $S \notin S$ and, contrapositively, $\mathcal{P}(S) \in \mathcal{P}(S)$, i.e., $\mathcal{P}(S) \subseteq S$. We also have $S \in \mathcal{P}(S)$. But since $X \subseteq Y$ iff for all $x : x \in X \Rightarrow x \in Y$, we thus obtain that $\mathcal{P}(S) \subseteq S$ and $S \in \mathcal{P}(S)$ imply $S \in S$, contradicting IH. □

Example 2.32
Show that the set L, Example 2.19.(1), equals the set $T = \{a^n b^n : n \in \mathbb{N}\}$.
(1) We show first the inclusion $L \subseteq T$ by structural induction on L.

BASIS :: $\epsilon = a^0 b^0 \in T$.

IND. :: Assume that $s \in L$ satisfies the property, i.e., $s = a^n b^n$ for some $n \in \mathbb{N}$. From this s, the rule will produce $asb = aa^n b^n b = a^{n+1} b^{n+1} \in T$. Hence all the strings of L have the required form and $L \subseteq \{a^n b^n : n \in \mathbb{N}\}$.

Notice again that the precise nature of the induced ordering is inessential for this proof, which merely follows the steps of the inductive definition of L. (This induced ordering $\prec$ is given by: $a^n b^n \prec a^m b^m$ iff $n < m$.)

(2) On the other hand, each element of $\{a^n b^n : n \in \mathbb{N}\}$ can be generated in n steps applying the L's construction rule. This is shown by induction on n, that is, on the natural ordering $\langle \mathbb{N}, < \rangle$:

BASIS :: For $n = 0$, we have $a^0 b^0 = \epsilon \in L$.
   IND. :: If IH : $a^n b^n \in L$, then $aa^n b^n b = a^{n+1} b^{n+1} \in L$ by the induction step of definition of L.

Hence $\{a^n b^n : n \in \mathbb{N}\} \subseteq L$. $\square$

Example 2.33
Recall the language L defined in Example 2.19.(2). Define its subset S:

BASIS :: $B = \{a, b, \neg c, (a \to (b \to c)), (c \to b)\} \subset S$
  IND. :: if $s \in S$ and $(s \to t) \in S$ then $t \in S$
         if $\neg t \in S$ and $(s \to t) \in S$ then $\neg s \in S$

We show that $S = B \cup \{(b \to c), \neg b, c\}$ constructing S step by step:

$S_0 = B = \{a, b, \neg c, (a \to (b \to c)), (c \to b)\}$
$S_1 = S_0 \cup \{(b \to c)\}$, with a for s and $(b \to c)$ for t in the first rule
    $= \{a, b, \neg c, (a \to (b \to c)), (c \to b), (b \to c)\}$
$S_2 = S_1 \cup \{\neg b, c\}$ taking b for s and c for t and applying both rules
    $= \{a, b, \neg c, (a \to (b \to c)), (c \to b), (b \to c), \neg b, c\}$
$\;\; S = $ Every application of a rule to some new combination of strings now yields a string which is already in S_2. Since no new strings can be produced, the process stops here and we obtain $S = S_2$. $\square$

The last example illustrates some more intricate points which often may appear in proofs by structural induction.

Example 2.34
Recall the definition of finite non-empty strings Σ^+ from Example 2.17. For the same alphabet Σ, define the set Σ' inductively as follows:

BASIS :: $x \in \Sigma'$ for all $x \in \Sigma$

IND-A :: if $x \in \Sigma$ and $s \in \Sigma'$ then $xs \in \Sigma'$

IND-B :: if $x \in \Sigma$ and $s \in \Sigma'$ then $sx \in \Sigma'$.

This definition allows to generate most finite non-empty strings in various ways – it is not 1-1. For instance, abc may be obtained from bc by prepending a according to rule A, or from ab by appending c according to rule B. To make sure that these operations yield the same result, we should augment the above definition with the equation:

$$x(sy) = (xs)y. \tag{2.35}$$

Intuitively, $\Sigma^+ = \Sigma'$ but this needs a proof. We show two inclusions.

$\Sigma^+ \subseteq \Sigma'$. This inclusion is trivial since anything which can be generated according to definition of Σ^+ can be generated by the same process following definition of Σ'. (Strictly speaking, we show it by structural induction following the definition of Σ^+: every element of its basis is also in the basis of Σ'; and whatever can be obtained by the rule from the definition of Σ^+ can be obtained by the first rule from the definition of Σ'.)

$\Sigma' \subseteq \Sigma^+$, i.e., for every $s : s \in \Sigma' \to s \in \Sigma^+$. We show the claim by structural induction on s, i.e., on the definition of Σ'.

BASIS :: if $x \in \Sigma$, then $x \in \Sigma'$ and also $x \in \Sigma^+$.

IH :: Assume the claim for s, i.e., $s \in \Sigma' \to s \in \Sigma^+$. According to Idea 2.27, we have to show now the claim *for each way* a "new" element may be obtained. The two different rules in the inductive definition of Σ', give two cases to be considered.

IND-A :: $xs \in \Sigma'$ – by IH, $s \in \Sigma^+$, and so by the induction step of the definition of $\Sigma^+ : xs \in \Sigma^+$.

IND-B :: $sx \in \Sigma'$ ...? There does not seem to be any helpful information to show that then $sx \in \Sigma^+$...

Let us therefore try *a new level* of induction, now for showing this particular case. We have the assumption of IH above, and proceed by sub-induction, again on s:

BASIS2 :: $s = y \in \Sigma$ and $x \in \Sigma$ – then $xy \in \Sigma^+$, by the induction step of its definition.

IH2 :: Basis has been shown for arbitrary x and $y = s$, and so we may assume that: for every $x \in \Sigma$ and for $s \in \Sigma' : sx \in \Sigma^+$. We have again two cases for s:

IND2-A :: $ys \in \Sigma'$, and $(ys)x \in \Sigma'$ – then $(ys)x \overset{(2.35)}{=} y(sx)$ and $sx \in \Sigma^+$ by IH2. This suffices to conclude $y(sx) \in \Sigma^+$.

IND2-B :: $sy \in \Sigma'$, and $(sy)x \in \Sigma'$ – by IH2 we have that $sy \in \Sigma^+$, but what can we then do with the whole term $(sy)x$? That $sy \in \Sigma^+$ means, by the definition of Σ^+, that sy can be actually written as zt for some $z \in \Sigma$ and $t \in \Sigma^+$, i.e., $sy = zt$. Can we conclude that $(sy)x = (zt)x \overset{(2.35)}{=} z(tx) \in \Sigma^+$? Not really, because this would require $tx \in \Sigma^+$, and that we do not know. We might, perhaps, try yet another level of induction – now on t – but this should start looking suspect.

What we know about tx is that i) $tx \in \Sigma'$ (since $zt \in \Sigma'$ so $t \in \Sigma'$) and that ii) $length(tx) = length(sy)$. If IH2 could be assumed for all such tx (and not only for the particular sy from which the actual $(sy)x$ is built), we would be done. And, as a matter of fact, we *are allowed to assume just that* in our (sub-)induction hypothesis IH2. Why? Because proving our induction step for the term of the form $(sy)x$ (or $(ys)x$) we can assume the claim for *all terms which are smaller* in the actual ordering. This is the same as the strong version of the induction principle, like the one used in Example 2.9, as compared to the weaker version used, for instance, in Example 2.10. The actual formulation of Idea 2.27 allows us only the weaker version of induction hypothesis – namely, the assumption of the claim only for the *immediate* predecessors of the element for which we are proving the claim in the induction step. However, if we inspect the actual ordering induced by the inductive definition according to Idea 2.20 and apply the general Theorem 2.7, then we see that also this stronger version will be correct. □

We will encounter many examples of inductive proofs in the rest of the book. In fact, the most important sets we will consider will be defined inductively and most proofs of their properties will use structural induction.

2.3: TRANSFINITE INDUCTION .... [optional]

As a final example, we give an inductive definition of ordinals (introduced by Cantor and called also "ordinal numbers" though, as a matter of fact, they are sets), and show inductively some properties of this set. The example shows that induction is by no means restricted to finitely generated elements. It can be carried over to sets of arbitrary cardinality. Although the principle is the same as before, emphasising the context, one calls it "transfinite induction".

Define the collection of ordinals, $\mathbb{O}$, inductively as follows:

BASIS :: $\varnothing \in \mathbb{O}$
IND. :: 1) if $x \in \mathbb{O}$ then $x^+ = x \cup \{x\} \in \mathbb{O}$
        2) for any index set I, if for all $i \in I : x_i \in \mathbb{O}$ then $\bigcup_{i \in I} x_i \in \mathbb{O}$

1) says that the "successor" of an ordinal is an ordinal while 2) that an arbitrary

(possibly infinite) union of ordinals is an ordinal. A small initial segment of $\mathbb{O}$ is shown below, with the standard, abbreviated notation indicated in the right column.

$$
\omega + \omega \left\{
\begin{array}{ll}
\omega \left\{
\begin{array}{ll}
\varnothing & 0 \\
\{\varnothing\} & 1 \\
\{\varnothing, \{\varnothing\}\} & 2 \\
\{\varnothing, \{\varnothing\}, \{\varnothing, \{\varnothing\}\}\} & 3 \\
\vdots & \vdots
\end{array}
\right. \\
\bigcup_{x<\omega} x & \omega \\
\{\omega, \{\omega\}\} & \omega + 1 \\
\{\omega, \{\omega\}, \{\omega, \{\omega\}\}\} & \omega + 2 \\
\vdots & \vdots
\end{array}
\right.
$$

$$
\begin{array}{ll}
2\omega & 2\omega \\
\{2\omega, \{2\omega\}\} & 2\omega + 1 \\
\vdots & \vdots
\end{array}
$$

The rule 2) allows us to form infinite ordinals by taking unions of ordinals. This gives the possibility of forming special kind of *limit* ordinals, which have no immediate predecessor. ω is the first such limit ordinal (the union of all finite ordinals), $\omega + \omega = 2\omega$ the second one, 3ω the third one, etc. – the sequence of ordinals continues indefinitely:

$$0, 1, 2, 3, 4, 5, \ldots \omega, \omega + 1, \omega + 2, \ldots \omega + \omega =$$
$$2\omega, 2\omega + 1, 2\omega + 2, \ldots 3\omega, 3\omega + 1, \ldots 4\omega, 4\omega + 1, \ldots \omega * \omega =$$
$$\omega^2, \omega^2 + 1, \ldots \omega^3, \omega^3 + 1, \ldots \omega^4, \omega^4 + 1, \ldots$$
$$\omega^\omega, \omega^\omega + 1, \ldots \omega^{2\omega}, \ldots \omega^{3\omega}, \ldots \omega^{\omega * \omega}, \ldots \omega^{(\omega^3)}, \ldots \omega^{(\omega^\omega)}, \ldots$$

We show a few simple facts about $\mathbb{O}$ using structural induction.

(A) For all $x \in \mathbb{O} : y \in x \Rightarrow y \in \mathbb{O}$

BASIS :: Since there is no $y \in \varnothing$, this follows trivially.
    IND :: Let $x \in \mathbb{O}$ and 1) assume IH : $y \in x \Rightarrow y \in \mathbb{O}$. If $y \in x \cup \{x\}$ then either $y \in x$ and by IH $y \in \mathbb{O}$, or else $y = x$ but $x \in \mathbb{O}$.
           2) IH : for all $x_i : y \in x_i \Rightarrow y \in \mathbb{O}$. If $y \in \bigcup x_i$, then there is some x_k for which $y \in x_k$. Then, by IH, $y \in \mathbb{O}$.

(B) For all $x \in \mathbb{O} : y \in x \Rightarrow y \subset x$.

BASIS :: $x = \varnothing$, and as there is no $y \in \varnothing$, the claim follows trivially.
    IND :: 1) From IH : $y \in x \Rightarrow y \subset x$ show that $y \in x^+ \Rightarrow y \subset x^+$. $y \in x^+$ means that either a) $y \in x$, from which, by IH, we get $y \subset x$ and thus $y \subset x \cup \{x\}$, or b) $y = x$, and then $y \subseteq x$ but $y \neq x \cup \{x\}$, i.e., $y \subset x \cup \{x\}$.
           2) IH : for all $x_i : y \in x_i \Rightarrow y \subset x_i$. If $y \in \bigcup x_i$, then there is some $k : y \in x_k$ and, by IH, $y \subset x_k$. But then $y \subset \bigcup x_i$.

Property (A) says that every member of an ordinal is itself an ordinal, while (B) that every ordinal x is a *transitive set*, in the sense that its members are also its subsets, i.e., a member of a member of an ordinal x is also a member of x. (An ordinal can be defined as a transitive set of transitive sets, or else, as a transitive set totally ordered by set inclusion.)

As noted, rule 2) of the definition of $\mathbb{O}$ enables one to construct elements of $\mathbb{O}$ without any immediate predecessor. In the above proofs, the case 1) is treated by a simple induction assuming the claim about the immediate predecessor. The case 2), however, requires the stronger version assuming the claim for all ordinals x_i involved in forming the new ordinal by union. This step amounts to transfinite induction, allowing to "pass through" the limit ordinals.

Ordinals play the central role in set theory, providing the paradigmatic well-orderings. As the concept includes ordering, which is abstracted away in the concept of a cardinal number, there may be several ordinals of the same cardinality. E.g., the ordinals ω, $\omega + 1$ and $1 + \omega$, are easily seen to have the same cardinality, denoted $\aleph_0$:

$$
\begin{aligned}
\omega = \quad & 0 < 1 < 2 < 3 < 4 < 5 < \dots \\
\omega + 1 = \quad & 0 < 1 < 2 < 3 < 4 < 5 < \dots < \omega \\
1 + \omega = \quad & \bullet < 0 < 1 < 2 < 3 < 4 < 5 < \dots
\end{aligned}
$$

The functions $f : \omega + 1 \to 1 + \omega$, defined by $f(\omega) = \bullet$ and $f(n) = n$, and $g : 1 + \omega \to \omega$, defined by $g(\bullet) = 0$ and $g(n) = n+1$, are obviously bijections. g is, in addition order-homomorphism, since for all $x, y \in 1 + \omega : x < y \Rightarrow g(x) < g(y)$. This means that these two sets represent the same ordering. However, f does not preserve the ordering, since it maps the greatest element ω of the ordering $\omega + 1$ onto the smallest element $\bullet$ of the ordering $1 + \omega$. These two ordinals represent different ways of well-ordering the same number, $\aleph_0$, of elements. Consequently, the inequality $\omega + 1 \neq 1 + \omega$ holds in the ordinal arithmetics, in spite of the cardinal equality $|\omega + 1| = \aleph_0 = |1 + \omega|$ . . . . [end optional]

EXERCISES 2.

EXERCISE 2.1 Given the following inductively defined set $S \subset \mathbb{N} \times \mathbb{N}$:

> BASIS :: $\langle 0, 0 \rangle \in S$
> IND. :: If $\langle n, m \rangle \in S$ then $\langle s(n), m \rangle \in S$ and $\langle s(n), s(m) \rangle \in S$

Determine a property P allowing to set $S = \{\langle n, m \rangle : P(n, m)\}$.
Describe those elements of S which can be derived in more than one way.

EXERCISE 2.2 Let P and R be well-founded sPOs on a set S. Prove the true statement(s) and provide counter-examples to the false one(s):
(1.a) $P \cap R$ is an sPO.   (2.a) $P \cup R$ is an sPO.
(1.b) $P \cap R$ is well-founded.   (2.b) $P \cup R$ is well-founded.

EXERCISE 2.3 Let $\Sigma = \{a, b, c\}$, $\Gamma = \{\neg, \rightarrow, (,)\}$ and define the language WFF^Σ over $\Sigma \cup \Gamma$ inductively as follows:

BASIS :: $\Sigma \subseteq \mathsf{WFF}^\Sigma$
IND. :: If $A, B \in \mathsf{WFF}^\Sigma$ then $\neg A \in \mathsf{WFF}^\Sigma$ and $(A \rightarrow B) \in \mathsf{WFF}^\Sigma$.

(1) Which of the following strings belong to WFF^Σ:
$(a \rightarrow \neg b) \rightarrow c$, $a \rightarrow b \rightarrow c$, $\neg(a \rightarrow b)$, $(\neg a \rightarrow b)$, $\neg a \rightarrow b$?
(2) Replace Σ with $\Delta = \{*, \#\}$ and use the analogous definition of WFF^Δ. Which strings are in WFF^Δ:
$*\#$, $\neg *$, $\neg(\#\#)$, $(* \rightarrow \#)$, $* \rightarrow \#$, $* \leftarrow \#$?
(3) Use appropriate induction to show that every string in WFF^Σ can be derived in only one way (i.e., the definition is 1-1).

EXERCISE 2.4 Describe the ordering induced on the set Σ^+ from Example 2.22 by following Idea 2.20.

EXERCISE 2.5 Use induction on $\mathbb{N}$ to prove for all $1 \leq n \in \mathbb{N}$:
$(1 + 2 + \ldots + n)^2 = 1^3 + 2^3 + \ldots + n^3$.
(Hint: In the induction step expand the expression $((1 + 2 + \ldots + n) + (n + 1))^2$ and use Example 2.28).

EXERCISE 2.6 Use induction to prove that a finite set with n elements has 2^n subsets (cf. Exercise 1.15).

EXERCISE 2.7 Let A, B range over non-empty, finite sets and A^B denote the set of all functions from B to A.

(1) Show by appropriate induction that the cardinality of this set equals $|A|$ to the power $|B|$, i.e., $|A^B| = |A|^{|B|}$.
(2) Show that there are $2^{(n^2)}$ binary relations over a set with $n \in \mathbb{N}$ elements. (One sentence, referring to the relevant definitions/results and with an adequate use of the previous point, can be enough.)

EXERCISE 2.8 Use induction to show the general de Morgan law, namely, for every finite collection of $n \geq 2$ sets:

$$\overline{(A_1 \cap A_2 \cap \ldots \cap A_n)} = \overline{A_1} \cup \overline{A_2} \cup \ldots \cup \overline{A_n},$$

where $\overline{X}$ denotes the complement of X (relatively to a given universe U). The basis case for $n = 2$ can be assumed as given.

EXERCISE 2.9 Show that every finite partial order $<$ (a PO on a finite set):
(1) is well-founded;
(2) has a linear extension, i.e., a total order $<^*$ such that $< \subseteq <^*$.

EXERCISE 2.10 For arbitrary sets A, B, let $f : \mathcal{P}(A) \to \mathcal{P}(B)$ be a monotone function, i.e., such that $X \subseteq Y$ implies $f(X) \subseteq F(Y)$. Show by induction that for every natural number $n > 0 : f^n(X) \cup f^n(Y) \subseteq f^n(X \cup Y)$, where $f^1(Z) = f(Z)$ and $f^{n+1}(Z) = f(f^n(Z))$.

EXERCISE 2.11 For a possibly infinite set A with a strict partial ordering $<_1$, define its "lexicographic extension" to pairs $A \times A$, $\langle A^2, <_2 \rangle$, as follows:
$$\langle x, y \rangle <_2 \langle x', y' \rangle \text{ iff either } x <_1 x' \text{ or } (x = x' \text{ and } y <_1 y').$$
Inductively, given the ordering on the n-tuples $\langle A^n, <_n \rangle$, we obtain the induced ordering on A^{n+1} by letting:
$$\langle x, \overline{y} \rangle <_{n+1} \langle x', \overline{y}' \rangle \text{ iff either } x <_1 x' \text{ or } (x = x' \text{ and } \overline{y} <_n \overline{y}').$$
Use induction to show that, for every $n \geq 1$:

(1) $\langle A^n, <_n \rangle$ is a strict partial ordering,
(2) if $\langle A, <_1 \rangle$ is total, then $\langle A^n, <_n \rangle$ is total,
(3) if $\langle A, <_1 \rangle$ is well-founded, then $\langle A^n, <_n \rangle$ is well-founded.

EXERCISE 2.12 Given a set S, define inductively $P \subseteq \mathcal{P}(S) \times \mathcal{P}(S)$:

BASIS :: $\langle \{s\}, \{s\} \rangle \in P$ – for every $s \in S$
IND. :: If $s \in S$ and $\langle A, B \rangle \in P$, then
$\langle A \cup \{s\}, B \rangle \in P$ and $\langle A, B \cup \{s\} \rangle \in P$.

Let SS denote the set $\{ \langle A, B \rangle \in \mathcal{P}(S) \times \mathcal{P}(S) : A \cap B \neq \varnothing \}$.

(1) Using structural induction, show that $P \subseteq SS$.
(2) Show that for every finite A, B: if $\langle A, B \rangle \in SS$ then $\langle A, B \rangle \in P$.
(3) Why can't we claim (2) for infinite A or B?

EXERCISE 2.13 Let X be a country with finitely many cities. Show that for any such X, if every two cities in X have (at least) one one-way direct road (not passing through other cities), then there is some starting city x_0 and a route from x_0 which passes through every city exactly once.

EXERCISE 2.14 Binary trees (over a set S), BT, are defined inductively:

BASIS :: $\langle \rangle \in BT$ – the empty tree is a binary tree
IND. :: If $s \in S$ and $L, R \in BT$, then $\langle L, s, R \rangle \in BT$.

(1) Define by induction the function $height : BT \to \mathbb{N}$, giving the length of the longest branch. (The empty tree has height 0.)
(2) Define by induction the function $numn : BT \to \mathbb{N}$, giving the number of nodes in the tree.
(3) Let $CBT \subset BT$ be the subset of *complete* binary trees, i.e., ones where for every node $\langle L, s, R \rangle$: $L, R \in CBT$ and $height(L) = height(R)$.

Show by structural induction that, for every $t \in CBT : numn(t) = 2^{height(t)} - 1$.

Exercise 2.15 (1) For the binary trees BT from Exercise 2.14, define by induction function $mir : BT \rightarrow BT$, returning a mirror image of the argument. E.g., for the tree $T = \langle 1, 2, \langle 4, 3, 5 \rangle \rangle$, $mir(T) = \langle \langle 5, 3, 4 \rangle, 2, 1 \rangle$, as illustrated below (dropping the empty subtrees):

(2) Show that mir is idempotent, i.e., for all $T \in BT : mir(mir(T)) = T$.
(3) For BT over $\mathbb{N}$, define by induction function $sum : BT \rightarrow \mathbb{N}$, giving the sum of all nodes $(sum(\langle \rangle) = 0, sum(T) = 15)$, and show by structural induction that for all $T \in BT : sum(T) = sum(mir(T))$.

Exercise 2.16 Using the definition of strings Σ^* from Example 2.16, let function $f : \{a, b, \bullet\}^* \rightarrow \{a, b, \bullet\}^*$ be defined as follows:

Basis :: $f(\epsilon) = \epsilon$
Ind. :: $f(xs) = \bullet f(s)$.

Show by structural induction that for every $s \in \{a, b, \bullet\}^* : f(s) = f(f(s))$.

Exercise 2.17 Using the definition of strings Σ^* and concatenation/length functions from Examples 2.16 and 2.24, show by structural induction that

(1) for every string $t \in \Sigma^* : \epsilon \cdot t = t \cdot \epsilon$,
(2) for all string $s, t \in \Sigma^* : length(s \cdot t) = length(s) + length(t)$.

Exercise 2.18 Using the definition of strings Σ^* from Example 2.16 as the schema for structural induction, show that the reverse operation from Example 2.25 satisfies the following equations:

(1) for all $y \in \Sigma, s \in \Sigma^* : (s \vdash y)^R = y(s^R)$,
(2) for all $s \in \Sigma^* : (s^R)^R = s$.

Exercise 2.19 Bit-strings are finite, non-empty strings Δ^+, over the alphabet $\Delta = \{0, 1\}$, as defined in Example 2.17.

(1) Define Δ^+ by induction using right append as the constructor (i.e., adding an element $x \in \Delta$ at the end of the string $s \in \Delta^+$, written sx).
(2) Interpreting bit-strings as binary numbers, define inductively function $dec : \Delta^+ \rightarrow \mathbb{N}$, which converts the binary number to its decimal form, e.g., $dec(110) = 6, dec(101) = 5, dec(010) = 2$, etc.

(3) Show by structural induction that for every $s \in \Delta^+$:
  (a) $dec(s) = dec(0s)$, and
  (b) $dec(s)$ is the correct decimal value of the binary number s.

EXERCISE 2.20 Show that the operation $+$ on $\mathbb{N}$, defined in Example 2.26, is associative, i.e., that the identity $a + (b + c) = (a + b) + c$ holds for all $a, b, c \in \mathbb{N}$.

EXERCISE 2.21 Show that the operation $+$ on $\mathbb{N}$, defined in Example 2.26, is commutative, i.e., that the identity $n + m = m + n$ holds for all $n, m \in \mathbb{N}$. This requires a *nested* induction, i.e.,

- the basis: for all $m \in \mathbb{N}$, $0 + m = m + 0$, and
- the induction: for all $m \in \mathbb{N}$, $s(n) + m = m + s(n)$, given that for all $m \in \mathbb{N}$, $n + m = m + n$

can themselves be proved only by the use of induction (on m).

EXERCISE 2.22 Give an argument showing that the relation $\prec$, induced according to Idea 2.20, is a strict partial order.

EXERCISE 2.23 We have seen two different proof rules for induction on natural numbers – one (2.11) in Example 2.9 and another (2.29) in Example 2.28. Show that each can be derived from the other. Begin by stating clearly what exactly you are asked to prove! (One part of this proof is actually the argument after Idea 2.27.)

Part II
TURING MACHINES

Chapter 3

COMPUTABILITY AND DECIDABILITY

The idea of "mechanical" reasoning underlies the whole development of logic, and this chapter presents its ultimate expression as a Turing machine or, in more modern language, a computer program. It captures the "mechanical" decidability of a problem in terms of purely symbolic transformations of the input following precisely defined rules. It relies on the idea of an uninterpreted language, so we begin with this concept.

3.1 ALPHABETS AND LANGUAGES

──────────────── a background story ────────────────

The languages we use for daily communication are what we call *natural languages*. They are acquired in childhood and are suited to just about any communication we may have. They have evolved along with the development of mankind and form a background for any culture and civilisation. *Formal languages*, on the other hand, are explicitly designed by people for a clear, particular purpose. A semi-formal language was used in the preceding chapters for the purpose of talking about sets, functions, relations, etc. It was only *semi*-formal because it was not fully defined. It was introduced along as we needed some notation for particular concepts.

Formal language is a fundament of formal logical system and we will later encounter several examples of formal languages. Its most striking feature is that, although designed for some purposes, it is an entity on its own. It is completely specified without necessarily referring to its possible meaning. It is a pure syntax. Similar distinction can be drawn with respect to natural languages. The syntax of a natural language is captured by the intuition of the *grammatically correct expressions.* At the basic (written) level, there is an alphabet, e.g., the Latin one with a,b,c,... from which words are formed. Yet, not all possible combinations of the letters form words; aabez is a syntactic possibility but there is no such word. Using only valid words, still does not guarantee correctness. But even grammatically correct sentences may remain meaningless. Quadrille drinks procrastination is a grammatically correct expression consisting of the subject quadrille, verb in the proper form drinks, and object procrastination. But the fact that it is grammatical, does not ensure that it is meaningful. The sentence does convey an idea of some strange event which, unless it is employed in a very particular context, does not make any sense.

Pure syntax may try to approach but does not guarantee meaning. Yet, on the positive side, syntax is much easier to define and control than its intended meaning and we will now start using such purely syntactic languages. Their basis is determined by some, arbitrarily chosen alphabet. Then, one may design various syntactic rules determining which expressions built from the alphabet's symbols, form valid, well-formed words and sentences. In this chapter, we will merely introduce the fundamental definition of a language and observe how the formal notion of computability relates necessarily to some formal language. In the subsequent chapters, we will study some particular formal languages forming the basis of most common logical systems. _____

Definition 3.1　An *alphabet* is a (typically finite) set, its members are called *symbols*. The set of all finite strings over an alphabet Σ is denoted Σ^*. A *language* L over an alphabet Σ is a subset $L \subseteq \Sigma^*$.

Example 3.2
The language with natural numbers as the only expressions is a subset of $N \subseteq \Sigma^*$ where $\Sigma = \{0, 1, 2, 3, 4, 5, 6, 7, 8, 9\}$.

Typically, languages are defined inductively. The language L of natural number arithmetic can be defined over the alphabet $\Sigma' = \Sigma \cup \{), (, +, -, *\}$

BASIS :: If $n \in N$ then $n \in L$
IND. :: If $m, n \in L$ then $(m + n), (m - n), (m * n) \in L.$　　　　□

Alphabets of particular importance are the ones with only two symbols. Any finite alphabet Σ with n distinct symbols can be encoded using an alphabet with only two symbols, e.g., $\Delta = \{0, 1\}$.

Example 3.3

To code any 2-symbol alphabet Σ, we just choose any bijection $\Sigma \leftrightarrow \Delta$.

To code an alphabet with 3 symbols, $\Sigma = \{a, b, c\}$, we can represent $\{a \mapsto 00, b \mapsto 01, c \mapsto 10\}$. The string $aacb$ will be represented by the bit-string 00000101. Notice that to decode this string, we have to know that any symbol from Σ is represented by a string of exactly two bits.   □

In general, to code an alphabet with n symbols, bit-strings of length $\lceil log_2 n \rceil$ are needed for each symbol, e.g., coding a 5-symbol alphabet, requires bit-strings of length at least 3 for each symbol.

Binary representation of natural numbers is a more specific coding utilizing in particular way the positions of bits. We will not use it but it is worth mentioning as the basis of a computer's operations.

Definition 3.4   A binary number b is an element of Δ^* where $\Delta = \{0, 1\}$. A binary number $b = b_n b_{n-1} \ldots b_2 b_1 b_0$ represents the natural number
$$b_n * 2^n + b_{n-1} * 2^{n-1} + b_{n-2} * 2^{n-2} + \ldots + b_1 * 2^1 + b_0 * 2^0.$$

For instance, 0001 and 01 both represent 1; 1101 represents the number $1 * 2^3 + 1 * 2^2 + 0 * 2 + 1 * 1 = 13$ (cf. Exercise 2.19).

3.2   Turing machines

———————————— a background story ————————————

Turing machine (after English mathematician Alan Turing, 1912-1954) was the first general model designed for the purpose of separating problems which can be solved automatically from those which cannot. Although the model was purely mathematical, it was easy to imagine that a corresponding physical device could be constructed. In fact, it was and is today known as the computer.

Many other models of computability have been designed but, as it turns out, all such models define the same concept of computability, i.e., the same problems are mechanically computable irrespectively of which model one is using. The fundamental results about Turing machines apply to all computations on even the most powerful computers. The limits of Turing computability are also the limits of the modern computers.

The rest of the story below is taken from Turing's seminal paper "On computable numbers, with an application to the Entscheidungsproblem" from 1936:

"Computing is normally done by writing certain symbols on paper. We may suppose this paper is divided into squares like a child's arithmetic book. In elementary arithmetic the two-dimensional character of the paper is sometimes used. But such a use is always avoidable, and I think that it will be agreed that the two-dimensional character of paper is no essential of computation. I assume then that the computation is carried out one one-dimensional paper, i.e., on a tape divided into squares. I shall also suppose that the number of symbols which may be printed is finite. If we were to allow an infinity of symbols, then there would be symbols differing to an arbitrary small extent. The effect of this restriction of the number of symbols is not very serious. It is always possible to use sequences of symbols in the place of single symbols. An Arabic numeral such as 17 or 9999999999 is normally treated as a single symbol. Similarly, in any European language words are treated as single symbols. (Chinese, however, attempts to have an enumerable infinity of symbols.) The differences from our point of view between the single and compound symbols is that the compound symbols, if they are too lengthy, cannot be observed at one glance. This is in accordance with experience. We cannot tell at a glance whether 99999999999999 and 9999999999999 are the same.

The behaviour of the computer at any moment is determined by the symbols which he is observing, and his "state of mind" at that moment. We may suppose that there is a bound B to the number of symbols of squares which the computer can observe at one moment. If he wishes to observe more, he must use successive observations. We will also suppose that the number of states of mind which need be taken into account is finite. The reasons for this are of the same character as those which restrict the number of symbols. If we admitted an infinity of states of mind, some of them will be "arbitrarily close" and will be confused. Again, the restriction is not one which seriously affects computation, since the use of more complicated states of mind can be avoided by writing more symbols on the tape.

Let us imagine the operations performed by the computer to be split up into "simple operations" which are so elementary that it is not easy to imagine them further divided. Every such operation consists of some change of the physical system consisting of the computer and his tape. We know the state of the system if we know the sequence of symbols on the tape, which of these are observed by the computer (possibly with a special order), and the state of mind of the computer.

We may suppose that in a simple operation not more than one symbol
is altered. Any other change can be split up into simple changes of
this kind. The situation in regard to the squares whose symbols may
be altered in this way is the same as in regard to the observed squares.
We may, therefore, without loss of generality, assume that the squares
whose symbols are changed are always "observed" squares.

Besides these changes of symbols, the simple operations must include changes of distribution of observed squares. The new squares
must be immediately recognisable by the computer. I think it is reasonable to suppose that they can only be squares whose distance from
the closest of the immediately previously observed squares does not exceed a certain fixed amount. Let us say that each of the new observed
squares is within L squares of an immediately previously observed
square.

In connection with "immediate recognizability", it may be thought
that there are other kinds of square which are immediately recognisable. In particular, squares marked by special symbols might be taken
as immediately recognisable. Now if these squares are marked only by
single symbols there can be only finite number of them, and we should
not upset our theory by adjoining these marked squares to the observed squares. If, on the other hand, they are marked by a sequence
of symbols, we cannot regard the process of recognition as a simple
process. This is a fundamental point and should be illustrated. In
most mathematical papers the equations and theorems are numbered.
Normally the numbers do not go beyond (say) 1000. It is, therefore,
possible to recognise a theorem at a glance by its number. But if
the paper was very long, we might reach Theorem 157767733443477;
then, further on in the paper, we might find "...hence (applying Theorem 157767733443477) we have...". In order to make sure which
was the relevant theorem we should have to compare the two numbers
figure by figure, possibly ticking the figures off in pencil to make sure
of their not being counted twice. If in spite of this it is still thought
that there are other "immediately recognisable" squares, it does not
upset my contention so long as these squares can be found by some
process of which my type of machine is capable. [...]

The simple operations must therefor include:

(a) Changes of the symbol on one of the observed squares.

(b) Changes of one of the squares observed to another square within
L squares of one of the previously observed squares.

It may be that some of these changes necessarily involve a change of state of mind. The most general single operation must therefore be taken to be one of the following:

(A) A possible change (a) of symbol together with a possible change of state of mind.

(B) A possible change (b) of observed squares, together with a possible change of mind.

The operation actually performed is determined, as has been suggested, by the state of mind of the computer and the observed symbols. In particular, they determine the state of mind of the computer after the operation is carried out.

   We may now construct a machine to do the work of this computer. To each state of mind of the computer corresponds an "m-configuration" of the machine. The machine scans B squares corresponding to the B squares observed by the computer. In any move the machine can change a symbol on a scanned square or can change any one of the scanned squares to another square distant not more than L squares from one of the other scanned squares. The move which is done, and the succeeding configuration, are determined by the scanned symbol and the m-configuration." _____

Definition 3.5   A *Turing machine* M is a quadruple $\langle K, \Sigma, q_0, \tau \rangle$ where

- K is a finite set of *states* of M
- Σ is a finite alphabet of M
- $q_0 \in K$ is the *initial state*
- $\tau : K \times \Sigma \to K \times (\Sigma \cup \{\mathsf{L}, \mathsf{R}\})$ is a (partial) *transition function* of M.

For convenience, we will assume that Σ always contains the 'space' symbol #. This definition deviates only slightly from the one sketched by Turing and does not deviate from it at all with respect to the computational power of the respective machines. The difference is that our machine reads only a single symbol (a single square, or position) at a time, $B = 1$, and that it moves at most one square at the time, $L = 1$.

Idea 3.6 [**Operation of TM**] Imagining M as a physical device with a "reading head" moving along an infinite "tape" divided into discrete positions, the transition function τ determines its operation as follows:

- M starts in the initial state q_0, with its "head" at some position on the input tape.

- When M is in a state q and "reads" a symbol a, and the pair $\langle q, a \rangle$ is in the domain of τ, then M "performs the action" $\langle q', a' \rangle = \tau(\langle q, a \rangle)$: "passes to the state" q' "doing" a' which may be one of the two things:
  - if $a' \in \Sigma$, M "prints" the symbol a' at the current position on the tape;
  - otherwise $a' = \mathsf{L}$ or $a' = \mathsf{R}$ – then M "moves its head" one step to the left or right, respectively.
- When M is in a state q and reads a symbol a, and the pair $\langle q, a \rangle$ is not in the domain of τ then M "stops its execution" – it *halts*.
- To "run M on the input string w" is to write w on an otherwise blank tape, placing the reading head, typically, on the first symbol of w (anywhere on the tape, if w is the empty string), and then starting the machine (from the state q_0).
- $M(w)$ denotes the tape's content after M has run on the input w.

τ may be written as a set of quadruples $\{\langle q_1, a_1, q_1', a_1' \rangle, \langle q_2, a_2, q_2', a_2' \rangle, \ldots\}$ called *instructions*. We often write a single instruction as $\langle q, a \rangle \mapsto \langle q', a' \rangle$. Initially, the tape contains the "input" for the computation, but it is also used by M for writing the "output".

Remark.
Allowing, as is sometimes done, τ to be a relation instead of a function, leads to *nondeterministic* Turing machines. Such machines are not more powerful than the machines from our definition, and we will not study them. Another variant, which does not increase the power and will not be discussed here either, allows TM to use several tapes. For instance, a machine with 2 tapes would have $\tau : K \times \Sigma \times \Sigma \to K \times (\Sigma \cup \{\mathsf{L}, \mathsf{R}\}) \times (\Sigma \cup \{\mathsf{L}, \mathsf{R}\})$. □

Turing machines embody the idea of *mechanically computable functions* or *algorithms*. The following three examples illustrate different flavours of such functions. The one in Example 3.7 disregards its input (for simplicity we made the input blank) and merely produces a constant value – it *computes a constant function*. The one in Example 3.8 does not modify the input but *recognizes* whether it belongs to a specific language. Starting on the leftmost 1 it halts if and only if the number of consecutive 1's is even – in this case we say that it *accepts the input string*. If the number is odd the machine goes forever. The machine in Example 3.9 *computes a function* of the input x. Using unary representation of natural numbers, it returns $\lceil x/2 \rceil$ – the least natural number greater than or equal to $x/2$.

Example 3.7
The following machine goes PING! Starting on an empty (filled with blanks)

tape, it writes "PING" and halts. The alphabet is $\{\#, P, I, N, G\}$, there are 4 states $q_0 \ldots q_3$ with the initial state q_0, and the transition function is as follows (graphically, on the right, state q_x is marked by (x) and the initial state q_z by $\boxed{z}$):

$\langle q_0, \# \rangle \mapsto \langle q_0, P \rangle$ $\langle q_0, P \rangle \mapsto \langle q_1, \mathsf{R} \rangle$

$\langle q_1, \# \rangle \mapsto \langle q_1, I \rangle$ $\langle q_1, I \rangle \mapsto \langle q_2, \mathsf{R} \rangle$

$\langle q_2, \# \rangle \mapsto \langle q_2, N \rangle$ $\langle q_2, N \rangle \mapsto \langle q_3, \mathsf{R} \rangle$

$\langle q_3, \# \rangle \mapsto \langle q_3, G \rangle$

$\square$

Example 3.8

The alphabet is $\{\#, 1\}$ and machine has 2 states. It starts on the leftmost 1 in state q_0 – the transition function is given on the left, while the corresponding graphical representation on the right:

$\langle q_0, 1 \rangle \mapsto \langle q_1, \mathsf{R} \rangle$

$\langle q_1, 1 \rangle \mapsto \langle q_0, \mathsf{R} \rangle$

$\langle q_1, \# \rangle \mapsto \langle q_1, \# \rangle$

Write the computations of this machine on the inputs 1, 11 and 111. $\square$

The machine in this last example was deliberately partial, entering an infinite loop whenever encountering blank in state 1. This could be easily remedied by introducing a new "halting" state h, extending the alphabet with Y, N and adding two instructions $\langle q_0, \# \rangle \mapsto \langle h, Y \rangle$ and $\langle q_1, \# \rangle \mapsto \langle h, N \rangle$ with the convention that output Y denotes acceptance and N rejection of the input. In general, such an operation may be impossible – there are partial functions which cannot be computed for all arguments by any known means, but we have to wait a little before showing that.

Example 3.9

To make a machine computing $\lceil x/2 \rceil$ – using the alphabet $\{\#, 1\}$ and unary representation of numbers – one might go through the input removing every second 1 (replacing it with $\#$) and then compact the result to obtain a contiguous sequence of 1's. We apply a different algorithm which keeps all 1's together all the time. The machine starts at the leftmost 1:

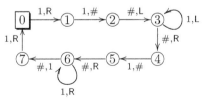

	#	1	Starting on the leftmost 1 in q_0
q_0		$\langle q_0,1\rangle \mapsto \langle q_1,\mathsf{R}\rangle$	$-$ jump over one 1; halt if no 1
q_1		$\langle q_1,1\rangle \mapsto \langle q_2,\#\rangle$	$-$ replace next 1 with #; halt if no 1
q_2	$\langle q_2,\#\rangle \mapsto \langle q_3,\mathsf{L}\rangle$		$-$ move left and
q_3	$\langle q_3,\#\rangle \mapsto \langle q_4,\mathsf{R}\rangle$	$\langle q_3,1\rangle \mapsto \langle q_3,\mathsf{L}\rangle$	return to the leftmost 1
q_4		$\langle q_4,1\rangle \mapsto \langle q_5,\#\rangle$	$-$ erase it
q_5	$\langle q_5,\#\rangle \mapsto \langle q_6,\mathsf{R}\rangle$		$-$ move right $-$ return to # inserted
q_6	$\langle q_6,\#\rangle \mapsto \langle q_7,1\rangle$	$\langle q_6,1\rangle \mapsto \langle q_6,\mathsf{R}\rangle$	in q_1 replacing it with 1 erased in q_4
q_7		$\langle q_7,1\rangle \mapsto \langle q_0,\mathsf{R}\rangle$	$-$ move right and continue from q_0

Try to follow the computations on the input tapes #, 11, 111 and 1111.□

3.2.1: COMPOSING TURING MACHINES .... [optional]
The following machine starts its computation anywhere within or just after a
string of 1's and adds a single 1 at the end of the string, M_{+1} : 1,R $\big(\!\!\big(\,0\,\big)\!\xrightarrow{\#,1}\!\big(\,1\,\big)$.
Now taking, for instance the machine $M_{/2}$ from Example 3.9, we should be able to
put the two together into a machine M_f computing the function $f(x) = \lceil x/2\rceil +1$,
by running first $M_{/2}$ and then M_{+1}.

To do this in general, one has to ensure that the configuration in which the
first machine halts (if it does) is of the form assumed by the second machine
when it starts. A closer look at the machine $M_{/2}$ shows that it halts with the
reading head just after the rightmost 1. This is an acceptable configuration for
starting M_{+1} and so we can write our machine M_f as $\boxed{M_{/2}}\xrightarrow{\#,\#}\boxed{M_{+1}}$. This
abbreviated notation says that: M_f starts by running $M_{/2}$ from its initial state
and then, *whenever* $M_{/2}$ halts *and* reads a blank #, it passes to the initial state
of M_{+1} writing #, and then runs M_{+1}.

We give a more elaborate example illustrating composition of Turing machines. A while-loop in a programming language is a command of the form while
B do F, where B is a boolean function and F is a command which we will assume
computes some function. Assuming availability of machines M_B and M_F, we construct a machine computing the function $G(y)$ which returns the least $x > 0$ such
that $B(x, F(x, y))$ is true (if no such x exists $G(y)$ is undefined). It is expressed
by the following while-loop:

```
G(y) = {  x:=1; z:=y;
          while not B(x,z) do {
            x:=x+1;
            z:=F(x,y);}
          return x; }
```

We consider the alphabet $\Sigma = \{\#, 1, Y, N\}$ and functions over positive natural
numbers (without 0) $\mathbb{N}$, with unary representation as strings of 1's. Let our given
functions be $F : \mathbb{N} \times \mathbb{N} \to \mathbb{N}$ and $B : \mathbb{N} \times \mathbb{N} \to \{Y, N\}$, and the corresponding
Turing machines, M_F, resp. M_B. More precisely, s_F is the initial state of M_F
which, starting in a configuration of the form $C1$, halts iff $z = F(x, y)$ is defined

in the final state e_F in the configuration of the form $C2$:

$$C1: \quad \cdots\#\ \overbrace{1\ \ldots\ 1}^{1^y}\ \#\ \overbrace{1\ \ldots\ 1}^{1^x}\ \#\ \cdots \qquad \uparrow s_F$$

$$M_F \downarrow$$

$$C2: \quad \cdots\#\ \overbrace{1\ \ldots\ 1}^{1^y}\ \#\ \overbrace{1\ \ldots\ 1}^{1^x}\ \#\ \overbrace{1\ \ldots\ 1}^{1^z}\ \#\ \cdots \qquad \uparrow e_F$$

If, for some pair x, y, $F(x, y)$ is undefined, $M_F(y, x)$ may go forever.

B, on the other hand, is total and M_B always halts in its final state e_B, when started from a configuration of the form $C2$ and initial state s_B, yielding a configuration of the form $C3$:

$$C2: \quad \cdots\#\ \overbrace{1\ \ldots\ 1}^{1^y}\ \#\ \overbrace{1\ \ldots\ 1}^{1^x}\ \#\ \overbrace{1\ \ldots\ 1}^{1^z}\ \#\ \cdots \qquad \uparrow s_B$$

$$M_B \downarrow$$

$$C3: \quad \cdots\#\ \overbrace{1\ \ldots\ 1}^{1^y}\ \#\ \overbrace{1\ \ldots\ 1}^{1^x}\ \#\ \overbrace{1\ \ldots\ 1}^{1^z}\ u\ \#\ \cdots \qquad \uparrow e_B$$

where $u = Y$ iff $B(x, z) = Y$ (true) and $u = N$ iff $B(x, z) = N$ (false).

Using M_F and M_B, we design a TM M_G which starts in its initial state s_G in a configuration $C0$, and halts in its final state e_G iff $G(y) = x$ is defined, with the tape as shown in T:

$$C0: \quad \cdots\#\ \overbrace{1\ \ldots\ 1}^{1^y}\ \#\ \cdots \qquad \uparrow s_G$$

$$M_G \downarrow$$

$$T: \quad \cdots\#\ \overbrace{1\ \ldots\ 1}^{1^y}\ \#\ \overbrace{1\ \ldots\ 1}^{1^x}\ \#\ \cdots$$

M_G adds a single 1 ($= x$) past the first $\#$ to the right of the input y, and runs M_F and M_B on these two numbers. If M_B halts with Y, it only cleans up $F(x, y)$. If M_B halts with N, M_G erases $F(x, y)$, extends x with a single 1 and continues:

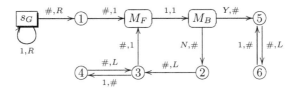

In case of success, M_B exits along the Y and states 5-6 erase the sequence of 1's representing $F(x, y)$. M_G stops in state 6 to the right of x. If M_B got N, this N is erased and states 3-4 erase the current $F(x, y)$. The first blank $\#$ encountered in the state 3 is the blank right to the right of the last x. This $\#$ is replaced with 1 – increasing x to $x + 1$ – and M_F is started in a configuration of the form $C1$.

$M_G(y)$ will go forever if no x exists such that $B(x, F(x, y)) = Y$. However, it may also go forever if such x exists but $F(x', y)$ is undefined for some $0 < x' < x$. Then the function $G(y)$ computed by M_G is undefined. In the theory of recursive functions, such a schema is called μ-recursion ("μ" for minimal) – here it is the function $G : \mathbb{N} \to \mathbb{N}$

$$G(y) \;=\; \text{the least } x \in \mathbb{N} \text{ such that } B(x, F(x, y)) = Y$$
$$\text{and } F(x', y) \text{ is defined for all } x' \leq x.$$

If $G(y) = x$ (i.e., when G is defined on y) then also $F(x', y)$ must be defined for all $x' \leq x$, and this reflects the mechanical, or blind, computability – M_G simply checks *all* consecutive values of x' until the correct one is found... [end optional]

3.3 UNIVERSAL TURING MACHINE

Informally, we might say that one Turing machine M' *simulates* another one M if M' is able to perform all the computations which can be performed by M or, more precisely, if any input w for M can be represented as an input w' for M' and the result $M'(w')$ represents the result $M(w)$.

This may happen in various ways, the most trivial one being the case when M' is strictly more powerful than M. If M is a multiplication machine (returning $n * m$ for any two natural numbers), while M' can do both multiplication and addition, then augmenting the input w for M with the indication of multiplication, we can use M' to do the same thing as M would do. Another possibility might be some encoding of the instructions of M in such a way that M', using this encoding as a part of its input, can act as if it was M. This is what happens in a computer since a computer program is a description of an algorithm, while an algorithm is just a mechanical procedure for performing computations of some specific type – i.e., it is a Turing machine. A program in a high level language is a Turing machine M – compiling it into a machine code amounts to constructing a machine M' which can simulate M. Execution of $M(w)$ proceeds by representing the high level input w as an input w' acceptable for M', running $M'(w')$ and converting the result back to the high level representation.

We won't define formally the notions of representation and simulation, relying instead on their intuitive understanding and the example of a Universal Turing machine, UTM, we will present. Such a machine is a Turing machine which can simulate any other Turing machine. It is a conceptual prototype and paradigm of the programmable computers as we know them.

Idea 3.10 [UTM] To build a UTM which can simulate an arbitrary TM M:

(1) choose a coding of Turing machines so that they can be represented on an input tape for UTM,
(2) represent the input of M on the input tape for UTM,
(3) choose a way of representing the state of the simulated machine M (the current state and position of the head) on the tape of UTM,
(4) design the set of instructions for the UTM.

To simplify the task, without losing generality, we will assume that the simulated machines work only on the default alphabet $\Sigma = \{*, \#\}$. At the same time, the UTM will use an extended alphabet with several symbols, which is the union of the following sets:

Σ – the alphabet of M
$\{S, N, \mathsf{R}, \mathsf{L}\}$ – additional symbols to represent instructions of M
$\{X, Y, 0, 1\}$ – symbols to keep track of the state and position of M
$\{(, A, B\}$ – auxiliary symbols for bookkeeping

We will code machine M together with its original input as follows:

(	instructions of M	current state	input and head position	(3.11)

1. A possible coding of TMs.

(1) Get the set of instructions from the description of a TM $M = \langle K, \Sigma, q_1, \tau \rangle$.
(2) Each instruction $t \in \tau$ is a four-tuple

$$t : \quad \langle q_i, a \rangle \mapsto \langle q_j, b \rangle$$

where $q_i, q_j \in K$, a is $\#$ or $*$, and $b \in \Sigma \cup \{\mathsf{L}, \mathsf{R}\}$. We assume that states are numbered from 1 up to $n > 0$. Represent t as C_t:

$$C_t : \quad \underset{1 \qquad i \qquad 1 \qquad j}{\boxed{S|\ldots|S|a|b|N|\ldots|N}}$$

i.e., first i S-symbols representing the current state q_i, then the read symbol a, so the action b – either the symbol to be written or R, L – and finally j N-symbols for the resulting state q_j.

(3) String the representations of all the instructions, with no extra spaces, in increasing order of state numbers. If for a state i there are two instructions, $t_i^\#$ for input symbol $\#$ and t_i^* for input symbol $*$, put t_i^* before $t_i^\#$.

(4) Put the "end" symbol '(' to the left:

(	C_{t_1}	C_{t_2}	$\cdots$	C_{t_z}	current state $\cdots$

Example 3.12

For $M = \langle \{q_1, q_2, q_3\}, \{*, \#\}, q_1, \tau \rangle$, the τ is given in the left part of the table, and its coding in the right part:

$\langle q_1, * \rangle \mapsto \langle q_1, \mathsf{R} \rangle$	$S * \mathsf{R}N$
$\langle q_1, \# \rangle \mapsto \langle q_2, * \rangle$	$S\# * NN$
$\langle q_2, * \rangle \mapsto \langle q_2, \mathsf{L} \rangle$	$SS * LNN$
$\langle q_2, \# \rangle \mapsto \langle q_3, \mathsf{R} \rangle$	$SS\#\mathsf{R}NNN$

The whole machine will be coded as:

(	S	$*$	R	N	S	$\#$	$*$	N	N	S	S	$*$	L	N	N	S	S	$\#$	R	N	N	N	$\cdots$

It is not necessary to perform the above conversion but – can you tell what M does? □

2. Input representation. We included the alphabet of the original machines $\Sigma = \{*, \#\}$ in the alphabet of the UTM. There is no need to code this part of the simulated machines.

3. Current state. After the instruction set of M, we reserve part of the tape for the representation of the current state. There are n states of M, so we reserve $n + 1$ fields for unary representation of the number of the current state. The i-th state is represented by i X's followed by $(n + 1 - i)$ Y's: if M is in the state i, this part of the tape will be:

instructions	X	$\cdots$	X	Y	$\cdots$	Y	input
	1		i		$n+1$		

We use $n + 1$ positions so that there is always at least one Y to the right of the sequence of X's representing the current state.

To "remember" the current position of the head, we will use the two extra symbols 0 and 1 corresponding, respectively, to $\#$ and $*$. The current symbol under the head will be always changed to 0, resp., 1. When the head is moved away, these symbols will be restored back to the original ones $\#$, resp., $*$. For instance, if M's head on the input tape $* * \#\# * \#*$ is in the 4-th place, the input part of the UTM tape will be $* * \#0 * \#*$.

4. Instructions for UTM. We let UTM start execution with its head at the rightmost X in the bookkeeping section of the tape. After completing the simulation of one step of M's computation, the head will again be placed at the rightmost X. The simulation of each step of computation of M will involve several things:

(4.1) Locate the instruction to be used next.

(4.2) Execute this instruction, i.e., either print a new symbol or move the head on M's tape.

(4.3) Write down the new state in the bookkeeping section.

(4.4) Prepare for the next step: clean up and go to the rightmost X.

In more detail, UTM performs these four steps as follows:

4.1. Find instruction. In a loop we erase one X at a time, replacing it by Y, and pass through all the instructions converting one S to A in each instruction. If there are too few S's in an instruction, we convert all the N's to B's in that instruction.

When all X's have been replaced by Y's, the instructions corresponding to the actual state have only A's instead of S. We deactivate the instructions which still contain S by going through all the instructions: if there is some S not converted to A, we replace all N's by B's in that instruction. Now, there remain at most 2 N-lists associated with the instruction(s) for the current state.

We go and read the current symbol on M's tape and replace N's by B's at the instruction (if any) which does not correspond to what we read.

The instruction to be executed has N's – others have only B's.

4.2. Execute instruction. UTM starts now looking for a sequence of N's. If none is found, then M – and UTM – stops. Otherwise, we check what to do looking at the symbol to the left of the leftmost N. If it is R or L, we go to the M's tape and move its head restoring the current symbol to its Σ form and replacing the new symbol by 1, resp. 0. If the instruction is to write a new symbol, we just write the appropriate thing.

4.3. Write new state. We find again the sequence of N's and write the same number of X's in the bookkeeping section indicating the next state.

4.4. Clean up. Finally, convert all A's and B's back to S and N's, and move the head to the rightmost X.

3.4 Undecidability

The definition of Turing machine captures formally the idea of *mechanical computability* – we are willing to say that a function $F : \mathbb{N} \to \mathbb{N}$ is computable iff some Turing machine computes $F(x)$ for all $x \in \mathbb{N}$. (Such functions are also called *recursive*. Considering only functions on $\mathbb{N}$ is no real restriction, as languages of interest can be encoded in $\mathbb{N}$.) Note that if $F(u)$ is undefined on some $u \in \mathbb{N}$, this would require assigning some special value to $F(u)$. In general, we may be unable to decide if a function is defined on a given argument. For such partial functions one uses therefore a weaker notion.

function F is		
computable (recursive)	iff	there is a TM which halts with $F(x)$ for all inputs $x \in \mathbb{N}$
semi-computable	iff	there is a TM which halts with $F(x)$ whenever F is defined on x but may not halt when F is undefined on x

A problem P of YES-NO type is a set L_P, of its instances, together with a subset $Y_P \subseteq L_P$, of the instances for which the answer is YES. For example, the problem if a given natural number is odd, consists of the instances $L_{Odd} = \mathbb{N}$ of natural numbers and its subset Y_{Odd} of odd numbers. A problem gives rise to a special case of function F_P which, for each instance $x \in L_P$, returns one of the only two values, YES or NO, depending on whether $x \in Y_P$ or not. We get here a third notion.

problem P is		
decidable	iff	F_P is computable – the machine computing F_P always halts returning correct answer YES or NO
semi-decidable	iff	F_P is semi-computable – the machine computing F_P halts with the correct answer YES, but may not halt when the answer is NO
co-semi-decidable	iff	not-F_P is semi-computable – the machine computing F_P halts with the correct answer NO, but may not halt when the answer is YES

Thus a problem is decidable iff it is both semi- and co-semi-decidable.

Set membership is a special case of YES-NO problem but for sets one uses a different terminology than for their membership problem:

set $S \subseteq \mathbb{N}$ is	iff	the membership problem $x \in S$ is
recursive	iff	decidable
recursively enumerable	iff	semi-decidable
co-recursively enumerable	iff	co-semi-decidable

A set is recursive iff it is both recursively and co-recursively enumerable.

One of the fundamental results about Turing machines, besides the existence of UTMs, concerns the undecidability of the halting problem. Following our strategy for encoding TMs and their inputs for simulation by a UTM, we assume that the encoding of the instruction set of a machine M is $E(M)$, while the encoding of input w for M is just w itself.

Problem 3.13 **[The halting problem]** Is there a Turing machine M_H such that for any machine M and input w, $M_H(E(M), w)$ always halts and

$$M_H(E(M), w) = \begin{cases} Y(es) \text{ if } M(w) \text{ halts} \\ N(o) \text{ if } M(w) \text{ does not halt} \end{cases} ?$$

The problem is trivially semi-decidable: given an M and w, simply run $M(w)$ and see what happens. If the computation halts, we get the correct YES answer. If it does not halt, then we may wait forever. Unfortunately, the following theorem shows that, in general, there is not much else to do than waiting to see what happens.

Theorem 3.14 **[Undecidability of the halting problem]** There is no Turing machine which decides the halting problem.

Proof. Assume, on the contrary, that there is such a machine M_H.

(1) We can easily design a machine M_N that is undefined (does not halt) on input Y and defined everywhere else, e.g., a machine with one state q_0 and instruction $\langle q_0, Y \rangle \mapsto \langle q_0, Y \rangle$.

(2) Now, construct machine M_1 which on the input $(E(M), w)$ gives $M_N(M_H(E(M), w))$. It has the property that $M_1(E(M), w)$ halts iff $M(w)$ does not halt. In particular:

$$M_1(E(M), E(M)) \text{ halts iff } M(E(M)) \text{ does not halt.}$$

(3) Let M^* be a machine which to an input w first computes (w, w) and then $M_1(w, w)$. In particular, $M^*(E(M^*)) = M_1(E(M^*), E(M^*))$. This one has the property that:

$$\begin{aligned} M^*(E(M^*)) \text{ halts} \quad &\text{iff} \quad M_1(E(M^*), E(M^*)) \text{ halts} \\ &\text{iff} \quad M^*(E(M^*)) \text{ does not halt.} \end{aligned}$$

This is clearly a contradiction, from which the theorem follows.

QED (3.14)

Thus the set $H = \{ \langle E(M), w \rangle : M \text{ halts on input } w \}$ is semi-recursive but not recursive. In terms of programming, the undecidability of halting problem means that it is impossible to write a program which could 1) take as input an arbitrary program M and its possible input w and 2) determine whether M run on w will terminate or not.

The theorem gives rise to a series of corollaries identifying other undecidable problems. The usual strategy for such proofs is to show that if a given problem were decidable then we could use it to decide another (e.g., halting) problem already known to be undecidable. As indicated before, a problem P (of YES-NO type) can be seen as a language $Y_P \subseteq L_P \subseteq \Sigma_P^*$, where Σ_P is the used alphabet, L_P contains all strings identifying instances of P and Y_P is the subset of instances with the YES answer. A (computable) *reduction* from a problem P to a problem Q is a computable function $r : L_P \to L_Q$ such that for every instance $p \in L_P : p \in Y_P$ iff $r(p) \in Y_Q$. Given such a reduction and an instance $p \in L_P$, we can compute the instance $r(p) \in L_Q$ and, if we can decide Q, the conclusion about $r(p) \in Y_Q$ can be applied to $p \in Y_P$ by the equivalence.

Corollary 3.15 There is no Turing machine

(1) M_D which, for any machine M, always halts with $M_D(E(M)) = Y$ iff M is total (always halts) and with N iff M is undefined for some input;
(2) M_E which, for given two machines M_1, M_2, always halts with Y iff the two halt on the same inputs and with N otherwise.

> **Proof.** (1) Assume that we have an M_D. Given an instance of the halting problem, an M and some input w, we can easily construct a machine M_w which, for any input x computes $M(w)$. In particular M_w is total iff $M(w)$ halts. Then we can use $M_D(E(M_w))$ to decide the halting problem. Hence there is no such M_D.
> (2) Assume that we have an M_E. Take as M_1 a machine which does nothing but halts immediately on any input. Then we can use M_E and M_1 to construct an M_D, which does not exist by the previous point.
> $\qquad$ QED (3.15)

For the further use of Turing machines, we observe that although they do provide an ultimately precise definition of computability and of (un)decidable problems, they are not necessarily easiest to use. Our (un)-decidability arguments will often rely on some informal intuitions arising from experience with programming. Correctness of such a simplification is ensured by the fact that each Turing machine can be expressed as a computer program while each computer program is a Turing machine.

EXERCISES 3.

EXERCISE 3.1 The questions at the end of Examples 3.8 and 3.9 (run the respective machines on the suggested inputs).

EXERCISE 3.2 Consider the alphabet $\Sigma = \{a, b\}$ and the language from Example 2.19.(1), i.e., $L = \{a^n b^n : n \in \mathbb{N}\}$.

(1) Build a TM M_1 which given a string s over Σ (possibly with additional blank symbol $\#$) halts iff $s \in L$ and goes forever iff $s \notin L$. If you find it necessary, you may allow M_1 to modify the input string.

(2) Modify M_1 to an M_2 which does a similar thing but always halts in the same state indicating the answer. For instance, the answer 'YES' may be indicated by M_2 just halting, and 'NO' by M_2 writing some specific string (e.g., 'NO') and halting.

EXERCISE 3.3 Let $\Sigma = \{a, b, c\}$ and $\Delta = \{0, 1\}$ (Example 3.3). Specify an encoding of Σ in Δ^* and build two Turing machines:

(1) M_c which, given a string over Σ, converts it to a string over Δ,

(2) M_d which, given a string over Δ, converts it to a string over Σ.

The two should act so that their composition gives identity, i.e., for all $s \in \Sigma^* : M_d(M_c(s)) = s$ and, for all $d \in \Delta^*$, if d codes a word over Σ then $M_c(M_d(d)) = d$. Choose the initial and final position of the head for both machines so that executing one after another will produce the same initial string. Run then one machine after another on some tapes to check if the input and output are the same.

EXERCISE 3.4 The language (S) is defined inductively (relatively to a given set S of other expressions, not containing '(' nor ')'):

BASIS :: Each $s \in S$ and the empty word, ϵ, belong to (S)
IND. :: If s and t belong to (S) then so do: (s) and st.

(1) Define inductively functions $l, r : (S) \to \mathbb{N}$, returning the number of left/right parentheses in the argument string.

(2) Show, using structural induction, that for every $e \in (S) : l(e) = r(e)$.

(3) Show, using structural induction, that for every $e \in (S)$ and every $1 \leq i \leq l(e)$ the i-th '(' comes before the i-th ')', i.e., the leftmost '(' comes before the leftmost ')', the second leftmost '(' before the second leftmost ')', etc.

(4) Show that a string satisfying (2) and (3) belongs to the language (S).

EXERCISE 3.5 The machine below decides the membership in the set (S) from Exercise 3.4. It starts reading the input on its leftmost symbol and

halts in state 7 iff the input belongs to (S) and in state 3 otherwise.
Its alphabet Σ consists of two disjoint sets $\Sigma_1 \cap \Sigma_2 = \varnothing$, where Σ_1 is some
set of symbols (for writing S-expressions) and $\Sigma_2 = \{X, Y, (,), \#\}$. In the
diagram we use the abbreviation '?' to indicate 'any other symbol from
Σ not mentioned explicitly among the transitions from this state'. For
instance, when in state 2 and reading $\#$ the machine goes to state 3 and
writes $\#$; reading) it writes Y and goes to state 4 – while reading any
other symbol ?, it moves the head to the right remaining in state 2.

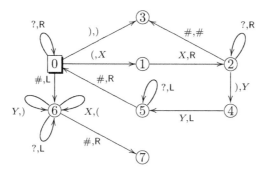

Using the claim from Exercise 3.4, justify that this machine *decides* membership in (S).

EXERCISE 3.6 The representation of the tape for the simulated TM, given
in (3.11), allows it to be infinite only in one direction, extending indefinitely to the right, but terminating on the left at the preceding block of
X's and Y's coding the current state. Explain why this is not any real
restriction, i.e., why everything computed by a TM with tape infinite in
both directions, can also be computed by a TM with tape infinite only in
one direction.

EXERCISE 3.7 Describe an algorithm showing that the set WFF^Σ from Exercise 2.3 is recursive.

EXERCISE 3.8 Define the set $TM = \{E(M) : M$ is a Turing machine$\}$,
assuming a uniform representation of Turing machines (i.e., all machines
work on the same finite alphabet and are coded in the same way, so that
each occurs exactly once in TM.)
(1) What is the cardinality of the set TM?
(2) What is the cardinality of the set $\mathcal{P}(\{1\}^*)$?
(3) Now, show that there exists an undecidable subset of $\{1\}^*$.

EXERCISE 3.9 After Theorem 3.14, we noted that the set $H = \{\langle E(M), w \rangle :$
M halts on input $w\}$ is not recursive. For this statement to be meaningful,

H must be a subset of $\mathbb{N}$, so show that it can be seen as such.

(Hint: Use (1) from the previous exercise and the idea of enumerating $\mathbb{N} \times \mathbb{N}$ from Example 1.26.)

Part III
PROPOSITIONAL LOGIC

Chapter 4

SYNTAX AND PROOF SYSTEMS

4.1 AXIOMATIC SYSTEMS

———————————— a background story ————————————

One of the fundamental goals of all scientific inquiry is to achieve precision and clarity of a body of knowledge. This "precision and clarity" means, among other things:

- all assumptions of a given theory are stated explicitly;
- the language is designed carefully by choosing some basic, primitive notions and defining others in terms of these ones;
- the theory contains some basic principles – all other claims of the theory follow from its basic principles by applications of definitions and some explicit laws.

Axiomatization in a formal system is the ultimate expression of these postulates. Axioms play the role of basic principles – explicitly stated fundamental assumptions which may be disputable but, once assumed, imply other claims, the theorems. These follow from the axioms not by some unclear arguments but by formal deductions according to well defined rules.

The most famous example of an axiomatization (and the one which, in more than one way gave the origin to the modern axiomatic systems) was Euclidean geometry. Euclid systematized geometry by showing how many geometrical statements could be logically derived from a small set of axioms and principles. The axioms he postulated were supposed to be intuitively obvious:

A1. Given two points, there is an interval that joins them.

A2. An interval can be prolonged indefinitely.

A3. A circle can be constructed when its center, and a point on it, are *given.*

A4. *All right angles are equal.*

There was also the famous fifth axiom – we will return to it shortly. Another part of the system were "common notions" which may be perhaps more adequately called inference rules about equality:

R1. *Things equal to the same thing are equal.*

R2. *If equals are added to equals, the wholes are equal.*

R3. *If equals are subtracted from equals, the remainders are equal.*

R4. *Things that coincide with one another are equal.*

R5. *The whole is greater than a part.*

Presenting a theory as an axiomatic system has tremendous advantages. For the first, it is economical – instead of long lists of facts and claims, we can store only axioms and deduction rules, since the rest is derivable from them. They, so to speak, "code" the knowledge of the whole field. More importantly, it systematizes knowledge by displaying the fundamental assumptions which form the logical basis of the field. In a sense, Euclid uncovered "the essence of geometry" by identifying axioms and rules which are sufficient and necessary for deriving all geometrical theorems. Finally, having such a compact presentation of a complicated field, makes it possible to relate not only to particular theorems but also to the whole field as such. This possibility is reflected in us speaking about Euclidean geometry vs. non-Euclidean ones. The differences between them concern precisely changes of some basic principles – inclusion or removal of the fifth postulate.

As an example of a proof in Euclid's system, we show how using the above axioms and rules he deduced the following proposition ("Elements", Book 1, Proposition 4):

Proposition 4.1 If two triangles have two sides equal to two sides respectively, and have the angles contained by the equal straight lines equal, then they also have the base equal to the base, the triangle equals the triangle, and the remaining angles equal the remaining angles respectively, namely those opposite the equal sides.

> **Proof.** Let ABC and DEF be two triangles having the two sides AB and AC equal to the two sides DE and DF respectively, namely AB equal to DE and AC equal to DF, and the angle BAC equal to the angle EDF.

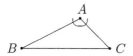 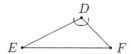

I say that the base BC also equals the base EF, the triangle ABC equals the triangle DEF, and the remaining angles equal the remaining angles respectively, namely those opposite the equal sides, that is, the angle ABC equals the angle DEF, and the angle ACB equals the angle DFE.

If the triangle ABC is superposed on the triangle DEF, and if the point A is placed on the point D and the straight line AB on DE, then the point B also coincides with E, because AB equals DE.

Again, AB coinciding with DE, the straight line AC also coincides with DF, because the angle BAC equals the angle EDF. Hence the point C also coincides with the point F, because AC again equals DF.

But B also coincides with E, hence the base BC coincides with the base EF and – by R4 – equals it. Thus the whole triangle ABC coincides with the whole triangle DEF and – by R4 – equals it.   QED (4.1)

The proof is allowed to use only the given assumptions, the axioms and the deduction rules. Yet, the Euclidean proofs are not exactly what we mean by a formal proof in an axiomatic system. Why? Because Euclid presupposed a particular *model*, namely, the abstract set of points, lines and figures in an infinite, homogeneous space. This presupposition need not be wrong (although, according to modern physics, it is), but it has important bearing on the notion of proof. For instance, it is intuitively obvious what Euclid means by "superposing one triangle on another". Yet, this operation hides some further assumptions, for instance, that length does not change during such a process. This implicit assumption comes most clearly forth in considering the language of Euclid's geometry. Here are just few definitions from *"Elements"*:

D1. A point is that which has no part.

D2. A line is breadthless length.

D3. The ends of a line are points.

D4. A straight line is a line which lies evenly with the points on itself.

D23. Parallel straight lines are straight lines which, being in the same plane and being produced indefinitely in both directions, do not meet one another in either direction.

These are certainly smart formulations but one can wonder if, say, D1 really defines anything or, perhaps, merely states a property of something intended by the name "point". Or else, does D2 define anything *if one does not presuppose* some intuition of what length is? To make a genuinely formal system, one would have to identify some basic notions as truly primitive – that is, with *no intended interpretation*. For these notions one may postulate some properties. For instance, one might say that we have the primitive notions of P, L and IL (for point, line and indefinitely prolonged line). P has no parts; L has two ends, both being P's; any two P's determine an L (whose ends they are – this reminds of A1); any L determines uniquely an IL (cf. A2.), and so on. Then, one may identify derived notions which are defined in terms of the primitive ones. Thus, for instance, the notion of parallel lines can be defined from the primitives as it was done in D23.

The difference may seem negligible but is of the utmost importance. By insisting on the *uninterpreted* character of the primitive notions, it opens an entirely new perspective. On the one hand, we have our primitive, uninterpreted notions. These can be manipulated according to the axioms and rules we have postulated. On the other hand, there are *various* possibilities of *interpreting* these primitive notions. All such interpretations will have to satisfy the axioms and conform to the rules, but otherwise they may be vastly different. This was the insight which led, first, to non-Euclidean geometry and, then, to the formal systems. We now illustrate this first stage of the development.

The strongest, and relatively simple formulation of the famous fifth axiom, the "Parallel Postulate", is as follows:

A5. Given a line L and a point p not on line L, exactly one line L' can be drawn through p parallel to L (i.e., not intersecting L no matter how far extended).

This axiom seems to be much less intuitive than the other ones and mathematicians had spent centuries trying to derive it from the other ones. Failing to do that, they started to wonder: "But, does this postulate have to be true? What if it isn't?"

Well, it may seem that it is true – but how can we check? It
may be hard to prolong any line indefinitely. Thus we encounter the
other aspect of formal systems, which we will study in the following
chapters, namely, what is the *meaning* or *semantics* of such a system.
Designing an axiomatic system, one has to specify precisely what are
its primitive terms and how these terms may interact in derivation
of the theorems. On the other hand, one specifies what these terms
are supposed to denote. In fact, terms of a formal system may denote
anything which conforms to the rules specified for their interaction.
Euclidean geometry was designed with a particular model in mind –
the abstract set of points, lines and figures that can be constructed with
compass and straightedge in an infinite space. But now, allowing for
the primitive character of the basic notions, we can consider other
interpretations. We can consider as our space a finite, open circle C,
interpret P as any point within C, L as any closed interval within C
and IL as an open-ended chord of the circle, i.e., a straight line within
the circle which approaches indefinitely closely, but never touches the
circumference. (Thus one can "prolong a line indefinitely" without
ever meeting the circumference.)

This interpretation does not satisfy the fifth postulate. Starting with
a line L and a point p not on L, we can choose two other points x
and y and, by A1, obtain two lines xp and yp which can be prolonged
indefinitely according to A2. As we see, neither of these indefinitely
prolonged lines intersects L. Thus, both are parallel to L according to
the very same, old definition D23.

Failing to satisfy the fifth postulate, this interpretation is not a
model of Euclidean geometry. But it is a model of the first non-
Euclidean geometry – the Bolyai-Lobachevsky geometry, which keeps
all the definitions, postulates and rules except the fifth postulate.
Later, many other non-Euclidean geometries have been developed –
perhaps the most famous one, by Hermann Minkowski as a four-
dimensional space-time universe of the relativity theory.

Now we can see another advantage of using axiomatic systems. Since non-Euclidean geometry preserves all Euclid's postulates except the fifth one, all the theorems derived *without* the use of the fifth postulate remain valid. For instance, Proposition 4.1 needs no new proof in the new geometries. Revealing which sets of postulates are needed to establish which consequences, allows their reuse. Studying some new phenomena, one can then start by asking which postulates are satisfied by them. An answer to this question yields then immediately all the theorems which have been proven from these postulates.

It should be constantly kept in mind that axiomatic systems, their primitive terms and proofs, are purely syntactic, that is, do *not presuppose any interpretation*. Of course, their eventual usefulness depends on whether we can find interesting interpretations for their terms and rules but this is another story. In this chapter, we study some fundamental axiomatic systems without considering such interpretations, which will be addressed later on. ──────────────

Recall that an inductive definition of a set consists of a BASIS, an INDUCTION part, and an implicit CLOSURE condition. When the set defined is a language, i.e., a set of strings, we often talk about an *axiomatic system*. In this case, the elements of the basis are called *axioms*, the induction part is given by a set of *proof rules*, and *theorems* are the members of so defined set. An axiomatic system is thus a finite representation of a typically infinite set of theorems. The symbol ⊢ denotes this set, but the membership $A \in \, \vdash$ is written ⊢ A. Usually, it is a subset of some other language $L \subseteq \Sigma^*$, thus ⊢ $\subseteq L \subseteq \Sigma^*$.

Definition 4.2 An axiomatic system ⊢ over $L \subseteq \Sigma^*$ has the form:

AXIOMS :: A set $Ax \subseteq \, \vdash \subseteq L$, and

PROOF
RULES :: the elements $R \in L^n \times L$, written $R : \dfrac{\vdash A_1; \ldots; \vdash A_n}{\vdash C}$.

A proof rule says that if the *premises* are theorems, ⊢ $A_1 \ldots$ ⊢ A_n, then so is the *conclusion* ⊢ C. The rules are always designed so that C is in L if $A_1, \ldots, A_n$ are, thus ⊢ is guaranteed to be a subset of L. A formula A is a *theorem* of the system iff there is a proof of A in the system.

Definition 4.3 A *proof* in an axiomatic system is a finite sequence $A_1, \ldots, A_n$ of strings from L, such that for each A_i

- either $A_i \in Ax$ or else
- there are $A_{i_1}, \ldots, A_{i_k}$ in the sequence with all $i_1, \ldots, i_k < i$, and an
  application of a proof rule $R : \dfrac{\vdash A_{i_1}; \ldots; \vdash A_{i_k}}{\vdash A_i}$.

A proof *of* A is a proof in which A is the final string.

Remark.
For a given language L there may be several axiomatic systems which all define the same subset of L, albeit, by means of very different axioms and rules. There are also variations which we will consider, where the predicate $\vdash$ is defined on various sets built over L, for instance, $\mathcal{P}(\mathsf{L}) \times \mathsf{L}$. $\square$

4.2 Propositional syntax

The basic logical system, originating with Boole's algebra, is propositional logic (PL). The name reflects the fact that the expressions of the language are "intended as" propositions. This interpretation will be part of the semantics of PL to be discussed in the following chapter. Here we introduce syntax and the associated axiomatic proof system for PL (cf. Exercise 2.3).

Definition 4.4   The language of well-formed formulae of PL is given by:

(1) An alphabet: a set of propositional variables $\Sigma = \{a, b, \ldots\}$ together with the (formula building) connectives, $\neg$ and $\rightarrow$, and auxiliary symbols $(,)$.

(2) The *well-formed formulae*, $\mathsf{WFF}_{\mathsf{PL}}^{\Sigma}$ are defined inductively:

   Basis :: $\Sigma \subseteq \mathsf{WFF}_{\mathsf{PL}}^{\Sigma}$;
   Ind :: (a) if $A \in \mathsf{WFF}_{\mathsf{PL}}^{\Sigma}$ then $\neg A \in \mathsf{WFF}_{\mathsf{PL}}^{\Sigma}$
           (b) if $A, B \in \mathsf{WFF}_{\mathsf{PL}}^{\Sigma}$ then $(A \rightarrow B) \in \mathsf{WFF}_{\mathsf{PL}}^{\Sigma}$.

(3) The propositional variables are called *atomic formulae*, the formulae of the form A or $\neg A$, where A is atomic are called *literals*.

Remark 4.5 [Some conventions]
1) The outermost pair of parentheses is often suppressed, hence $A \rightarrow (B \rightarrow C)$ stands for $(A \rightarrow (B \rightarrow C))$ while $(A \rightarrow B) \rightarrow C$ for $((A \rightarrow B) \rightarrow C)$.

2) Formulae are strings over Σ extended with additional symbols, $\mathsf{WFF}_{\mathsf{PL}}^{\Sigma} \subseteq (\Sigma \cup \{), (, \rightarrow, \neg\})^*$. We use lower case letters for the propositional variables from Σ, while upper case letters for the schematic variables, standing for arbitrary formulae. The definition yields a different set of formulae for different Σ's, e.g., the sets $\mathsf{WFF}_{\mathsf{PL}}^{\Sigma}$ and $\mathsf{WFF}_{\mathsf{PL}}^{\Sigma_1}$, for $\Sigma = \{a, b\}$ and $\Sigma_1 = \{c, d\}$,

are disjoint (though bijective). Writing $\mathsf{WFF}_{\mathsf{PL}}$ we mean well-formed PL formulae over an arbitrary alphabet and most of our discussion is concerned with this general case irrespectively of the particular Σ.

3) It is always implicitly assumed that $\Sigma \neq \varnothing$.

4) For the reasons which we will explain later, occasionally, we may use the abbreviations $\bot$ for $\neg(B \to B)$ and $\top$ for $B \to B$, for arbitrary B. $\square$

In the following we will always – unless explicitly stated otherwise – assume that the considered formulae are well-formed.

4.3 HILBERT'S AXIOMATIC SYSTEM $\mathcal{H}$

Hilbert's system $\mathcal{H}$ for PL is defined as the predicate $\vdash_{\mathcal{H}} \subseteq \mathsf{WFF}_{\mathsf{PL}}$, and $\vdash_{\mathcal{H}} B$ is read as B *is provable in* $\mathcal{H}$.

Definition 4.6   The predicate $\vdash_{\mathcal{H}}$ of *Hilbert's system* for PL is defined inductively by:

AXIOMS :: A1: $\vdash_{\mathcal{H}} A \to (B \to A)$;
   A2: $\vdash_{\mathcal{H}} (A \to (B \to C)) \to ((A \to B) \to (A \to C))$;
   A3: $\vdash_{\mathcal{H}} (\neg B \to \neg A) \to (A \to B)$;

PROOF
RULE :: called *Modus Ponens:* $\dfrac{\vdash_{\mathcal{H}} A \ ; \ \vdash_{\mathcal{H}} A \to B}{\vdash_{\mathcal{H}} B}$.

Remark 4.7 [Axioms vs. axiom schemata]
$A1$–$A3$ are in fact axiom *schemata*; the actual axioms comprise all formulae of the indicated form with the letters A, B, C instantiated to arbitrary formulae. For each particular alphabet Σ, there will be a different (infinite) collection of actual axioms. Similar instantiations are performed in the proof rule. For instance, for $\Sigma = \{a, b, c, d\}$, all the following formulae are instances of axiom schemata:
   $A1 : b \to (a \to b)$, $(b \to d) \to (\neg a \to (b \to d))$, $a \to (a \to a)$,
   $A3 : (\neg\neg d \to \neg b) \to (b \to \neg d)$.
The following formulae are *not* (instances of the) axioms:
   $a \to (b \to b)$, $(\neg b \to a) \to (\neg a \to b)$.
Hence, an axiom schema, like $A1$, is actually a predicate giving, for any Σ, the set of Σ-instances $A1^{\Sigma} = \{x \to (y \to x) : x, y \in \mathsf{WFF}_{\mathsf{PL}}^{\Sigma}\} \subset \mathsf{WFF}_{\mathsf{PL}}^{\Sigma}$.

   Also every proof rule R with n premises (in an axiomatic system over a language L) is typically given as a *schema* – a relation $R \subseteq \mathsf{L}^n \times \mathsf{L}$. Each substitution for the schematic variables $A, B, \ldots$ of some formulae over the actual alphabet Σ, yields a particular *application of the rule* $r \in R$. For instance, Hilbert's Modus Ponens schema yields an infinite set of its applications $MP^\Sigma = \{\langle x, x \to y, y \rangle : x, y \in \mathsf{WFF}^\Sigma_{\mathsf{PL}}\} \subset \mathsf{WFF}^\Sigma_{\mathsf{PL}} \times \mathsf{WFF}^\Sigma_{\mathsf{PL}} \times \mathsf{WFF}^\Sigma_{\mathsf{PL}}$. The following are examples of such applications for $\{a, b, c\} \subseteq \Sigma$:

$$\frac{\vdash_\mathcal{H} a \; ; \; \vdash_\mathcal{H} a \to b}{\vdash_\mathcal{H} b} \qquad \frac{\vdash_\mathcal{H} a \to \neg b \; ; \; \vdash_\mathcal{H} (a \to \neg b) \to (b \to c)}{\vdash_\mathcal{H} b \to c} \qquad \cdots$$

In $\mathcal{H}$, the sets Ai^Σ and MP^Σ are *recursive*, provided that Σ is (which it always is by assumption). Recursivity of MP^Σ means that we can always *decide* whether a given triple of formulae is an application of the rule. Recursivity of the set of axioms means that we can always *decide* whether a given formula is an axiom or not. Axiomatic systems which do not satisfy these conditions are of lesser interest and we will not encounter them. $\quad\square$

That both sets Ai^Σ and MP^Σ of $\mathcal{H}$ are recursive does not imply that so is $\vdash_\mathcal{H}$. This only means that *given* a sequence of formulae, we can *decide* whether it is a proof or not. To decide if a given formula belongs to $\vdash_\mathcal{H}$ would require a procedure for deciding if such a proof exists – probably, a procedure for *constructing* a proof. We will see several examples illustrating that, even if such a procedure for $\vdash_\mathcal{H}$ exists, it is by no means simple.

Lemma 4.8 For an arbitrary $B \in \mathsf{WFF}_{\mathsf{PL}} : \vdash_\mathcal{H} B \to B$.

Proof.

$1 : \vdash_\mathcal{H} (B \to ((B \to B) \to B)) \; \to \; ((B \to (B \to B)) \to (B \to B)) \quad A2$

$2 : \vdash_\mathcal{H} B \to ((B \to B) \to B) \quad A1$

$3 : \vdash_\mathcal{H} (B \to (B \to B)) \to (B \to B) \quad MP : 2, 1$

$4 : \vdash_\mathcal{H} B \to (B \to B) \quad A1$

$5 : \vdash_\mathcal{H} B \to B \quad MP : 4, 3$

$\hfill$ QED (4.8)

The phrase "for an arbitrary $B \in \mathsf{WFF}_{\mathsf{PL}}$" indicates that any formula of the above form (with any well-formed formula over any actual alphabet Σ substituted for B) is derivable, e.g. $\vdash_\mathcal{H} a \to a$, $\vdash_\mathcal{H} (a \to \neg b) \to (a \to \neg b)$, etc. All the results about PL will be stated in such a schematic way.

But we cannot substitute different formulae for the two occurrences of B. If we try to apply the above proof to deduce $\vdash_\mathcal{H} A \to B$ it will fail – identify the place(s) where it would require invalid transitions.

The above lemma means that we can always, for arbitrary formula B, use $\vdash_\mathcal{H} B \to B$ as a step in a proof. Also derivations can be reused, when they are "stored" as admissible rules.

Definition 4.9 Let C be an axiomatic system. A rule $\dfrac{\vdash_C A_1 \; ; \; \ldots \; ; \vdash_C A_n}{\vdash_C C}$ is

admissible in C if whenever there are proofs in C of all the premises, i.e., $\vdash_C A_i$ for all $1 \leq i \leq n$, then there is a proof in C of the conclusion $\vdash_C C$.

Lemma 4.10 The following rules are admissible in $\mathcal{H}$:

(1) $\dfrac{\vdash_{\mathcal{H}} A \to B \; ; \; \vdash_{\mathcal{H}} B \to C}{\vdash_{\mathcal{H}} A \to C}$ (2) $\dfrac{\vdash_{\mathcal{H}} B}{\vdash_{\mathcal{H}} A \to B}$

Proof.

(1) $1 : \vdash_{\mathcal{H}} (A \to (B \to C)) \to ((A \to B) \to (A \to C))$ $A2$
   $2 : \vdash_{\mathcal{H}} (B \to C) \to (A \to (B \to C))$ $A1$
   $3 : \vdash_{\mathcal{H}} B \to C$ *assumption*
   $4 : \vdash_{\mathcal{H}} A \to (B \to C)$ $MP : 3, 2$
   $5 : \vdash_{\mathcal{H}} (A \to B) \to (A \to C)$ $MP : 4, 1$
   $6 : \vdash_{\mathcal{H}} A \to B$ *assumption*
   $7 : \vdash_{\mathcal{H}} A \to C$ $MP : 6, 5$

(2) $1 : \vdash_{\mathcal{H}} B$ *assumption*
   $2 : \vdash_{\mathcal{H}} B \to (A \to B)$ $A1$
   $3 : \vdash_{\mathcal{H}} A \to B$ $MP : 1, 2$

QED (4.10)

The proof of the following lemma illustrates the usefulness of admissible rules in constructing new proofs.

Lemma 4.11 (1) $\vdash_{\mathcal{H}} \neg\neg B \to B$ (2) $\vdash_{\mathcal{H}} B \to \neg\neg B$.

Proof.

(1) $1 : \vdash_{\mathcal{H}} \neg\neg B \to (\neg\neg\neg\neg B \to \neg\neg B)$ $A1$
   $2 : \vdash_{\mathcal{H}} (\neg\neg\neg\neg B \to \neg\neg B) \to (\neg B \to \neg\neg\neg B)$ $A3$
   $3 : \vdash_{\mathcal{H}} \neg\neg B \to (\neg B \to \neg\neg\neg B)$ $L.4.10.(1) : 1, 2$
   $4 : \vdash_{\mathcal{H}} (\neg B \to \neg\neg\neg B) \to (\neg\neg B \to B)$ $A3$
   $5 : \vdash_{\mathcal{H}} \neg\neg B \to (\neg\neg B \to B)$ $L.4.10.(1) : 3, 4$
   $6 : \vdash_{\mathcal{H}} (\neg\neg B \to (\neg\neg B \to B)) \to$
          $\qquad\qquad ((\neg\neg B \to \neg\neg B) \to (\neg\neg B \to B))$ $A2$
   $7 : \vdash_{\mathcal{H}} (\neg\neg B \to \neg\neg B) \to (\neg\neg B \to B)$ $MP : 5, 6$
   $8 : \vdash_{\mathcal{H}} \neg\neg B \to B$ $MP : 7, L.4.8$

(2) $1 : \vdash_{\mathcal{H}} \neg\neg\neg B \to \neg B$ *point* (1)
   $2 : \vdash_{\mathcal{H}} (\neg\neg\neg B \to \neg B) \to (B \to \neg\neg B)$ $A3$
   $3 : \vdash_{\mathcal{H}} B \to \neg\neg B$ $MP : 1, 2$ QED (4.11)

4.4 The axiomatic system $\mathcal{N}$

In the system $\mathcal{N}$, instead of the unary predicate $\vdash_{\mathcal{H}}$ we use a binary relation $\vdash_{\mathcal{N}} \subseteq \mathcal{P}(\mathsf{WFF_{PL}}) \times \mathsf{WFF_{PL}}$, written as $\Gamma \vdash_{\mathcal{N}} B$. It reads B *is provable in* $\mathcal{N}$ *from the assumptions* Γ.

Definition 4.12 The axioms and rules of $\mathcal{N}$ are as in Hilbert's system with the additional axiom schema A0:

Axioms :: A0: $\Gamma \vdash_{\mathcal{N}} B$, whenever $B \in \Gamma$;
     A1: $\Gamma \vdash_{\mathcal{N}} A \rightarrow (B \rightarrow A)$;
     A2: $\Gamma \vdash_{\mathcal{N}} (A \rightarrow (B \rightarrow C)) \rightarrow ((A \rightarrow B) \rightarrow (A \rightarrow C))$;
     A3: $\Gamma \vdash_{\mathcal{N}} (\neg B \rightarrow \neg A) \rightarrow (A \rightarrow B)$;
Proof
Rule :: Modus Ponens: $\dfrac{\Gamma \vdash_{\mathcal{N}} A \;\; ; \;\; \Gamma \vdash_{\mathcal{N}} A \rightarrow B}{\Gamma \vdash_{\mathcal{N}} B}$.

As for $\mathcal{H}$, the "axioms" are actually *schemata*. The real set of axioms is the infinite set of actual formulae obtained from these schemata by substituting actual formulae for the upper case variables. Similarly for the proof rule.

The next lemma corresponds exactly to Lemma 4.10. In fact, the proof of that lemma (and most others) can be taken over line for line from $\mathcal{H}$, without any modification (just replace $\vdash_{\mathcal{H}}$ by $\Gamma \vdash_{\mathcal{N}}$), giving a proof in $\vdash_{\mathcal{N}}$.

Lemma 4.13 The following rules are admissible in $\mathcal{N}$:

(1) $\dfrac{\Gamma \vdash_{\mathcal{N}} A \rightarrow B \;\; ; \;\; \Gamma \vdash_{\mathcal{N}} B \rightarrow C}{\Gamma \vdash_{\mathcal{N}} A \rightarrow C}$
     (2) $\dfrac{\Gamma \vdash_{\mathcal{N}} B}{\Gamma \vdash_{\mathcal{N}} A \rightarrow B}$.

The name $\mathcal{N}$ reflects the intended association with the so-called "natural deduction" reasoning. This system is not exactly what is usually so called and we have adopted $\mathcal{N}$ because it corresponds so closely to $\mathcal{H}$. But while $\mathcal{H}$ derives only single formulae, tautologies, $\mathcal{N}$ provides also means for *reasoning from the assumptions* Γ. This is the central feature which it shares with natural deduction systems: they both satisfy the following Deduction Theorem. (The expression "Γ, A" is short for "$\Gamma \cup \{A\}$.")

Theorem 4.14 [Deduction Theorem] If $\Gamma, A \vdash_{\mathcal{N}} B$, then $\Gamma \vdash_{\mathcal{N}} A \rightarrow B$.

Proof. By induction on the length l of a proof of $\Gamma, A \vdash_{\mathcal{N}} B$. Basis, $l = 1$, means that the proof consists merely of an instance of an axiom and it has two cases depending on which axiom was involved:

A1-A3 :: If B is one of these axioms, then we also have $\Gamma \vdash_{\mathcal{N}} B$
     and Lemma 4.13.(2) gives the conclusion.

A0 :: If B results from this axiom, we have two subcases:

(1) If $B = A$, then Lemma 4.8 gives that $\Gamma \vdash_{\mathcal{N}} B \to B$.

(2) If $B \neq A$, then $B \in \Gamma$, and so $\Gamma \vdash_{\mathcal{N}} B$. Lemma 4.13.(2) gives $\Gamma \vdash_{\mathcal{N}} A \to B$.

MP :: B follows by MP: $\dfrac{\Gamma, A \vdash_{\mathcal{N}} C \ ; \ \Gamma, A \vdash_{\mathcal{N}} C \to B}{\Gamma, A \vdash_{\mathcal{N}} B}$.

Induction hypothesis gives lines 1.-2. of the following

$$
\begin{array}{lll}
1 : \Gamma \vdash_{\mathcal{N}} A \to C & & IH \\
2 : \Gamma \vdash_{\mathcal{N}} A \to (C \to B) & & IH \\
3 : \Gamma \vdash_{\mathcal{N}} (A \to (C \to B)) \ \to \ ((A \to C) \to (A \to B)) & A2 \\
4 : \Gamma \vdash_{\mathcal{N}} (A \to C) \ \to \ (A \to B) & & MP : 2, 3 \\
5 : \Gamma \vdash_{\mathcal{N}} A \to B & & MP : 1, 4
\end{array}
$$

QED (4.14)

Example 4.15

Deduction Theorem significantly shortens the proofs. The tedious proof of Lemma 4.8 takes now two lines: from the axiom A0, $B \vdash_{\mathcal{N}} B$, DT gives $\vdash_{\mathcal{N}} B \to B$. □

Lemma 4.16 $\vdash_{\mathcal{N}} (A \to B) \to (\neg B \to \neg A)$.

Proof.
$$
\begin{array}{lll}
1 : A \to B \vdash_{\mathcal{N}} (\neg\neg A \to \neg\neg B) \to (\neg B \to \neg A) & A3 \\
2 : A \to B \vdash_{\mathcal{N}} \neg\neg A \to A & L.4.11.(1) \\
3 : A \to B \vdash_{\mathcal{N}} A \to B & A0 \\
4 : A \to B \vdash_{\mathcal{N}} \neg\neg A \to B & L.4.13.(1) : 2, 3 \\
5 : A \to B \vdash_{\mathcal{N}} B \to \neg\neg B & L.4.11.(2) \\
6 : A \to B \vdash_{\mathcal{N}} \neg\neg A \to \neg\neg B & L.4.13.(1) : 4, 5 \\
7 : A \to B \vdash_{\mathcal{N}} \neg B \to \neg A & MP : 6, 1 \\
8 : \vdash_{\mathcal{N}} (A \to B) \to (\neg B \to \neg A) & DT \qquad \text{QED (4.16)}
\end{array}
$$

The empty set of assumptions in the above lemma – as $\vdash_{\mathcal{N}} C$ in any other context – means that $\Gamma \vdash_{\mathcal{N}} C$ holds for arbitrary Γ. Often, however, like in Deduction Theorem or the following corollary, we write Γ explicitly, because the proof of $\vdash_{\mathcal{N}} A \to B$ or $A \vdash_{\mathcal{N}} B$ may actually depend on it. Still, the claim holds then for arbitrary Γ satisfying the assumptions (in the corollary below, for which either side is provable). According to this corollary, Deduction Theorem is a kind of dual to MP: each gives one implication of the following equivalence.

Corollary 4.17 $\Gamma, A \vdash_{\mathcal{N}} B$ iff $\Gamma \vdash_{\mathcal{N}} A \to B$.

Proof. $\Rightarrow$) is Deduction Theorem 4.14. $\Leftarrow$) By Exercise 4.5, the assumption can be strengthened to $\Gamma, A \vdash_{\mathcal{N}} A \to B$. But then, also $\Gamma, A \vdash_{\mathcal{N}} A$, and by MP $\Gamma, A \vdash_{\mathcal{N}} B$.   QED (4.17)

Corollary 4.18 The rule $\dfrac{\Gamma \vdash_{\mathcal{N}} A \to (B \to C)}{\Gamma \vdash_{\mathcal{N}} B \to (A \to C)}$ is admissible in $\mathcal{N}$.

Proof. Follows easily from Corollary 4.17: $\Gamma \vdash_{\mathcal{N}} A \to (B \to C)$ iff $\Gamma, A \vdash_{\mathcal{N}} B \to C$ iff $\Gamma, A, B \vdash_{\mathcal{N}} C$. As Γ, A, B abbreviates the set $\Gamma \cup \{A, B\}$, this is also equivalent to $\Gamma, B \vdash_{\mathcal{N}} A \to C$, and then to $\Gamma \vdash_{\mathcal{N}} B \to (A \to C)$.   QED (4.18)

4.5 PROVABLE EQUIVALENCE

Equational reasoning is based on the simple principle of substitution of equals for equals. E.g., having the arithmetical expression $2 + (7 + 3)$ and knowing that $7 + 3 = 10$, we also obtain $2 + (7 + 3) = 2 + 10$. The rule applied in such cases may be written as $\dfrac{a = b}{F[a] = F[b]}$ where $F[\_]$ is an expression "with a hole" (a variable or a place-holder) into which we may substitute other expressions. We now illustrate a logical counterpart of this idea.

Lemma 4.8 showed that any formula of the form $(B \to B)$ is derivable in $\mathcal{H}$ and, by Lemma 4.29, in $\mathcal{N}$. It allows us to use, for instance, 1) $\vdash_{\mathcal{N}} a \to a$, 2) $\vdash_{\mathcal{N}} (a \to b) \to (a \to b)$, 3) ... as a step in any proof. Putting it a bit differently, the lemma says that 1) is provable iff 2) is provable iff ... Recall the abbreviation $\top$ for an *arbitrary* formula of this form introduced in Remark 4.5. It also introduced the abbreviation $\bot$ for an *arbitrary* formula of the form $\neg(B \to B)$. These abbreviations indicate that all the formulae of the respective form are equivalent in the following sense.

Definition 4.19 Formulae A and B are *provably equivalent* in a system $\mathcal{C}$ for PL, if both $\vdash_{\mathcal{C}} A \to B$ and $\vdash_{\mathcal{C}} B \to A$ (denoted $\vdash_{\mathcal{C}} A \leftrightarrow B$).

In view of Lemma 4.8 and 4.10.(1), generalized to $\mathcal{N}$, the relation $Im \subseteq \mathsf{WFF}_{\mathsf{PL}}^{\Sigma} \times \mathsf{WFF}_{\mathsf{PL}}^{\Sigma}$ defined by $Im(A, B) \Leftrightarrow \vdash_{\mathcal{N}} A \to B$ is reflexive and transitive. Definition 4.19 adds the requirement of symmetricity making $\leftrightarrow$ the greatest equivalence on the set $\mathsf{WFF}_{\mathsf{PL}}^{\Sigma}$ contained in Im.

Lemma 4.11 provides an example of provable equivalence, namely

$$\vdash_{\mathcal{H}} B \leftrightarrow \neg\neg B. \tag{4.20}$$

Another example follows from axiom A3 and Lemma 4.16:

$$\vdash_{\mathcal{N}} (A \to B) \leftrightarrow (\neg B \to \neg A). \tag{4.21}$$

In Exercise 4.2, you are asked to show that all formulae $\top$ are provably equivalent, i.e., (the proof can start with $\vdash_{\mathcal{N}} B \to B$)

$$\vdash_{\mathcal{N}} (A \to A) \leftrightarrow (B \to B). \tag{4.22}$$

To show the analogous equivalence of all $\bot$ formulae,

$$\vdash_{\mathcal{N}} \neg(A \to A) \leftrightarrow \neg(B \to B), \tag{4.23}$$

we have to proceed differently, since we do not have $\vdash_{\mathcal{N}} \neg(B \to B)$. (This is not true, as we will see later on.) We can use (4.22) and Lemma 4.16:

$$
\begin{array}{lll}
1: & \vdash_{\mathcal{N}} (A \to A) \to (B \to B) & \\
2: & \vdash_{\mathcal{N}} ((A \to A) \to (B \to B)) \to (\neg(B \to B) \to \neg(A \to A)) & L.4.16 \\
3: & \vdash_{\mathcal{N}} \neg(B \to B) \to \neg(A \to A) & MP: 1,2
\end{array}
$$

and the opposite implication is again an instance of this one.

The provable equivalence $A \leftrightarrow B$ means – and this is its main importance – that the two formulae are interchangeable. Whenever we have a proof of a formula $F[A]$ containing A (as a subformula, possibly with several occurrences), we can replace A by B – the result will be provable too. This fact is a powerful tool in simplifying proofs and is expressed in the following theorem. (The analogous version holds for $\mathcal{H}$.)

Theorem 4.24 For any formula $F[A]$, the following rule is admissible in $\mathcal{N}$:

$$\frac{\Gamma \vdash_{\mathcal{N}} A \leftrightarrow B}{\Gamma \vdash_{\mathcal{N}} F[A] \leftrightarrow F[B]}.$$

Proof. By induction on the complexity of $F[\_]$ viewed as a formula "with a hole", i.e., with $[\_]$ added to the basis of the inductive definition 4.4.(2). (There may be several occurrences of the "hole", i.e., $F[\_]$ may be $\neg[\_]$ or $[\_] \to (\neg G \to [\_])$, etc.). We drop Γ in the notation.

$a \in \Sigma$ :: i.e., the premise $a \leftrightarrow a$ remains unchanged in the conclusion.

$\quad [\_]$ :: i.e., the premise $A \leftrightarrow B$ remains unchanged in the conclusion.

$\neg G[\_]$ :: IH gives the claim for $G[\_]$: $\dfrac{\vdash_{\mathcal{N}} A \leftrightarrow B}{\vdash_{\mathcal{N}} G[A] \leftrightarrow G[B]}$. Then

$$
\begin{array}{lll}
1: & \vdash_{\mathcal{N}} A \to B & assumption \\
2: & \vdash_{\mathcal{N}} G[A] \to G[B] & IH \\
3: & \vdash_{\mathcal{N}} (G[A] \to G[B]) \to (\neg G[B] \to \neg G[A]) & L.4.16 \\
4: & \vdash_{\mathcal{N}} \neg G[B] \to \neg G[A] & MP: 2,3
\end{array}
$$

Similarly, starting with $\vdash_{\mathcal{N}} B \to A$ gives $\vdash_{\mathcal{N}} \neg G[A] \to \neg G[B]$.

$G \to H$ :: Given $\vdash_N A \leftrightarrow B$, the IH gives the following four assumptions:
$\vdash_N G[A] \to G[B]$ and $\vdash_N G[B] \to G[A]$, and
$\vdash_N H[A] \to H[B]$ and $\vdash_N H[B] \to H[A]$.
We show only one implication, namely, $\vdash_N F[A] \to F[B]$:

$$
\begin{array}{lll}
1 : \vdash_N H[A] \to H[B] & & IH \\
2 : G[A] \to H[A] \vdash_N H[A] \to H[B] & & Ex.4.5 \\
3 : G[A] \to H[A] \vdash_N G[A] \to H[A] & & A0 \\
4 : G[A] \to H[A] \vdash_N G[A] \to H[B] & & L.4.13.(1) : 3, 2 \\
5 : \vdash_N G[B] \to G[A] & & IH \\
6 : G[A] \to H[A] \vdash_N G[B] \to G[A] & & Ex.4.5 \\
7 : G[A] \to H[A] \vdash_N G[B] \to H[B] & & L.4.13.(1) : 6, 4 \\
8 : \vdash_N (G[A] \to H[A]) \to (G[B] \to H[B]) & & DT : 7
\end{array}
$$

The other implication follows by a symmetric proof.

QED (4.24)

The theorem, together with the preceding observations about the equivalence of all $\top$, respectively, all $\bot$ formulae justify the use of these abbreviations: in a proof, any formula of the form $\top$, respectively $\bot$, can be replaced by any other formula of the same form. As a simple consequence of the theorem, we obtain:

Corollary 4.25 For any formula $F[A]$, the following rule is admissible:

$$
\frac{\Gamma \vdash_N F[A] \ ; \ \Gamma \vdash_N A \leftrightarrow B}{\Gamma \vdash_N F[B]} .
$$

Proof. If $\Gamma \vdash_N A \leftrightarrow B$, Theorem 4.24 gives us $\Gamma \vdash_N F[A] \leftrightarrow F[B]$ which, in particular, implies $\Gamma \vdash_N F[A] \to F[B]$. MP applied to this and the premise $\Gamma \vdash_N F[A]$, gives $\Gamma \vdash_N F[B]$. QED (4.25)

4.6 CONSISTENCY

Lemma 4.8, and the discussion of provable equivalence above, show that for any Γ (also for $\Gamma = \varnothing$) we have $\Gamma \vdash_N \top$, where $\top$ is an arbitrary instance of $B \to B$. The following notion indicates that the similar fact for $\bot$, namely $\Gamma \vdash_N \bot$, need not always hold.

Definition 4.26 A set of formulae Γ is *consistent* iff $\Gamma \nvdash_N \bot$.

An equivalent formulation says that Γ is consistent iff there is a formula A such that $\Gamma \nvdash_N A$. In fact, if $\Gamma \vdash_N A$ for all A then, in particular, $\Gamma \vdash_N \bot$. Equivalence follows then by the next lemma.

Lemma 4.27 The following rule, called *Ex Falso Quodlibet*, is admissible in the system $\mathcal{N}$ (and $\mathcal{H}$): $\dfrac{\Gamma \vdash_{\mathcal{N}} \bot}{\Gamma \vdash_{\mathcal{N}} A}$.

Proof. (Observe how Corollary 4.25 simplifies the proof.)

$$
\begin{array}{lll}
1 : \Gamma \vdash_{\mathcal{N}} \neg(B \to B) & assumption \\
2 : \Gamma \vdash_{\mathcal{N}} B \to B & L.4.8 \\
3 : \Gamma \vdash_{\mathcal{N}} \neg A \to (B \to B) & 2 + L.4.13.(2) \\
4 : \Gamma \vdash_{\mathcal{N}} \neg(B \to B) \to \neg\neg A & C.4.25 \ (4.21) \\
5 : \Gamma \vdash_{\mathcal{N}} \neg\neg A & MP : 1, 4 \\
6 : \Gamma \vdash_{\mathcal{N}} A & C.4.25 \ (4.20) & \text{QED } (4.27)
\end{array}
$$

The formula A in the conclusion of the rule is arbitrary. This is the syntactic reason for why inconsistent *theories*, sets of "assumptions" Γ, are uninteresting. Such a set makes the machinery of the proof system irrelevant for checking whether something is a theorem or not – every well-formed formula is a theorem. Similarly, an axiomatic system, like $\mathcal{H}$, is inconsistent if its rules and axioms allow us to derive $\vdash_{\mathcal{H}} \bot$.

Notice that the definition of consistency requires that $\bot$ is *not* derivable. Thus, to decide if Γ is consistent it does not suffice to run enough proofs and see what can be derived from Γ. One must show that, no matter what, one will never be able to derive $\bot$. This, in general, may be an infinite task requiring searching through all the proofs. If $\bot$ is derivable, we will eventually construct a proof of it, but if it is not, we will never reach any conclusion. In general, consistency of a given theory may be only co-semi-decidable. In $\mathcal{N}$, consistency of any finite Γ is decidable, since "being a theorem" is decidable for finite Γ (see Subsection 4.8.1). But for infinite Γ, it is only co-semi-decidable. In some cases, the following theorem may be used to ease the process of determining the (in)consistency of an infinite Γ.

Theorem 4.28 [Compactness] Γ is consistent iff each finite subset $\Delta \subseteq \Gamma$ is consistent.

Proof. $\Rightarrow$) If $\Gamma \nvdash_{\mathcal{N}} \bot$ then no subset of Γ proves $\bot$.

$\Leftarrow$) Contrapositively, assume that Γ is inconsistent. The proof of $\bot$ must be finite and, in particular, uses only a finite number of assumptions $\Delta \subseteq \Gamma$. This means that the proof $\Gamma \vdash_{\mathcal{N}} \bot$ can be carried from a finite subset Δ of Γ, i.e., $\Delta \vdash_{\mathcal{N}} \bot$.   QED (4.28)

4.7 $\mathcal{H}$ VERSUS $\mathcal{N}$

In $\mathcal{H}$ we prove only single formulae, while in $\mathcal{N}$ we work "from assumptions Γ" proving their consequences. Since the axiom schemata and rules of $\mathcal{H}$ are special cases of their counterparts in $\mathcal{N}$, it is obvious that for any formula B, if $\vdash_{\mathcal{H}} B$ then $\varnothing \vdash_{\mathcal{N}} B$. In fact this can be strengthened to an equivalence. (We follow the convention of writing $\vdash_{\mathcal{N}} B$ for $\varnothing \vdash_{\mathcal{N}} B$.)

Lemma 4.29 For any formula B we have: $\vdash_{\mathcal{H}} B$ iff $\vdash_{\mathcal{N}} B$.

Proof. One direction is noted above. In fact, any proof of $\vdash_{\mathcal{H}} B$ itself qualifies as a proof of $\vdash_{\mathcal{N}} B$. The other direction is almost as obvious, since there is no way to make any real use of A0 in a proof of $\vdash_{\mathcal{N}} B$. More precisely, take any proof of $\vdash_{\mathcal{N}} B$ and delete all lines (if any) of the form $\Gamma \vdash_{\mathcal{N}} A$ for $\Gamma \neq \varnothing$. The result is still a proof of $\vdash_{\mathcal{N}} B$, and now also of $\vdash_{\mathcal{H}} B$.

More formally, the lemma can be proved by induction on the length of a proof of $\vdash_{\mathcal{N}} B$: Since $\Gamma = \varnothing$ the last step of the proof could have used either an axiom A1, A2, A3 or MP. The same step can be then done in $\mathcal{H}$ – for MP, the proofs of $\vdash_{\mathcal{N}} A$ and $\vdash_{\mathcal{N}} A \to B$ for the appropriate A are shorter and hence by the IH have counterparts in $\mathcal{H}$. QED (4.29)

The next lemma is a further generalization of this equivalence..

Lemma 4.30 $\vdash_{\mathcal{H}} G_1 \to (G_2 \to ...(G_n \to B)...)$ iff $\{G_1, G_2, \ldots, G_n\} \vdash_{\mathcal{N}} B$.

Proof. We prove the lemma by induction on n:

BASIS :: This case, corresponding to $n = 0$, is just the previous lemma.

IND. :: Suppose the IH, i.e., for any B:

$\vdash_{\mathcal{H}} G_1 \to (G_2 \to ...(G_n \to B)...)$ iff $\{G_1, G_2...G_n\} \vdash_{\mathcal{N}} B$.

Then, taking $(G_{n+1} \to B)$ for B, we obtain

$\vdash_{\mathcal{H}} G_1 \to (G_2 \to ...(G_n \to (G_{n+1} \to B))..)$

(by IH)   iff $\{G_1, G_2...G_n\} \vdash_{\mathcal{N}} (G_{n+1} \to B)$

(by Corollary 4.17)   iff $\{G_1, G_2, \ldots, G_n, G_{n+1}\} \vdash_{\mathcal{N}} B$.

                        QED (4.30)

Lemma 4.29 stated the equivalence of $\mathcal{N}$ and $\mathcal{H}$ with respect to the simple formulae of $\mathcal{H}$. This lemma states a more general equivalence of these two systems: for every finite Γ, B there is a corresponding $\mathcal{H}$-expression H such that $\Gamma \vdash_{\mathcal{N}} B$ iff $\vdash_{\mathcal{H}} H$, and vice versa.

Observe, however, that this equivalence is restricted to finite Γ. The significant difference between the two systems consists in that $\mathcal{N}$ allows to consider also consequences of infinite sets of assumptions, for which there are no corresponding formulae in $\mathcal{H}$, since every formula must be finite.

4.8   GENTZEN'S AXIOMATIC SYSTEM $\mathcal{G}$

By now you should be convinced that it is rather cumbersome to design proofs in $\mathcal{H}$ or $\mathcal{N}$. From the mere form of the axioms and rules of these systems it is by no means clear that they define recursive sets of formulae. (As usual, it is easy to see (a bit more tedious to prove) that these sets are recursively enumerable.)

We give yet another axiomatic system for PL in which proofs can be constructed mechanically. The relation $\vdash_{\mathcal{G}} \subseteq \mathcal{P}(\mathsf{WFF}_{\mathsf{PL}}) \times \mathcal{P}(\mathsf{WFF}_{\mathsf{PL}})$, contains expressions, called *sequents*, of the form $\Gamma \vdash_{\mathcal{G}} \Delta$, where $\Gamma, \Delta \subseteq \mathsf{WFF}_{\mathsf{PL}}$ are finite sets of formulae. It is defined inductively as follows:

AXIOMS :: $\Gamma \vdash_{\mathcal{G}} \Delta$, whenever $\Gamma \cap \Delta \neq \varnothing$
  RULES ::

$$(\neg\vdash) \ \frac{\Gamma \vdash_{\mathcal{G}} \Delta, A}{\Gamma, \neg A \vdash_{\mathcal{G}} \Delta} \qquad\qquad (\vdash\neg) \ \frac{\Gamma, A \vdash_{\mathcal{G}} \Delta}{\Gamma \vdash_{\mathcal{G}} \Delta, \neg A}$$

$$(\rightarrow\vdash) \ \frac{\Gamma \vdash_{\mathcal{G}} \Delta, A \ ; \ \Gamma, B \vdash_{\mathcal{G}} \Delta}{\Gamma, A \rightarrow B \vdash_{\mathcal{G}} \Delta} \qquad\qquad (\vdash\rightarrow) \ \frac{\Gamma, A \vdash_{\mathcal{G}} \Delta, B}{\Gamma \vdash_{\mathcal{G}} \Delta, A \rightarrow B}$$

The power of the system is the same whether we allow Γ's and Δ's in the axioms to contain arbitrary formulae or only atomic ones. We comment now on the "mechanical" character of $\mathcal{G}$ and the way one can use it.

4.8.1   DECIDABILITY OF PL

Gentzen's system defines a set $\vdash_{\mathcal{G}} \subseteq \mathcal{P}(\mathsf{WFF}_{\mathsf{PL}}) \times \mathcal{P}(\mathsf{WFF}_{\mathsf{PL}})$. Unlike for $\mathcal{H}$ or $\mathcal{N}$, it is (almost) obvious that this set is recursive – we do not give a formal proof but indicate its main steps.

Theorem 4.31   Relation $\vdash_{\mathcal{G}}$ is decidable.

Proof. Given any sequent $\Gamma \vdash_{\mathcal{G}} \Delta = G_1, ..., G_n \vdash_{\mathcal{G}} D_1, ..., D_m$, we view both Γ and Δ as sequences, rather than as sets. We can start processing the formulae in an arbitrary order, for instance, from left to right, applying relevant rules *bottom-up!* For instance, $B \rightarrow A, \neg A \vdash_{\mathcal{G}} \neg B$ is shown by building the proof starting at the bottom line:

$$
\begin{array}{c}
(\vdash\neg) \dfrac{B \vdash_{\mathcal{G}} B, A}{\vdash_{\mathcal{G}} \neg B, B, A} \qquad A \vdash_{\mathcal{G}} \neg B, A \\[2mm]
(\to\vdash) \dfrac{\quad}{\dfrac{B \to A \vdash_{\mathcal{G}} \neg B, A}{(\neg\vdash)\dfrac{\quad}{B \to A, \neg A \vdash_{\mathcal{G}} \neg B}}}
\end{array}.
$$

In general, the proof in $\mathcal{G}$ proceeds as follows:

- If G_i is atomic, we continue with G_{1+i}, and then with D's.
- If G_i/D_i is not atomic, it is either $\neg X$ or $X \to Y$. In either case there is *only one* rule which can be applied (bottom-up). Premise(s) of this rule are uniquely determined by the conclusion (G_i/D_i processed at the moment) and its application will remove the main connective, i.e., *reduce the number of* $\neg$, *resp.* $\to$!
- Thus, eventually, we will arrive at a sequent $\Gamma' \vdash_{\mathcal{G}} \Delta'$ which contains only atomic formulae. We then only have to check whether $\Gamma' \cap \Delta' = \varnothing$, which is obviously a decidable problem since both sets are finite.

Notice that the rule $(\to\vdash)$ "splits" the proof into two branches, but each of them contains fewer connectives. We have to process both branches but, again, for each we will eventually arrive at sequents with only atomic formulae. The initial sequent is derivable in $\mathcal{G}$ iff all such branches terminate with axioms. And it is not derivable iff at least one terminates with a non-axiom (i.e., $\Gamma' \vdash_{\mathcal{G}} \Delta'$ where $\Gamma' \cap \Delta' = \varnothing$). Since all branches are guaranteed to terminate $\vdash_{\mathcal{G}}$ is decidable.   QED (4.31)

Now, notice that the expressions used in $\mathcal{N}$ are special cases of sequents, namely, the ones with exactly one formula on the right of $\vdash_{\mathcal{N}}$. If we restrict our attention in $\mathcal{G}$ to such sequents, the above theorem still tells us that the respective restriction of $\vdash_{\mathcal{G}}$ is decidable. We now indicate the main steps involved in showing that this restricted relation is the same as $\vdash_{\mathcal{N}}$. As a consequence, $\vdash_{\mathcal{N}}$ is decidable, too. That is, we want to show that

$$
\Gamma \vdash_{\mathcal{N}} B \quad \text{iff} \quad \Gamma \vdash_{\mathcal{G}} B. \tag{4.32}
$$

(1) In Exercise 4.9, you are asked to prove a part of the implication "if $\vdash_{\mathcal{N}} B$ then $\vdash_{\mathcal{G}} B$", by showing that all axioms of $\mathcal{N}$ are derivable in $\mathcal{G}$. It is tedious, but not too difficult, to show that also the MP rule is admissible in $\mathcal{G}$. It is there called the (cut)-rule, whose instance can be formulated as:

$$
\dfrac{\Gamma \vdash_{\mathcal{G}} A \;\; ; \;\; \Gamma, A \vdash_{\mathcal{G}} B}{\Gamma \vdash_{\mathcal{G}} B} \;\; (cut) \tag{4.33}
$$

and MP is easily derivable from it. (If $\Gamma \vdash_\mathcal{G} A \to B$, then it must have been derived using the rule ($\vdash \to$), i.e., we must have had earlier ("above") in the proof of the right premise $\Gamma, A \vdash_\mathcal{G} B$. Thus we could have applied (cut) at this earlier stage and obtain $\Gamma \vdash_\mathcal{G} B$, without deriving $\Gamma \vdash_\mathcal{G} A \to B$.)

(2) To complete the proof we would have to show also that $\mathcal{G}$ does not prove more formulae than $\mathcal{N}$ does, namely, "if $\vdash_\mathcal{G} B$ then $\vdash_\mathcal{N} B$". (Otherwise, the problem would be still open, since we would have a decision procedure for $\vdash_\mathcal{G}$ but not for $\vdash_\mathcal{N} \subset \vdash_\mathcal{G}$. I.e., for some formula $B \notin \vdash_\mathcal{N}$ we might still get the positive answer, which would merely mean that $B \in \vdash_\mathcal{G}$.) This part is more involved since Gentzen's rules for $\neg$ do not produce $\mathcal{N}$-expressions, i.e., a proof in $\mathcal{G}$ may go through intermediary steps involving expressions not existing ("illegal") in $\mathcal{N}$. (Exercise 6.14 gives another proof of (4.32).)

(3) Finally, if $\mathcal{N}$ is decidable, then so is $\mathcal{H}$ by Lemma 4.29 – according to it, to decide if $\vdash_\mathcal{H} B$ is the same as deciding if $\vdash_\mathcal{N} B$.

4.8.2   RULES FOR ABBREVIATED CONNECTIVES

Gentzen's rules form a very well-structured system. For each connective, $\to$, $\neg$ there are two rules – one treating its occurrence on the left, and one on the right of $\vdash_\mathcal{G}$. As we will soon see, it makes often things easier if one is allowed to work with some abbreviations for frequently occurring combinations of symbols. For instance, assume that in the course of some proof, we run again and again in the sequence of the form $\neg A \to B$. Processing it requires application of at least two rules. One may be therefore tempted to define a new connective $A \vee B \overset{\text{def}}{=} \neg A \to B$, and a new rule for its treatment. In fact, in Gentzen's system we should obtain two rules for the occurrence of this new symbol on the left, resp. on the right of $\vdash_\mathcal{G}$. Looking back at the original rules from the beginning of this section, we can see how such a connective should be treated:

$$\frac{\dfrac{\dfrac{\Gamma \vdash_\mathcal{G} A, B, \Delta}{\Gamma, \neg A \vdash_\mathcal{G} B, \Delta}\;(\neg\vdash)}{\Gamma \vdash_\mathcal{G} \neg A \to B, \Delta}\;(\vdash\to)}{\Gamma \vdash_\mathcal{G} A \vee B, \Delta}\;\text{def.} \qquad (\neg\vdash)\;\frac{\dfrac{\dfrac{\Gamma, A \vdash_\mathcal{G} \Delta}{\Gamma \vdash_\mathcal{G} \neg A, \Delta} \qquad \Gamma, B \vdash_\mathcal{G} \Delta}{\Gamma, \neg A \to B \vdash_\mathcal{G} \Delta}\;(\to\vdash)}{\Gamma, A \vee B \vdash_\mathcal{G} \Delta}\;\text{def.}$$

Abbreviating these two derivations yields the following two rules:

$$(\vdash\vee)\;\frac{\Gamma \vdash_\mathcal{G} A, B, \Delta}{\Gamma \vdash_\mathcal{G} A \vee B, \Delta} \qquad (\vee\vdash)\;\frac{\Gamma, A \vdash_\mathcal{G} \Delta \;\;;\;\; \Gamma, B \vdash_\mathcal{G} \Delta}{\Gamma, A \vee B \vdash_\mathcal{G} \Delta}.$$

In a similar fashion, we may construct the rules for another, very common abbreviation, $A \wedge B \stackrel{\text{def}}{=} \neg(A \to \neg B)$:

$$(\vdash\wedge) \quad \frac{\Gamma \vdash_{\mathcal{G}} A, \Delta \quad ; \quad \Gamma \vdash_{\mathcal{G}} B, \Delta}{\Gamma \vdash_{\mathcal{G}} A \wedge B, \Delta} \qquad\qquad (\wedge\vdash) \quad \frac{\Gamma, A, B \vdash_{\mathcal{G}} \Delta}{\Gamma, A \wedge B \vdash_{\mathcal{G}} \Delta}.$$

It is hard to imagine how to perform a similar construction in the systems $\mathcal{H}$ or $\mathcal{N}$. We will meet the above abbreviations in the following chapters.

4.9 SOME PROOF TECHNIQUES

In the next chapter we will see that the formulae of PL may be interpreted as propositions – statements possessing boolean value true or false. The connective $\neg$ may be then interpreted as negation of the argument proposition, while $\to$ as (a kind of) implication. With this intuition, we may recognize some of the provable facts (either formulae or admissible rules) as giving rise to particular strategies of proof which can be – and are – utilized at all levels, in fact, throughout the whole of mathematics, as well as in much of informal reasoning. Most results from PL can be viewed in this way, and we give only a few common examples.

• As a trivial example, the provable equivalence $\vdash_{\mathcal{N}} B \leftrightarrow \neg\neg B$ from (4.20), means that in order to show double negation $\neg\neg B$, it suffices to show B. One will hardly try to say "I am *not unmarried.*" – "I am married." is both more convenient and natural.

• Let G, D stand, respectively, for the statements '$\Gamma \vdash_{\mathcal{N}} \bot$' and '$\Delta \vdash_{\mathcal{N}} \bot$ for some $\Delta \subseteq \Gamma$' from the proof of Theorem 4.28. In the second point, we showed $\neg D \to \neg G$ contrapositively, i.e., by showing $G \to D$. That this is a legal and sufficient way of proving the first statement can be justified by appealing to (4.21) – $\vdash_{\mathcal{N}} (A \to B) \leftrightarrow (\neg B \to \neg A)$ says precisely that proving one is the same as (equivalent to) proving the other.

• Another proof technique is expressed in Corollary 4.17: $A \vdash_{\mathcal{N}} B$ iff $\vdash_{\mathcal{N}} A \to B$. Treating formulae on the left of $\vdash_{\mathcal{N}}$ as assumptions, this tells us that in order to prove that A implies B, $A \to B$, we may prove $A \vdash_{\mathcal{N}} B$, i.e., *assume* that A is true and show that then also B must be true.

• In Exercise 4.1.(6) you are asked to show admissibility of the rule $\dfrac{A \vdash_{\mathcal{N}} \bot}{\vdash_{\mathcal{N}} \neg A}$.
Interpreting $\bot$ as something which can never be true, a contradiction or an absurdity, this rule expresses *reductio ad absurdum* (see Zeno's argument about Achilles and tortoise in Section A.1 on the history of logic): if A can

be used to derive an absurdity, then A can not be true i.e., applying the law of excluded middle, its negation must be.

EXERCISES 4.

EXERCISE 4.1 Prove the following statements in $\mathcal{N}$:

(1) $\vdash_{\mathcal{N}} \neg A \rightarrow (A \rightarrow B)$
(Hint: Complete the following proof:

$1 : \vdash_{\mathcal{N}} \neg A \rightarrow (\neg B \rightarrow \neg A)$ $A1$
$2 : \neg A \vdash_{\mathcal{N}} \neg B \rightarrow \neg A$ $C.4.17$
$3 :$ $A3$
$4 :$ $MP : 2, 3$
$5 :$ $DT : 4$)

(2) $\neg B, A \vdash_{\mathcal{N}} \neg (A \rightarrow B)$
(Hint: Start as follows, then apply DT, Lemma 4.16 and, possibly, Corollary 4.17:

$1 : A, A \rightarrow B \vdash_{\mathcal{N}} A$ $A0$
$2 : A, A \rightarrow B \vdash_{\mathcal{N}} A \rightarrow B$ $A0$
$3 : A, A \rightarrow B \vdash_{\mathcal{N}} B$ $MP : 1, 2$
$\quad \vdots$)

(3) $\vdash_{\mathcal{N}} (A \rightarrow B) \rightarrow (A \rightarrow (A \rightarrow B))$
(4) $\vdash_{\mathcal{N}} A \rightarrow (\neg B \rightarrow \neg (A \rightarrow B))$
(5) $\vdash_{\mathcal{N}} (A \rightarrow \bot) \rightarrow \neg A$
(6) Show now admissibility in $\mathcal{N}$ of the rules

$$\text{(a)} \quad \frac{\vdash_{\mathcal{N}} A \rightarrow \bot}{\vdash_{\mathcal{N}} \neg A} \qquad\qquad \text{(b)} \quad \frac{A \vdash_{\mathcal{N}} \bot}{\vdash_{\mathcal{N}} \neg A}$$

(Hint: for (a) use (5) and MP; for (b) use Deduction Theorem and (a).)

(7) Prove the first formula in $\mathcal{H}$, i.e., show $(1') : \vdash_{\mathcal{H}} \neg A \rightarrow (A \rightarrow B)$.

EXERCISE 4.2 Show the claim (4.22), i.e., $\vdash_{\mathcal{N}} (A \rightarrow A) \leftrightarrow (B \rightarrow B)$.
(Hint: use Lemma 4.8 and then Lemma 4.13.(2).)

EXERCISE 4.3 Show that for an arbitrary formula A:

(1) $\vdash_{\mathcal{N}} A \rightarrow \top$.
(2) $\vdash_{\mathcal{N}} A$ iff $\vdash_{\mathcal{N}} \top \rightarrow A$.

EXERCISE 4.4 Show the provable equivalence (Corollary 4.18 and Deduction Theorem may be useful):

$$\vdash_{\mathcal{N}} (A \rightarrow (B \rightarrow C)) \leftrightarrow (B \rightarrow (A \rightarrow C)).$$

EXERCISE 4.5 Lemma 4.13 generalized Lemma 4.10 to the expressions involving assumptions $\Gamma \vdash_{\mathcal{N}} \ldots$ We can, however, reformulate the rules in a different way, namely, by placing the antecedents of $\to$ to the left of $\vdash_{\mathcal{N}}$. Show the admissibility in $\mathcal{N}$ of the rules:

$$(1)\quad \frac{\Gamma \vdash_{\mathcal{N}} B}{\Gamma, A \vdash_{\mathcal{N}} B} \qquad\qquad (2)\quad \frac{\Gamma, A \vdash_{\mathcal{N}} B \;\; ; \;\; \Gamma, B \vdash_{\mathcal{N}} C}{\Gamma, A \vdash_{\mathcal{N}} C}$$

((1) must be shown directly, for instance, by induction on the length of the proof of $\Gamma \vdash_{\mathcal{N}} B$, without using Corollary 4.17. Why? But this Corollary can be used for showing (2). Why?)

EXERCISE 4.6 Show that the following two statements are equivalent (for any fixed formula A):

(1.a) for every $\Gamma : \Gamma \vdash_{\mathcal{N}} A$

(1.b) $\vdash_{\mathcal{N}} A$, i.e., for Γ being the empty theory: $\varnothing \vdash_{\mathcal{N}} A$.

Give an example of Γ, A and B showing the difference between (non-equivalence of) the following two equivalences:

(2.a) for every $\Gamma : \Gamma \vdash_{\mathcal{N}} A$ iff $\Gamma \vdash_{\mathcal{N}} B$.

(2.b) $\vdash_{\mathcal{N}} A$ iff $\vdash_{\mathcal{N}} B$

EXERCISE 4.7 Show that Definition 4.26 is equivalent to the following one:

Γ is consistent iff there is no formula A such that both $\Gamma \vdash A$ and $\Gamma \vdash \neg A$.

(Hint: You should show that for arbitrary Γ one has that:

$\Gamma \not\vdash_{\mathcal{N}} \bot$ iff   for no A : $\Gamma \vdash_{\mathcal{N}} A$ and $\Gamma \vdash_{\mathcal{N}} \neg A$, which is the same as:

$\Gamma \vdash_{\mathcal{N}} \bot$ iff   for some A : $\Gamma \vdash_{\mathcal{N}} A$ and $\Gamma \vdash_{\mathcal{N}} \neg A$.

The implication $\Rightarrow$ follows from the assumption $\Gamma \vdash_{\mathcal{N}} \neg(B \to B)$ and Lemma 4.8. For the opposite, start as follows (use Corollary 4.25 on 3, and then MP):

$1 : \Gamma \vdash_{\mathcal{N}} A$ assumption

$2 : \Gamma \vdash_{\mathcal{N}} \neg A$ assumption

$3 : \Gamma \vdash_{\mathcal{N}} \neg\neg(A \to A) \to \neg A$ $L.4.13.(2)$: 2

$\vdots$)

EXERCISE 4.8 Show that for every formula A, A and $(\neg A \to A)$ are provably equivalent in $\mathcal{N}$ (Definition 4.19), i.e., $\vdash_{\mathcal{N}} (\neg A \to A) \leftrightarrow A$.

[Hint: One way is trivial and needs only a concise justification. For the other, the so-called *Consequentia Mirabilis* or *Clavius' Law*, instead of constructing an actual proof in $\mathcal{N}$, it may be easier to show its existence using various auxiliary results. E.g., showing $\neg A \to A, \neg A \vdash_{\mathcal{N}} A$ and $\neg A \to A, \neg A \vdash_{\mathcal{N}} \neg A$, the claim will follow using Corollaries 4.17 and 4.25, Exercises 4.1 and 4.7, and Deduction Theorem.]

EXERCISE 4.9 Consider Gentzen's system $\mathcal{G}$ from Section 4.8.

(1) Show that all axioms of the $\mathcal{N}$ system are derivable in $\mathcal{G}$.

(2) Show derivability in $\mathcal{G}$ of the formulae (1), (2) and (4) from Exercise 4.1.

EXERCISE 4.10 Show that replacing the axiom schema in Gentzen system from page 136 by the schema $\Gamma, a \vdash_{\mathcal{G}} \Delta, a$, for every atomic formula $a \in \Sigma$ and arbitrary finite sets Γ, Δ of WFF^{Σ}, still allows one to prove every sequent of the form $\Gamma, F \vdash_{\mathcal{G}} \Delta, F$ for arbitrary formula $F \in \mathsf{WFF}^{\Sigma}$.

EXERCISE 4.11 Sketch an algorithm for showing the claim before Theorem 4.28, namely, that consistency in $\mathcal{N}$ of (countably) infinite theories is co-semi-decidable. (Note that this is the same as the claim that inconsistency of such theories is semi-decidable.)

Chapter 5

SEMANTICS OF PL

In this chapter we are leaving the proofs and axioms aside. For the time being, *none* of the concepts below should be referred to any earlier results on axiomatic systems. (Such connections will be studied in the following chapters.) Here, we are studying exclusively *the language* of PL – Definition 4.4 – and the standard way of assigning *meaning* to its expressions.

5.1 THE BOOLEAN SEMANTICS

―――――――――――― a background story ――――――――――

There is a huge field of Proof Theory which studies axiomatic systems per se, i.e., without reference to their possible meanings. This was the kind of study we were carrying out in the preceding chapter. As we emphasised at the beginning of that chapter, an axiomatic system may be given different interpretations and we will in this chapter see a few possibilities for interpreting the system of propositional logic. Yet, axiomatic systems are typically introduced for the purpose of studying particular areas or particular phenomena therein. They provide syntactic means for such a study: a language for referring to objects and their properties and a proof calculus capturing, hopefully, some of the essential relationships between various aspects of the domain.

As described in the history of logic, its original intention was to capture the patterns of correct reasoning which we otherwise carry out in natural language. Propositional logic, in particular, emerged as a logic of statements: propositional variables may be interpreted as arbitrary statements, while the connectives as the means of constructing new statements from others. For instance, consider the following argument:

	If it is raining, we will go to cinema.
and	If we go to cinema, we will see a Kurosawa film.
hence	If it is raining, we will see a Kurosawa film.

If we agree to represent the implication if ... then ... by the syntactic symbol $\rightarrow$, this reasoning is represented by interpreting A as It will rain, B as We will go to cinema, C as We will see a Kurosawa film and by the deduction $\dfrac{A \rightarrow B \;\; ; \;\; B \rightarrow C}{A \rightarrow C}$. As we have seen in Lemma 4.10, this is a valid rule in the system $\vdash_{\mathcal{H}}$. Thus, we might say that the system $\vdash_{\mathcal{H}}$ (as well as $\vdash_{\mathcal{N}}$) captures this aspect of our natural reasoning.

However, one has to be extremely careful with this kinds of analogies. They are never complete and any formal system runs, sooner or later, into problems when confronted with the richness and sophistication of natural language. Consider the following argument:

	If I am in Paris then I am in France.
and	If I am in Rome then I am in Italy.
hence	If I am in Paris then I am in Italy or else
	if I am in Rome then I am in France.

It does not look plausible, does it? Now, let us translate it into statement logic: P for being in Paris, F for being in France, R in Rome and I in Italy. Using Gentzen's rules with the standard reading of $\wedge$ as 'and' and $\vee$ as 'or', we obtain:

$$\frac{\dfrac{R \rightarrow I, P, R \vdash_{\mathcal{G}} I, F, P \;\; ; \;\; F, R \rightarrow I, P, R \vdash_{\mathcal{G}} I, F}{\dfrac{P \rightarrow F, R \rightarrow I, P, R \;\; \vdash_{\mathcal{G}} \;\; I, F}{\dfrac{P \rightarrow F, R \rightarrow I, P \;\; \vdash_{\mathcal{G}} \;\; I, R \rightarrow F}{\dfrac{P \rightarrow F, R \rightarrow I \;\; \vdash_{\mathcal{G}} \;\; P \rightarrow I, R \rightarrow F}{\dfrac{P \rightarrow F, R \rightarrow I \;\; \vdash_{\mathcal{G}} \;\; P \rightarrow I \vee R \rightarrow F}{\dfrac{P \rightarrow F \wedge R \rightarrow I \;\; \vdash_{\mathcal{G}} \;\; P \rightarrow I \vee R \rightarrow F}{\vdash_{\mathcal{G}} \;\; (P \rightarrow F \wedge R \rightarrow I) \rightarrow (P \rightarrow I \vee R \rightarrow F)}}}}}}$$

Our argument – the implication from ($P \rightarrow F$ and $R \rightarrow I$) to ($P \rightarrow I$ or $R \rightarrow F$) turns out to be provable in $\vdash_{\mathcal{G}}$. (It is so in the other systems as well.) Logicians happen to have an answer to this particular problem (we will return to it in Exercise 6.1). But there are other strange things which cannot be easily answered. Typically, any formal system attempting to capture some area of discourse, will capture only *some part* of it. Attempting to apply it beyond this area, leads inevitably to counterintuitive phenomena.

Propositional logic attempts to capture some simple patterns of reasoning at the level of propositions. A proposition can be thought of as a declarative sentence which may be assigned a unique truth value. The sentence "It is raining" is either true or false. Thus, the intended and possible *meanings* of propositions are truth values: true or false. Now, the meaning of the proposition If it rains, we will go to a cinema, $A \rightarrow B$, can be construed as: *if* 'it is true that it will rain' *then* 'it is true that we will go to a cinema'. The implication $A \rightarrow B$ says that if A is true then B must be true as well.

Now, since this implication is itself a proposition, it will have to be given a truth value as its meaning. And this truth value will depend on the truth value of its constituents A and B. If A is true (it is raining) but B is false (we are not going to a cinema), the whole implication $A \rightarrow B$ is false.

And now comes the question: what if A is false? Did the implication $A \rightarrow B$ assert anything about this situation? No, it did not. If A is false (it is not raining), we may go to a cinema or we may stay at home – nothing has been said about that case. Yet, the proposition *has to* have a meaning for all possible values of its parts. In this case – when the antecedent A is false – the whole implication $A \rightarrow B$ is declared true irrespectively of the truth value of B. You should notice that here something special is happening which does not necessarily correspond so closely to our intuition. And indeed, it is something very strange! If I am a woman, then you are Dalai Lama. Since I am not a woman, the implication happens to be true! But, as you know, this does not mean that you are Dalai Lama. This example, too, can be explained by the same argument as the above one (to be indicated in Exercise 6.1). However, the following implication is true, too, and there is no formal way of excusing it being so or explaining it away: If it is not true that when I am a man then I am a man, then you are Dalai Lama, $\neg(M \rightarrow M) \rightarrow D$. It is correct, it is true and ... it seems to be entirely meaningless.

In short, formal correctness and accuracy does not always correspond to something meaningful in natural language, even if such a correspondence was the original motivation. A possible discrepancy indicated above concerned, primarily, the discrepancy between our intuition about the meaning of sentences and their representation in a syntactic system. But the same problem occurs at yet another level – analogous discrepancies occur between our intuitive understanding of the world and its formal *semantic* model. Thinking about axiomatic systems as tools for modeling the world, we might be tempted to look at the relation as illustrated on the left side of the following figure: an axiomatic system modeling the world. In truth, however, the relation is more complicated as illustrated on the right of the figure.

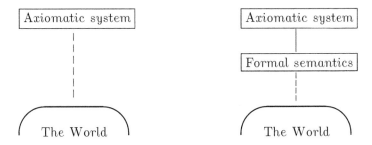

An axiomatic system *never* addresses the world directly. It addresses a possible semantic model which tries to give a formal representation of the world. As we have repeatedly said, an axiomatic system may be given various interpretations, each providing a possible formal semantic model of the system. To what extent these models capture our intuition about the world is a different question – about the "correctness" or "incorrectness" of modeling. An axiomatic system in itself is neither, because it can be endowed with different interpretations. The problems indicated above were really the problems with the semantic model of natural language which was implicitly introduced by assuming that statements are to be interpreted as truth values.

We will now endeavour to study the semantics – meaning – of the syntactic expressions from WFF_{PL}. We will see some alternative semantics starting with the standard one based on the so-called "truth functions" (which we will call "boolean functions"). To avoid confusion and surprises, one should always keep in mind that we are not talking about the world but are defining a *formal* model of PL which, at best, can provide an imperfect link between the syntax of PL and the world. The formality of the model, as always, will introduce some discrepancies as those described above and many things may turn out not exactly as we would expect them to be in the real world. _____

Let $\mathbb{B}$ be a set with two elements. *Any* such set would do but, for convenience, we will typically let $\mathbb{B} = \{1, 0\}$. Whenever one tries to capture the meaning of propositions as their truth value, and uses propositional logic with this intention, one interprets $\mathbb{B}$ as the set $\{\mathsf{true}, \mathsf{false}\}$. Since this gives too strong associations and leads often to incorrect intuitions without improving anything, we avoid the words $\mathsf{true}/\mathsf{false}$ and talk, instead, about "boolean values" ($\mathbf{1}$ and $\mathbf{0}$), "boolean functions", "boolean tables", etc..

For any $n \geq 0$, there are various functions mapping $\mathbb{B}^n \to \mathbb{B}$. For instance, for $n = 2$, a function $f : \mathbb{B} \times \mathbb{B} \to \mathbb{B}$ can be defined by $f(1, 1) \overset{\text{def}}{=} 1$, $f(1, 0) \overset{\text{def}}{=} 0$, $f(0, 1) \overset{\text{def}}{=} 1$ and $f(0, 0) \overset{\text{def}}{=} 1$. It can be written more concisely as the boolean table:

$$
\begin{array}{cc||c}
x & y & f(x,y) \\
\hline
1 & 1 & 1 \\
1 & 0 & 0 \\
0 & 1 & 1 \\
0 & 0 & 1
\end{array}
\tag{5.1}
$$

The first n-columns contain all the possible combinations of the arguments (giving 2^n distinct rows), and the last column specifies the value of the function for this combination of the arguments. For each of the 2^n rows a function takes one of the two possible values, so for any n there are exactly $2^{(2^n)}$ different functions $\mathbb{B}^n \to \mathbb{B}$. For $n = 0$, there are only two (constant) functions, for $n = 1$ there will be four distinct functions (which ones?) and so on. Surprisingly, the language of PL describes exactly such functions!

Definition 5.2 A PL *structure* consists of:

(1) A domain with two *boolean values*, $\mathbb{B} = \{1, 0\}$
(2) Interpretation of the connectives as boolean functions:
    $\underline{\neg} : \mathbb{B} \to \mathbb{B}$, given by $\underline{\neg}(1) = 0$ and $\underline{\neg}(0) = 1$, and
    $\underline{\to} : \mathbb{B}^2 \to \mathbb{B}$, given by the boolean table (5.1).

Given an alphabet Σ, a PL structure for Σ is a PL structure with
(3) an assignment of boolean values to all propositional variables, i.e., a function $V : \Sigma \to \{1, 0\}$ (also called a *valuation* of Σ).

Connectives are thus interpreted as functions on the set $\{1, 0\}$. To distinguish the two, we use the simple symbols $\neg$ and $\to$ when talking about syntax, and the underlined ones $\underline{\neg}$ and $\underline{\to}$ when we are talking about the semantic interpretation as boolean functions. $\neg$ is interpreted as the function $\underline{\neg} : \{1, 0\} \to \{1, 0\}$, defined by $\underline{\neg}(1) \overset{\text{def}}{=} 0$ and $\underline{\neg}(0) \overset{\text{def}}{=} 1$. $\to$ is binary and represents one of the functions from $\{1, 0\}^2$ into $\{1, 0\}$.

Example 5.3

Let $\Sigma = \{a, b\}$. $V = \{a \mapsto 1, b \mapsto 1\}$ is a Σ-structure (i.e., a structure interpreting all symbols from Σ) assigning 1 (true) to both variables. Another Σ-structure is, for instance, $V' = \{a \mapsto 1, b \mapsto 0\}$.

Let $\Sigma = \{\text{`Jon smokes', `Ann sings'}\}$ – two propositional variables with rather lengthy names. We can view $V = \{\text{`Jon smokes'} \mapsto 1, \text{`Ann sings'} \mapsto 0\}$ as a Σ-structure where "Jon smokes" and "Ann does not sing". $\square$

The domain of interpretation has two boolean values 1 and 0, and so we can imagine various functions, in addition to those interpreting the connectives. As remarked above, for arbitrary $n \geq 0$ there are $2^{(2^n)}$ distinct functions mapping $\{1, 0\}^n$ into $\{1, 0\}$.

Example 5.4

Here is a (somewhat involved) boolean function $F : \{1, 0\}^3 \to \{1, 0\}$:

x	y	z	$F(x, y, z)$
1	1	1	1
1	1	0	1
1	0	1	1
1	0	0	0
0	1	1	1
0	1	0	1
0	0	1	1
0	0	0	1

$\square$

According to Definition 5.2, the interpretation of the connectives is fixed once for all and only the valuation differs from one Σ-structure to another. Consequently, any valuation V can be extended in a canonical way to the interpretation of all formulae – a valuation of propositional variables induces a valuation of all well-formed formulae. We sometimes write $\widehat{V}$ for this extended valuation. This is given in the following definition which, intuitively, corresponds to the fact that if we know that 'John smokes' and 'Mary does not sing', then we also know that 'John smokes *and* Mary does not sing', or else that it is not true that 'John does not smoke'.

Definition 5.5 Any valuation of propositional variables $V : \Sigma \to \{1, 0\}$ induces a *unique* valuation of all formulae $\widehat{V} : \mathsf{WFF}_{\mathsf{PL}}^{\Sigma} \to \{1, 0\}$ as follows:

$$\widehat{V}(A) = V(A), \quad \text{for } A \in \Sigma$$
$$\widehat{V}(\neg B) = \underline{\neg}(\widehat{V}(B))$$
$$\widehat{V}(B \to C) = \widehat{V}(B) \underline{\to} \widehat{V}(C).$$

For the purposes of this section it is convenient to assume that some total ordering has been selected for the propositional variables, so that for instance a "comes before" b, which again "comes before" c.

Example 5.6

Given the alphabet $\Sigma = \{a, b, c\}$, we use the fixed interpretation of the connectives to determine the boolean value of, for instance, the formula $a \to (\neg b \to c)$ as follows:

a	b	c	$\neg\, b$	$\neg b \to c$	$a \to (\neg b \to c)$
1	1	1	0	1	1
1	1	0	0	1	1
1	0	1	1	1	1
1	0	0	1	0	0
0	1	1	0	1	1
0	1	0	0	1	1
0	0	1	1	1	1
0	0	0	1	0	1

$\square$

Ignoring the intermediary columns, this table displays exactly the same dependence of the entries in the last column on the entries in the first three ones as the function F from Example 5.4. We say that the formula $a \to (\neg b \to c)$ *determines* the function F. The general definition is given below.

Definition 5.7 For any formula B, let $\{b_1, \ldots, b_n\}$ be the propositional variables in B, listed in increasing order. Each assignment $V : \{b_1, \ldots, b_n\} \to \{1, 0\}$ determines a unique boolean value $\widehat{V}(B)$. Hence, each formula B determines a function $\underline{B} : \{1, 0\}^n \to \{1, 0\}$, given by the equation
$$\underline{B}(x_1, \ldots, x_n) = \{b_1 \mapsto x_1, \overbrace{\ldots, b_n \mapsto x_n}\}(B).$$

Example 5.8

Suppose a and b are in Σ, and a comes before b in the ordering. Then $(a \to b)$ determines the function $\to$, while $(b \to a)$ determines the function $\leftarrow$ with the boolean table shown below.

x	y	$x \to y$	$x \leftarrow y$
1	1	1	1
1	0	0	1
0	1	1	0
0	0	1	1

$\square$

Observe that although for a given n there are exactly $2^{(2^n)}$ boolean functions, there are infinitely many formulae over n propositional variables. Thus, different formulae will often determine the same boolean function. Deciding which formulae determine the same functions is an important problem which we will soon encounter.

5.1.1 Syntactic abbreviations

Intuitively, $\neg$ is supposed to express negation and we read $\neg B$ as "not B". $\rightarrow$ corresponds to implication: $A \rightarrow B$ is similar to "if A then B". These formal symbols and their semantics are not exact counterparts of the natural language expressions but they do try to *mimic* the latter as far as possible. In natural language there are several other connectives but, as we will see in Chapter 6, the two we have introduced for PL are all that are needed. We will, however, try to make our formulae shorter – and more readable – by using the following abbreviations:

Definition 5.9 Define the following abbreviations:

- $A \vee B \overset{\text{def}}{=} \neg A \rightarrow B$, read as "$A$ or B"
- $A \wedge B \overset{\text{def}}{=} \neg(A \rightarrow \neg B)$, read as "$A$ and B"
- $A \leftrightarrow B \overset{\text{def}}{=} (A \rightarrow B) \wedge (B \rightarrow A)$, read as "$A$ if and only if B".

The abbreviation $\leftrightarrow$ uses the same symbol as the provable equivalence from Definition 4.19. There is a connection between the two, but beware of confusing them. The one defined here introduces only a syntactic abbreviation, while provable equivalence concerns a proof system, and occurs only in contexts with $\vdash$, e.g., as $\vdash_{\mathcal{N}} A \leftrightarrow B$.

Example 5.10
Some intuitive justification for the reading of these abbreviations comes from the boolean tables for the functions they denote. For instance, the table for $\wedge$ is constructed according to the definition $x \wedge y \overset{\text{def}}{=} \neg(x \rightarrow \neg y)$:

x	y	$\neg y$	$x \rightarrow \neg y$	$\neg(x \rightarrow \neg y)$
1	1	0	0	1
1	0	1	1	0
0	1	0	1	0
0	0	1	1	0

Thus $A \wedge B$ evaluates to **1** (true) iff both components are true and, in this sense, corresponds to 'and' used in natural language. (In Exercise 5.1 you are asked to do the analogous thing for $\vee$.)   □

5.2 SEMANTIC PROPERTIES

A formula determines a boolean function and we now list some semantic properties of formulae, i.e., properties which are actually the properties of such induced functions.

Definition 5.11 Let $A, B \in \mathsf{WFF_{PL}}$, and V be a valuation.

A is	iff	condition holds	notation:
satisfied in V	iff	$\widehat{V}(A) = 1$	$V \models A$
not satisfied in V	iff	$\widehat{V}(A) = 0$	$V \not\models A$
valid/tautology	iff	for all $V : V \models A$	$\models A$
falsifiable	iff	there is a $V : V \not\models A$	$\not\models A$
satisfiable	iff	there is a $V : V \models A$	
unsatisfiable/contradiction	iff	for all $V : V \not\models A$	
(tauto)logical consequence of B	iff	$B \to A$ is valid	$B \Rightarrow A$
(tauto)logically equivalent to B	iff	$A \Rightarrow B$ and $B \Rightarrow A$	$A \Leftrightarrow B$

The first line makes **1** the *designated element*, representing the truth in the structure. If A is satisfied in V, we say that V *satisfies* A, or is a *model* of A. Otherwise V *falsifies* A. (Sometimes, one also says that A is valid *in* V, when A is satisfied in V. But notice that validity of A *in* V does not mean or imply that A is valid (in general), only that it is satisfiable.) Valid formulae – those satisfied in all structures – are also called *tautologies* and the unsatisfiable ones *contradictions*. Those which are both falsifiable and satisfiable, i.e., which are neither tautologies nor contradictions, are called *contingent*.

Sets of formulae are usually called *theories*. Many of the properties defined for formulae are defined for theories as well. Thus a valuation is said to satisfy a theory iff it satisfies every formula in the theory. Such a valuation is also said to be a *model* of the theory. The class of all models of a theory Γ is denoted $Mod(\Gamma)$. Like a single formula, a set of formulae Γ is *satisfiable* iff it has a model, i.e., iff $Mod(\Gamma) \neq \varnothing$. A theory which has no model is called *unsatisfiable* (rather than contradictory).

Example 5.12
$a \to b$ is not a tautology – assign $V(a) = \mathbf{1}$ and $V(b) = \mathbf{0}$. Hence $a \Rightarrow b$ does not hold. However, it is satisfiable, since it is true, for instance, under the valuation $\{a \mapsto \mathbf{1}, b \mapsto \mathbf{1}\}$. The formula is contingent.

$B \to B$ evaluates to **1** for any valuation (and any $B \in \mathsf{WFF_{PL}}$), and so

$B \Rightarrow B$. As a last example, we have that $B \Leftrightarrow \neg\neg B$.

B	B	$B \to B$
0	0	1
1	1	1

B	$\neg B$	$\neg\neg B$	$B \to \neg\neg B$	and	$\neg\neg B \to B$
1	0	1	1	and	1
0	1	0	1	and	1

$\square$

The operators $\Rightarrow$ and $\Leftrightarrow$ are *meta*-connectives stating that a corresponding relation ($\to$ and $\leftrightarrow$, respectively) between the two formulae holds *for all* boolean assignments. These operators are therefore used only at the outermost level, like for $A \Rightarrow B$ – we avoid something like $A \Leftrightarrow (A \Rightarrow B)$ or $A \to (A \Leftrightarrow B)$. Often, $\Leftrightarrow/\Rightarrow$ are used also in the statements of theorems instead of "iff"/"only if".

Fact 5.13 We have the obvious relations between the sets of Sat(isfiable), Fal(sifiable), Taut(ological), Contr(adictory) and All formulae:

- $Contr \subset Fal$
- $Fal \cap Sat \neq \varnothing$
- $Taut \subset Sat$
- $All = Taut \cup Contr \cup (Fal \cap Sat)$

5.2.1 SOME PROPOSITIONAL LAWS

The logical equivalence $A \Leftrightarrow B$ means that for all valuations (of the propositional variables in A and B) both formulae have the same boolean value. This means almost that they determine the same function, with one restriction which is discussed in Exercise 5.14. The definitions of semantics of the connectives together with the introduced abbreviations entitle us to conclude validity of some logical equivalences.

(1) Neutral elements
$$A \lor \bot \Leftrightarrow A$$
$$A \land \top \Leftrightarrow A$$

(2) Associativity
$$(A \lor B) \lor C \Leftrightarrow A \lor (B \lor C)$$
$$(A \land B) \land C \Leftrightarrow A \land (B \land C)$$

(3) Commutativity
$$A \lor B \Leftrightarrow B \lor A$$
$$A \land B \Leftrightarrow B \land A$$

(4) Distributivity
$$A \lor (B \land C) \Leftrightarrow (A \lor B) \land (A \lor C)$$
$$A \land (B \lor C) \Leftrightarrow (A \land B) \lor (A \land C)$$

(5) Complement
$$A \land \neg A \Leftrightarrow \bot$$
$$A \lor \neg A \Leftrightarrow \top$$

(6) de Morgan
$$\neg(A \lor B) \Leftrightarrow \neg A \land \neg B$$
$$\neg(A \land B) \Leftrightarrow \neg A \lor \neg B$$

(7) Conditional
$$A \to B \Leftrightarrow \neg A \lor B \qquad\qquad A \to B \Leftrightarrow \neg B \to \neg A$$

As shown in Example 5.12, $\top$ evaluates always to $\mathbf{1}$ and $\bot$ to $\mathbf{0}$. Thus, for instance, the second law of neutral element can be verified as follows:

A	$\top$	$\neg\top$	$A{\to}\;\neg\top$	$\neg(A{\to}\;\neg\top) \overset{\text{def}}{=} A{\wedge}\top$
1	1	0	0	1
0	1	0	1	0

Also the first complement law is verified directly from the definition of $\wedge$:

A	$\neg\neg A$	$A{\to}\;\neg\neg A$	$A{\wedge}\neg A = \bot$
1	1	1	0
0	0	1	0

For any A, B, C the two formulae $(A \wedge B) \wedge C$ and $A \wedge (B \wedge C)$ are distinct. However, as they are tautologically equivalent it is not a very urgent matter to distinguish between them. In general, there are a great many ways to insert the missing parentheses in an expression like $A_1 \wedge A_2 \wedge \ldots \wedge A_n$, but since they all yield equivalent formulae we usually do not care where these parentheses go. Hence for a sequence $A_1, A_2, \ldots, A_n$ of formulae we may just talk about their *conjunction* and mean any formula obtained by supplying missing parentheses to the expression $A_1 \wedge A_2 \wedge \ldots \wedge A_n$. Analogously, the *disjunction* of $A_1, A_2, \ldots, A_n$ is any formula obtained by supplying missing parentheses to the expression $A_1 \vee A_2 \vee \ldots \vee A_n$.

Moreover, the laws of commutativity and idempotency (which can be easily verified) tell us that order and repetition don't matter either. Hence we may talk about the conjunction of the formulae in some finite set, and mean any conjunction formed by the elements in some order or other. Similarly for disjunction.

The elements $A_1, \ldots, A_n$ of a conjunction $A_1 \wedge \ldots \wedge A_n$ are called the *conjuncts*. The term *disjunct* is used analogously.

5.3 SET-BASED SEMANTICS

Semantics of propositional logic was defined by interpreting the formulae as elements of the two-element set $\mathbb{B}$ and connectives as functions over $\mathbb{B}$. Some consequences, in form of the laws following from this definition, were listed in Subsection 5.2.1. Below, we note a close relationship to the laws obeyed by the set operations from Section 1.1 and define an alternative semantics of the propositional language based on set interpretation. The optional Subsection 5.3.2 gathers these similarities in the common concept of boolean algebra.

5.3.1 SETS AND PROPOSITIONS

Compare the set laws (1)-(6) from Section 1.1, p. 48, with the tautological
equivalences (1)-(6) from Subsection 5.2.1. It is easy to see that they have
"corresponding form" and can be obtained from each other by the following
translations.

set-expression		proposition
set variable $a, b...$	$-$	propositional variable $a, b...$
$\overline{}$	$-$	$\neg$
$\cap$	$-$	$\wedge$
$\cup$	$-$	$\vee$
$=$	$-$	$\Leftrightarrow$

One also translates :
$$U = \overline{\varnothing} \quad - \quad \top$$
$$\varnothing \quad - \quad \bot$$

Remark 5.14 [Formula- vs. set-operations]
Although there is some sense of connection between the subset $\subseteq$ and implication $\rightarrow$, the two have very different functions. The latter allows us to
construct new propositions. The former, $\subseteq$, is not however a set building
operation: $A \subseteq B$ does not denote any set but states a relation between
two sets. The consistency principles are not translated because they are
not so much laws as definitions introducing a new relation $\subseteq$ which holds
only under the specified conditions. In order to find a set operation corresponding to $\rightarrow$, we should reformulate the syntactic Definition 5.9 and
verify that $A \rightarrow B \Leftrightarrow \neg A \vee B$. The corresponding set building operation
$\triangleright$, would be then defined by $A \triangleright B \overset{\text{def}}{=} \overline{A} \cup B$.
   The set difference abbreviation (7).b from p. 48, $A \setminus B = A \cap \overline{B}$, has
no standard propositional form. But we can simply use the expression
$A \wedge \neg B$ corresponding to $A \cap \overline{B}$. We may translate the remaining set laws,
e.g., $A \cap \overline{A} = \varnothing$ as $A \wedge \neg A \Leftrightarrow \bot$, etc., and verify them as illustrated in
Subsection 5.2.1.   □

Let us see if we can discover the reason for this exact match of laws. For the
time being let us ignore the superficial differences of syntax, and settle for
the logical symbols on the right. Expressions built up from Σ with the use
of these, we call *boolean expressions*, BE^Σ. As an alternative to a valuation
$V : \Sigma \rightarrow \{0, 1\}$ we may consider a *set valuation* $SV : \Sigma \rightarrow \mathcal{P}(U)$, where U
is any non-empty set. Thus, instead of the boolean-value semantics in the
set $\mathbb{B}$, we are defining a set valued semantics in an arbitrary set U. Such

SV can be extended to $\widehat{SV} : \mathsf{BE}^\Sigma \to \mathcal{P}(U)$ according to the rules:

$$\begin{array}{rcl}
\widehat{SV}(a) & = & SV(a) \quad \text{for all } a \in \Sigma \\
\widehat{SV}(\top) & = & U \\
\widehat{SV}(\bot) & = & \varnothing \\
\widehat{SV}(\neg A) & = & U \setminus \widehat{SV}(A) \\
\widehat{SV}(A \wedge B) & = & \widehat{SV}(A) \cap \widehat{SV}(B) \\
\widehat{SV}(A \vee B) & = & \widehat{SV}(A) \cup \widehat{SV}(B).
\end{array}$$

The following lemma gives the first step for establishing the relation between the two semantics.

Lemma 5.15 Let $x \in U$ be arbitrary, $V : \Sigma \to \{1, 0\}$ and $SV : \Sigma \to \mathcal{P}(U)$ be such that for all $a \in \Sigma$ we have $x \in SV(a)$ iff $V(a) = 1$. Then for all $A \in \mathsf{BE}^\Sigma$ we have $x \in \widehat{SV}(A)$ iff $\hat{V}(A) = 1$.

Proof. By induction on the complexity of $A \in \mathsf{BE}^\Sigma$. For the basis of the induction, atomic a, the equivalence is stated in the assumption: $x \in SV(a)$ iff $V(a) = 1$. In the induction step for $\neg A$, the IH says that $\hat{V}(A) = 1$ iff $x \in \widehat{SV}(A)$, from which it follows that $\hat{V}(A) \neq 1$ iff $x \notin \widehat{SV}(A)$, i.e., $\hat{V}(\neg A) = 1$ iff $x \in U \setminus \widehat{SV}(A)$. All inductive steps follow in this way from the boolean tables of $\top, \bot, \neg, \wedge, \vee$ and the observation that:

$$\begin{array}{rcl}
x \in U & & \text{always} \\
x \in \varnothing & & \text{never} \\
x \in \overline{P} & \text{iff} & x \notin P \\
x \in P \cap Q & \text{iff} & x \in P \text{ and } x \in Q \\
x \in P \cup Q & \text{iff} & x \in P \text{ or } x \in Q.
\end{array}$$

QED (5.15)

The designated element x, satisfying the condition of the lemma, plays thus the role of the truth indicator. When it belongs to the set interpreting a given expression, the expression counts as true, while when it does not belong there, the expression counts as false.

Example 5.16

Let $\Sigma = \{a, b, c\}$, $U = \{4, 5, 6, 7\}$ and let the designated element be $4 \in U$. The upper part of the table shows an example of a valuation and set valuation satisfying the conditions of the lemma, and the lower part the values of some formulae (boolean expressions) under these valuations.

$$\begin{array}{ccccc}
\{1,0\} \overset{V}{\leftarrow} & \Sigma & \overset{SV}{\rightarrow} & \mathcal{P}(\{4,5,6,7\}) \\
\hline
1 & \leftarrow & a & \rightarrow & \{4,5\} \\
1 & \leftarrow & b & \rightarrow & \{4,6\} \\
0 & \leftarrow & c & \rightarrow & \{5,7\} \\
\hline
\{1,0\} \overset{\widehat{V}}{\leftarrow} & \mathsf{BE}^{\Sigma} & \overset{\widehat{SV}}{\rightarrow} & \mathcal{P}(\{4,5,6,7\}) \\
\hline
1 & \leftarrow & a \wedge b & \rightarrow & \{4\} \\
0 & \leftarrow & \neg a & \rightarrow & \{6,7\} \\
1 & \leftarrow & a \vee c & \rightarrow & \{4,5,7\} \\
0 & \leftarrow & \neg(a \vee c) & \rightarrow & \{6\} \\
\end{array}$$

The four formulae illustrate the general fact that for any $A \in \mathsf{BE}^{\Sigma}$ we have $\widehat{V}(A) = \mathbf{1} \Leftrightarrow 4 \in \widehat{SV}(A)$. □

The set identities from Section 1.1 say that the BE's on each side are interpreted identically by every set valuation. Hence the following theorem expresses the correspondence between the set identities and tautological equivalences.

Theorem 5.17 For every $A, B \in \mathsf{BE}^{\Sigma}$:

$$\left(\widehat{SV}(A) = \widehat{SV}(B) \text{ for all set valuations } SV, \text{ into all sets } U\right) \text{ iff } (A \Leftrightarrow B).$$

Proof. The idea is to show that for every set valuation that interprets A and B differently, there is some valuation that interprets them differently, and conversely.

$\Leftarrow$) First suppose $\widehat{SV}(A) \neq \widehat{SV}(B)$. Then there is some $x \in U$ that is contained in one but not the other. Let V_x be the valuation such that for all $a \in \Sigma$,

$$V_x(a) = \mathbf{1} \text{ iff } x \in SV(a).$$

Then $\widehat{V_x}(A) \neq \widehat{V_x}(B)$ follows from Lemma 5.15.

$\Rightarrow$) Now suppose $\widehat{V}(A) \neq \widehat{V}(B)$. Let SV be the set valuation into $\mathcal{P}(\{\bullet\}) = \{\varnothing, \{\bullet\}\}$ such that for all $a \in \Sigma$,

$$\bullet \in SV(a) \text{ iff } V(a) = \mathbf{1}.$$

Again Lemma 5.15 applies, and $\widehat{SV}(A) \neq \widehat{SV}(B)$ follows. QED (5.17)

This theorem provides an explanation for the validity of exactly the same laws for propositional logic and for sets. The set laws were universal equations, i.e., they stated equality of some set expressions *for all possible sets*.

The propositional laws were logical equivalences between corresponding logical formulae, i.e., equality of their boolean values *under all possible valuations*. We can now rewrite any valid equality $A = B$ between set expressions as $A' \Leftrightarrow B'$, where the primed symbols indicate the corresponding logical formulae; and vice versa. The theorem says that one is valid if and only if the other one is.

Let us reflect briefly over this result which is quite significant. For the first, observe that the semantics with which we started, namely, the one interpreting connectives and formulae over the set $\mathbb{B}$, turns out to be a special case of the set based semantics. We said that $\mathbb{B}$ may be an arbitrary two-element set. Now, take $U = \{\bullet\}$; then $\mathcal{P}(U) = \{\varnothing, \{\bullet\}\}$ has two elements. Using $\bullet$ as the designated element (x from Lemma 5.15), the set based semantics over this set will coincide with the propositional semantics which identifies $\varnothing$ with $\mathbf{0}$ and $\{\bullet\}$ with $\mathbf{1}$. Reinterpreting corollary with this in mind, i.e., substituting $\mathcal{P}(\{\bullet\})$ for $\mathbb{B}$, tells us that $A = B$ is valid (in all possible $\mathcal{P}(U)$ for all possible assignments) iff it is valid in $\mathcal{P}(\{\bullet\})$! In other words, to check if some set equality holds under all possible interpretations of the involved set variables, it is enough to check if it holds under all possible interpretations of these variables in the structure $\mathcal{P}(\{\bullet\})$. In this sense, this structure is a *canonical representative* of all such set based interpretations of propositional logic. It reduces the problem which might seem to involve infinitely many possibilities (all possible sets standing for each variable), to the simple task of checking the solutions with all combinations of $\{\bullet\}$ and $\varnothing$ substituted for the involved variables.

5.3.2: BOOLEAN ALGEBRAS   .... [optional]

The discussion in Subsection 5.3.1 shows the concrete connection between the set interpretation and the standard interpretation of the language of PL. The fact that both set operations and (functions interpreting the) propositional connectives obey essentially the same laws can be, however, stated more abstractly – they are both examples of yet other, general structures called *"boolean algebras"*.

Definition 5.18   The language of boolean algebra is given by 1) the set of *boolean expressions*, BE^Σ, relatively to a given alphabet Σ of variables:

BASIS :: $\perp, \top \in \mathsf{BE}^\Sigma$ and $\Sigma \subset \mathsf{BE}^\Sigma$
  IND. :: If $t \in \mathsf{BE}^\Sigma$ then $\neg t \in \mathsf{BE}^\Sigma$
      :: If $s, t \in \mathsf{BE}^\Sigma$ then $s \vee t \in \mathsf{BE}^\Sigma$ and $s \wedge t \in \mathsf{BE}^\Sigma$

and by 2) the formulae which are equations $s \equiv t$ for arbitrary $s, t \in \mathsf{BE}^\Sigma$.

A *boolean algebra* is any non-empty set X with an interpretation
  • of $\perp, \top$ as elements $\underline{\perp}, \underline{\top} \in X$ ("bottom" and "top");

- of $\neg$ as a unary operation $\underset{\neg}{} : X \to X$ ("complement"), and
- of $\vee, \wedge$ as binary operations $\underline{\vee}, \underline{\wedge} : X^2 \to X$ ("join" and "meet"),
- of $\equiv$ as identity, $=$,

satisfying the axioms (1)-(5) from Section 5.2.1, page 152 (with $\equiv$ replacing $\Leftrightarrow$).

Be wary of confusing the symbols, $\wedge, \neg$, etc. used here with the logical connectives we have been using. The latter denote special instances of the former, but the boolean connectives can be interpreted much more freely, as we have just seen on the example of set interpretation. Also, the constant symbol $\top$ (similarly for $\bot$) used here is only analogous to that introduced as an abbreviation for $B \to B$. Although the latter evaluates always to the designated value $\mathbf{1}$, it strictly speaking does not denote $\mathbf{1}$ but only abbreviates constant functions returning $\mathbf{1}$.

Roughly speaking, the word "algebra", stands here for the fact that the formulae are equalities and that one uses *equational reasoning* based on the properties of equality: reflexivity – $x \equiv x$, symmetry – $\frac{x \equiv y}{y \equiv x}$, transitivity – $\frac{x \equiv y \ ; \ y \equiv z}{x \equiv z}$, and "substitution of equals for equals", according to the rule:

$$\frac{g[x] \equiv z \ ; \ x \equiv y}{g[y] \equiv z.} \tag{5.19}$$

(Compare this to the provable equivalence from Theorem 4.24, in particular, the rule from Corollary 4.25.) For instance, in this manner, and using only the axioms (1)-(5) (for sets), we have shown the idempotency of $\wedge$

$$x \equiv x \wedge x \qquad\qquad x \equiv x \vee x \tag{5.20}$$

already in Section 1.1, on page 49.

- Another fact is a form of **absorption**:

$$\bot \wedge x \equiv \bot \qquad\qquad x \vee \top \equiv \top \tag{5.21}$$

$: \bot \wedge x \overset{(5)}{\equiv} (x \wedge \neg x) \wedge x \overset{(3)}{\equiv} (\neg x \wedge x) \wedge x \overset{(2)}{\equiv} \neg x \wedge (x \wedge x) \overset{(5.20)}{\equiv} \neg x \wedge x \overset{(5)}{\equiv} \bot.$

$: x \vee \top \overset{(5)}{\equiv} x \vee (x \vee \neg x) \overset{(3),(2)}{\equiv} (x \vee x) \vee \neg x \overset{(5.20)}{\equiv} x \vee \neg x \overset{(5)}{\equiv} \top.$

- **Complement of any** x **is determined uniquely** by the two properties from (5), namely, any y satisfying both these properties is necessarily x's complement:

$$\text{if a) } x \vee y \equiv \top \text{ and b) } y \wedge x \equiv \bot \text{ then } y \equiv \neg x \tag{5.22}$$

$: y \overset{(1)}{\equiv} y \wedge \top \overset{(5)}{\equiv} y \wedge (x \vee \neg x) \overset{(4)}{\equiv} (y \wedge x) \vee (y \wedge \neg x) \overset{b)}{\equiv} \bot \vee (y \wedge \neg x) \overset{(5)}{\equiv} (x \wedge \neg x) \vee (y \wedge \neg x) \overset{(4)}{\equiv}$
$(x \vee y) \wedge \neg x \overset{a)}{\equiv} \top \wedge \neg x \overset{(3),(1)}{\equiv} \neg x.$

- **Involution**,

$$\neg\neg x \equiv x \tag{5.23}$$

follows from (5.22). By (5) we have $x \wedge \neg x \equiv \bot$ and $x \vee \neg x \equiv \top$ which, by (5.22) imply that $x \equiv \neg\neg x$.

The fact that any set $\mathcal{P}(U)$ obeys the set laws from page 48, and that the set $\mathbb{B} = \{1, 0\}$ obeys the PL-laws from Subsection 5.2.1 amounts to the statement that these structures are, in fact, boolean algebras under the described interpretation of boolean operations. (We have not verified all the axioms of boolean algebras but this is an easy task.) Consequently, all the above formulae (5.20)–(5.23) will be valid in these structures:

$\mathcal{P}(U)$-law	$\leftarrow$	boolean algebra law	$\rightarrow$	PL-law	
$A \cap A = A$	$\leftarrow$	$x \wedge x \equiv x$	$\rightarrow$	$A \wedge A \Leftrightarrow A$	(5.20)
$\varnothing \cap A = \varnothing$	$\leftarrow$	$\bot \wedge x \equiv \bot$	$\rightarrow$	$\bot \wedge A \Leftrightarrow \bot$	(5.21)
$U \cup A = U$	$\leftarrow$	$\top \vee x \equiv \top$	$\rightarrow$	$\top \vee A \Leftrightarrow \top$	(5.21)
$\overline{(\overline{A})} = A$	$\leftarrow$	$\neg\neg x \equiv x$	$\rightarrow$	$\neg(\neg A) \Leftrightarrow A$	(5.23)

For instance, the last fact for PL was verified in Example 5.12.

Now, boolean algebras come with the reasoning system – equational logic – allowing one to prove equations $A \equiv B$, for $A, B \in$ BE. On the other hand, Hilbert's axiomatic systems for PL proves only simple boolean expressions: $\vdash_\mathcal{H} A$. Are these two reasoning systems related in some way? They are, indeed, but we will not study precise relationship in detail. At this point we only observe that, in boolean algebras, $\top$ plays the role of the designated element in the sense that: if $\vdash_\mathcal{H} A$ then also the equation $A' \equiv \top$ is provable in equational logic, where A' is obtained by replacing all subformulae $x \rightarrow y$ of A by the respective expressions $\neg x \vee y$ (recall $x \rightarrow y \Leftrightarrow \neg x \vee y$). For instance, the equation corresponding to the first axiom of $\vdash_\mathcal{H} A \rightarrow (B \rightarrow A)$ is obtained by translating $\rightarrow$ to the equivalent boolean expression: $(\neg A \vee \neg B \vee A) \equiv \top$. You may easily verify provability of this equation from axioms (1)-(5), as well as that it holds under set interpretation – for any set U and any of its subsets $A, B \subseteq U : \overline{A} \cup \overline{B} \cup A = U$.... [end optional]

EXERCISES 5.

EXERCISE 5.1 Recall Example 5.10 and set up the boolean table for the formula $a \vee b$ with $\vee$ trying to represent "or". Use your definition to represent the following statements, or explain why it can not be done:

(1) $x < y$ or $x = y$.
(2) John is ill or Paul is ill.

EXERCISE 5.2 Decide to which among the four classes from Fact 5.13 (Definition 5.11) the following formulae belong:

(1) $a \rightarrow (b \rightarrow a)$
(2) $(a \rightarrow (b \rightarrow c)) \rightarrow ((a \rightarrow b) \rightarrow (a \rightarrow c))$
(3) $(\neg b \rightarrow \neg a) \rightarrow (a \rightarrow b)$
(4) $a \rightarrow ((a \wedge \neg b) \vee b)$
(5) $(a \vee (c \rightarrow (a \wedge d))) \vee (a \rightarrow c)$

EXERCISE 5.3 Verify whether $(a \to b) \to b$ is a tautology. Is the following proof correct? If not, what is wrong with it?

$$1 : A, A \to B \vdash_N A \to B \quad A0$$
$$2 : A, A \to B \vdash_N A \qquad\quad A0$$
$$3 : A \to B \vdash_N B \qquad\quad MP : 2, 1$$
$$4 : \vdash_N (A \to B) \to B \quad DT$$

EXERCISE 5.4 Using boolean tables, show that the following formulae are contradictions:

(1) $\neg(B \to B)$

(2) $\neg(B \vee C) \wedge C$

(3) $(B \wedge C) \wedge (\neg(D \to B) \vee (C \to \neg C))$

Determine now what sets are denoted by these expressions – for the set interpretation of $\to$ recall Remark 5.14.

EXERCISE 5.5 Is it possible to find propositional formulae A and B such that both $A \to B$ and $B \to A$ are contradictions? Justify your answer.

EXERCISE 5.6 Which of the following sets of propositions can be satisfied simultanesoulsy (by one and the same valuation)?

(1) $\{p, q \vee r, \neg r\}$ (3) $\{p, p \to q, q\}$ (5) $\{p \vee q, \neg p \vee \neg q\}$

(2) $\{p, q \wedge r, \neg r\}$ (4) $\{p, p \to q, \neg q\}$ (6) $\{p \vee q, \neg p \wedge \neg q\}$

EXERCISE 5.7 For each of the following formulae, find a shortest possible, logically equivalent formula (using freely all connectives $\{\neg, \to, \vee, \wedge\}$):

(1) $\neg(\neg\neg P \to \neg\neg\neg Q)$ (3) $\neg P \wedge (Q \to P)$

(2) $\neg(\neg P \vee (\neg Q \to \neg R))$ (4) $Q \wedge (\neg\neg Q \to \neg Q)$

EXERCISE 5.8 Verify the following facts:

(1) $(A \wedge (A \to B)) \Rightarrow B$

(2) $A_1 \to (A_2 \to B) \Leftrightarrow (A_1 \wedge A_2) \to B.$

(3) Show by appropriate induction that for every $n \geq 0$:
$$(A_1 \wedge A_2 \wedge \ldots \wedge A_n) \to B \Leftrightarrow A_1 \to (A_2 \to (\ldots \to (A_n \to B)\ldots)).$$

EXERCISE 5.9 Which of the relations $\Rightarrow, \Leftarrow$ hold between the following pairs of formulae:

$$(1) \quad A \to (B \to C) \ ? \ (A \to B) \to C$$
$$(2) \quad A \to (B \to C) \ ? \ B \to (A \to C)$$
$$(3) \qquad\quad A \wedge \neg B \ ? \ \neg(A \to B)$$
$$(4) \quad A \to (B \vee C) \ ? \ (A \to B) \vee (A \to C)$$
$$(5) \quad (A \vee B) \to C \ ? \ (A \to C) \vee (B \to C)$$
$$(6) \quad (A \vee B) \to C \ ? \ (A \to C) \wedge (B \to C)$$

EXERCISE 5.10 Use Exercise 5.9.(3) to verify the logical equivalence:
$(C \wedge D) \to (A \wedge \neg B) \Leftrightarrow (C \wedge D) \to \neg(A \to B)$.
How does this earlier exercise simplify the work here?

EXERCISE 5.11 Let P^Σ denote the propositional formulae (over some alphabet Σ) using only $\wedge, \vee$ and $\to$. Show that if V is a valuation $V(x) = 1$ for all $x \in \Sigma$, then for every $A \in P^\Sigma : \widehat{V}(A) = 1$.

EXERCISE 5.12 Complete the proof of Lemma 5.15, by verifying the induction step for the cases of the formula A being $B \vee C$, $B \wedge C$ and $B \to C$.

EXERCISE 5.13 The biconditional is given by Definition 5.9 as $A \leftrightarrow B = (A \to B) \wedge (B \to A)$. Show the folowing equivalence for every $n \geq 1$:

$$Q_n \leftrightarrow \Big(Q_{n-1} \leftrightarrow \big(\ldots \leftrightarrow (Q_1 \leftrightarrow Q_0)\ldots\big)\Big)$$
$$\Longleftrightarrow \quad \neg Q_n \leftrightarrow \Big(Q_{n-1} \leftrightarrow \big(\ldots \leftrightarrow (Q_1 \leftrightarrow \neg Q_0)\ldots\big)\Big).$$

(Hint: Induction on n may be a good idea.)

EXERCISE 5.14 Tautological equivalence $A \Leftrightarrow B$ amounts *almost* to the fact that A and B have the same interpretation. We have to make the meaning of this "almost" more precise.

(1) Show that neither of the two relations $A \Leftrightarrow B$ and $\underline{A} = \underline{B}$ imply the other, i.e., give examples of A and B such that (a) $A \Leftrightarrow B$ but $\underline{A} \neq \underline{B}$ and (b) $\underline{A} = \underline{B}$ but not $A \Leftrightarrow B$.
   (Hint: Use extra/different propositional variables not affecting the truth of the formula.)

(2) Explain why the two relations are the same whenever A and B contain the same variables.

(3) Finally explain that if $\underline{A} = \underline{B}$ then there exists some formula C obtained from B by "renaming" the propositional variables, such that $A \Leftrightarrow C$.

EXERCISE 5.15 (Compositionality and substitutivity)
Let $F[\_]$ be a formula with (one or more) "holes" and A, B be arbitrary formulae. Assuming that for all valuations V, $\widehat{V}(A) = \widehat{V}(B)$, use induction on the complexity of $F[\_]$ to show that then also for all valuations $V : \widehat{V}(F[A]) = \widehat{V}(F[B])$.

(Hint: The structure of the proof will be similar to that of Theorem 4.24. Observe, however, that here you are proving a completely different fact concerning not the provability relation but the semantic interpretation of the formulae – not their provable but tautological equivalence.)

EXERCISE 5.16 Let Φ be an arbitrary, possibly infinite, set of formulae. The following conventions generalize the notion of (satisfaction of) binary conjunction/disjunction to such arbitrary sets, under a valuation V:

- conjunction: $\widehat{V}(\bigwedge \Phi) = \mathbf{1}$ iff for all A : if $A \in \Phi$ then $\widehat{V}(A) = \mathbf{1}$.
- disjunction: $\widehat{V}(\bigvee \Phi) = \mathbf{1}$ iff there is an $A \in \Phi$ such that $\widehat{V}(A) = \mathbf{1}$.

Let now Φ be a set containing zero or one formulae. What are then the interpretations of the expressions "the conjunction of formulae in Φ" and "the disjunction of formulae in Φ"?

EXERCISE 5.17 Using only equational reasoning and the axioms of boolean algebra, (1)-(5), p. 152, derive the following equalities:

(1) $x \vee \top \equiv \top$ and $x \wedge \bot \equiv \bot$.
(2) $(x \vee y) \vee (\neg x \wedge \neg y) \equiv \top$.
(3) $(x \vee y) \wedge (\neg x \wedge \neg y) \equiv \bot$.
(4) Conclude, using (5.22), derivability from the axioms of boolean algebra of de Morgan's law $\neg(x \vee y) \equiv \neg x \wedge \neg y$.

Chapter 6

SOUNDNESS AND COMPLETENESS

This chapter studies the relations between the syntax and the axiomatic systems of PL, and their semantic counterparts. Before we discuss the central concepts of soundness and completeness of the proof systems, we will ask about the expressive completeness of the language of PL, which can be identified with the possibilities it provides for defining various boolean functions. In Section 6.1 we show that *all* boolean functions can be defined by the formulae in our language. Section 6.2 explores a useful consequence of this fact showing that each formula can be written equivalently in a special normal form. The rest of the chapter shows then soundness and completeness of our axiomatic systems.

6.1 EXPRESSIVE COMPLETENESS

This and the next section study the relation between formulae of PL and boolean functions established by Definition 5.7, according to which every PL formula defines a boolean function. The question now is the opposite: Can every boolean function be defined by some formula of PL?

Introducing abbreviations ∧, ∨ and others in Section 5.1.1, we remarked that they are not necessary but merely convenient. Their being "not necessary" means that any function which can be defined by a formula containing these connectives, can also be defined by a formula which does not contain them. E.g., a function defined using ∨ can be also defined using ¬ and →.

Concerning our main question we need a stronger notion of an *expressively complete* set of connectives, namely, one allowing to define all boolean functions.

Definition 6.1 A set S of connectives is *expressively complete* if for every

$n > 0$ and function $f : \{1,0\}^n \to \{1,0\}$, there is a formula D, containing only the connectives from S, such that $\underline{D} = f$ (cf. Definition 5.7).

It is easy to see that, for example, the set with only negation $\{\neg\}$, is not expressively complete. It is a unary operation, so that it will never give rise to, for instance, a function with two arguments. But it can not even define all unary functions. It can be used to define only two functions $\mathbb{B} \to \mathbb{B}$ – inverse (i.e., $\neg$ itself) and identity ($\neg\neg(x) = x$). (The proof-theoretic counterpart of this last fact was Lemma 4.11, showing provable equivalence of B and $\neg\neg B$.) Any further applications of $\neg$ will yield one of these two functions. The constant functions ($f(x) = 1$ or $f(x) = 0$) can not be defined using exclusively this single connective. The following theorem identifies the first expressively complete set.

Theorem 6.2 The set $\{\neg, \wedge, \vee\}$ is expressively complete.

Proof. Let $f : \{1,0\}^n \to \{1,0\}$ be an arbitrary boolean function of n arguments (for some $n > 0$) with a given boolean table. If f always equals 0 then the contradiction $(a_1 \wedge \neg a_1) \vee \ldots \vee (a_n \wedge \neg a_n)$ determines f. For the case when f equals 1 for at least one row of arguments, we write the proof to the left illustrating it with an example to the right.

Proof	Example
Let $a_1, a_2, \ldots, a_n$ be distinct proposi-tional variables listed in increasing order. The boolean table for f has 2^n rows. Let $\underline{a}_c^r$ denote the entry in the c-th column and r-th row.	$\begin{array}{cc\|c} a_1 & a_2 & f(a_1, a_2) \\ \hline 1 & 1 & 0 \\ 1 & 0 & 1 \\ 0 & 1 & 1 \\ 0 & 0 & 0 \end{array}$
For each $1 \leq r \leq 2^n$, $1 \leq c \leq n$ let $$L_c^r = \begin{cases} a_c & \text{if } \underline{a}_c^r = 1 \\ \neg a_c & \text{if } \underline{a}_c^r = 0 \end{cases}$$	$L_1^1 = a_1,\ L_2^1 = a_2$ $L_1^2 = a_1,\ L_2^2 = \neg a_2$ $L_1^3 = \neg a_1,\ L_2^3 = a_2$ $L_1^4 = \neg a_1,\ L_2^4 = \neg a_2$
For each row $1 \leq r \leq 2^n$ form the conjunction: $C^r = L_1^r \wedge L_2^r \wedge \ldots \wedge L_n^r$. Then for all rows r and $p \neq r$: $\underline{C}^r(\underline{a}_1^r, \ldots, \underline{a}_n^r) = 1$ and $\underline{C}^r(\underline{a}_1^p, \ldots, \underline{a}_n^p) = 0$. Let D be the disjunction of those C^r for which $f(\underline{a}_1^r, \ldots, \underline{a}_n^r) = 1$.	$C^1 = a_1 \wedge a_2$ $C^2 = a_1 \wedge \neg a_2$ $C^3 = \neg a_1 \wedge a_2$ $C^4 = \neg a_1 \wedge \neg a_2$ $D = C^2 \vee C^3$ $\quad = (a_1 \wedge \neg a_2) \vee (\neg a_1 \wedge a_2)$

The claim is that D determines f, i.e., $\underline{D} = f$. If $f(\underline{a}_1^r, \ldots, \underline{a}_n^r) = 1$ then D contains the corresponding disjunct C^r which makes $\underline{D}(\underline{a}_1^r, \ldots, \underline{a}_n^r) = 1$, because $\underline{C}^r(\underline{a}_1^r, \ldots, \underline{a}_n^r) = 1$. If $f(\underline{a}_1^r, \ldots, \underline{a}_n^r) = 0$, then D does not contain the corresponding disjunct C^r. But for all $p \neq r$ we have $\underline{C}^p(\underline{a}_1^r, \ldots, \underline{a}_n^r) = \mathbf{0}$, so none of the disjuncts in D will be $\mathbf{1}$ for these arguments, and hence $\underline{D}(\underline{a}_1^r, \ldots, \underline{a}_n^r) = \mathbf{0}$.   QED (6.2)

Having some expressively complete set S, one shows expressive completeness of another set T by demonstrating how every connective $s \in S$ can be defined using only connectives from T.

Corollary 6.3 The following sets of connectives are expressively complete:

(1) $\{\neg, \vee\}$ (2) $\{\neg, \wedge\}$ (3) $\{\neg, \rightarrow\}$

Proof. (1) By de Morgan's law $A \wedge B \Leftrightarrow \neg(\neg A \vee \neg B)$, we can express each conjunction by negation and disjunction. Using distributive and associative laws, this allows us to rewrite the formula obtained in the proof of Theorem 6.2 to an equivalent one without conjunction.
(2) The same argument as above, using $A \vee B \Leftrightarrow \neg(\neg A \wedge \neg B)$.
(3) According to Definition 5.9, $A \vee B \overset{\text{def}}{=} \neg A \rightarrow B$. This, however, was a merely syntactic definition of a new symbol '$\vee$'. Here we have to show that the boolean functions $\underline{\neg}$ and $\underline{\rightarrow}$ can be used to define the boolean function $\underline{\vee}$. But this was done in Exercise 5.2.(1) where the semantics (boolean table) for $\underline{\vee}$ was given according to Definition 5.9, i.e., where $A \vee B \Leftrightarrow \neg A \rightarrow B$, required here, was shown. So the claim follows from point (1).   QED (6.3)

Remark 6.4

Our definition of "expressively complete" does not require that any formula determines the functions from $\{\mathbf{1}, \mathbf{0}\}^0$ into $\{\mathbf{1}, \mathbf{0}\}$. $\{\mathbf{1}, \mathbf{0}\}^0$ is the singleton set $\{\epsilon\}$ and there are two functions from it into $\{\mathbf{1}, \mathbf{0}\}$, namely $\{\epsilon \mapsto \mathbf{1}\}$ and $\{\epsilon \mapsto \mathbf{0}\}$. These functions are not determined by any formula in the connectives $\wedge, \vee, \rightarrow, \neg$. The best approximations are tautologies and contradictions like $(a \rightarrow a)$ and $\neg(a \rightarrow a)$, which we in fact took to be the special formulae $\top$ and $\bot$. However, these determine the constant functions $\{\mathbf{1} \mapsto \mathbf{1}, \mathbf{0} \mapsto \mathbf{1}\}$ and $\{\mathbf{1} \mapsto \mathbf{0}, \mathbf{0} \mapsto \mathbf{0}\}$, which in a strict set theoretic sense are distinct from the functions above. To obtain a set of connectives that is expressively complete in a stricter sense, one would have to introduce $\top$ or $\bot$ as a special formula (in fact, a 0-argument connective) that does not contain any propositional variables.   □

6.2 Disjunctive and conjunctive normal forms

The fact that, for instance, $\{\neg, \rightarrow\}$ is an expressively complete set, vastly reduces the need for elaborate syntax when studying propositional logic. We can (as we indeed have done) restrict the syntax of $\mathsf{WFF}^{\mathsf{PL}}$ to the necessary minimum. This simplifies many proofs concerned with the syntax and the axiomatic systems since such proofs involve often induction on the syntactic definitions (of WFF, of $\vdash$, etc.). Expressive completeness of a set of connectives means that any entity (any function defined by a formula) has some specific, "normal" form using only the connectives from the set.

Now we will show that even more "normalization" is possible. Not only every boolean function can be defined by *some* formula using only the connectives from one expressively complete set – every such a function can be defined by such a formula which, in addition, has a very specific form.

Definition 6.5 A formula B is in

(1) *disjunctive normal form*, DNF, iff $B = C_1 \vee \ldots \vee C_n$, where each C_i is a conjunction of literals.
(2) *conjunctive normal form*, CNF, iff $B = D_1 \wedge \ldots \wedge D_n$, where each D_i is a disjunction of literals.

Example 6.6
Let $\Sigma = \{a, b, c\}$.
- $(a \wedge b) \vee (\neg a \wedge \neg b)$ and $(a \wedge b \wedge \neg c) \vee (\neg a \wedge c)$ are both in DNF
- $a \vee b$ and $a \wedge b$ are both in DNF and CNF
- $(a \vee (b \wedge c)) \wedge (\neg b \vee a)$ is neither in DNF nor in CNF
- $(a \vee b) \wedge c \wedge (\neg a \vee \neg b \vee \neg c)$ is in CNF but not in DNF
- $(a \wedge b) \vee (\neg a \vee \neg b)$ is in DNF but not in CNF.

The last formula can be transformed into CNF using the laws like those from Subsection 5.2.1, p. 152. The distributivity and associativity laws yield:
$$(a \wedge b) \vee (\neg a \vee \neg b) \Leftrightarrow (a \vee \neg a \vee \neg b) \wedge (b \vee \neg a \vee \neg b)$$
and the formula on the right hand side is in CNF.   □

Recall the form of the formula constructed in the proof of Theorem 6.2 – it was in DNF! Thus, this proof tells us not only that the set $\{\neg, \wedge, \vee\}$ is expressively complete but also

Corollary 6.7 Each formula is logically equivalent to a formula in DNF.

Proof. For any B there is a D in DNF such that $\underline{B} = \underline{D}$. By "renaming" the propositional variables of D (Exercise 5.14), one obtains a new formula B_D in DNF, such that $B \Leftrightarrow B_D$.   QED (6.7)

Corollary 6.7 implies also the existence of conjunctive normal form.

Corollary 6.8   Each formula is logically equivalent to a formula in CNF.

Proof. Assuming, by Corollary 6.3, that the only connectives in B are $\neg$ and $\wedge$, we proceed by induction on B's complexity:

a ::   A propositional variable is a conjunction over one literal, and hence is in CNF.

$\neg A$ ::   By Corollary 6.7, A is equivalent to a formula A_D in DNF. Exercise 6.10 allows us to conclude that B is equivalent to B_C in CNF.

$C \wedge A$ ::   By IH, both C and A have CNF : C_C, A_C. Then $C_C \wedge A_C$ is easily transformed into CNF (using associative laws), i.e., we obtain an equivalent B_C in CNF.   QED (6.8)

The concepts of disjunctive and conjunctive normal forms may be ambiguous, as the definitions do not determine uniquely the form. If 3 variables a, b, c are involved, the CNF of $\neg(a \wedge \neg b)$ can be naturally seen as $\neg a \vee b$. But one could also require all variables to be present, in which case the CNF would be $(\neg a \vee b \vee c) \wedge (\neg a \vee b \vee \neg c)$. Applying distributivity to the following formula in DNF: $(b \wedge a) \vee (c \wedge a) \vee (b \wedge \neg a) \vee (c \wedge \neg a)$, we obtain CNF $(b \vee c) \wedge (a \vee \neg a)$. The last, tautological conjunct can be dropped, so that also $b \vee c$ can be considered as the CNF of this formula.

6.2.1: CNF, CLAUSES AND SAT   .... [optional]
An algorithm for constructing a DNF follows from the proof of Theorem 6.2. CNF can be constructed by performing the dual moves from a boolean table for a function. If it is always **1**, take a tautology with appropriate variables, $(a_1 \vee \neg a_1) \wedge \ldots \wedge (a_n \vee \neg a_n)$. Otherwise, pick the rows with value **0** and for each such row form the conjunction of its literals, combining the conjunctions with the disjunction as in the proof of Theorem 6.2. Take now the negation of the whole formula – pushing the negation inwards using de Morgan, will yield the conjunction of disjunctions of the negations of all literals. CNF results now from removing double negations wherever they occur. For the function from the proof of Theorem 6.2, we obtain $\neg(a_1 \wedge a_2) \wedge \neg(\neg a_1 \wedge \neg a_2)$ which, after de Morgan and elimination of double negation, becomes $(\neg a_1 \vee \neg a_2) \wedge (a_1 \vee a_2)$.

A disjunction of literals is called a *clause* and thus CNF is a conjunction of clauses. It plays a crucial role in many applications and the problem of deciding if a given CNF formula is satisfiable, SAT, is the paradigmatic NP-complete problem. (Deciding if a given DNF formula is satisfiable is trivial – it suffices to check if it contains any conjunction without any pair of complementary literals.)

For a given set V of $n = |V|$ variables, a V-clause or (abstracting from the actual names of the variables and considering only their number) n-clause is one with n literals. There are 2^n distinct n-clauses, and the set containing them all is denoted $C(V)$. The following fact may seem at first surprising, claiming that every subset of such n-clauses, except the whole $C(V)$, is satisfiable.

Fact 6.9 For a $T \subseteq C(V) : Mod(T) = \varnothing \Leftrightarrow T = C(V)$.

Proof. Proceeding by induction on the number $|V| = n$ of variables, the claim is trivially verified for $n = 1$ or $n = 2$. For any $n + 1 > 2$, in any subset of $2^n - 1 = 2^{n-1} + 2^{n-1} - 1$ clauses, at least one of the literals, say x, occurs 2^{n-1} times. These clauses are satisfied by making $x = \mathbf{1}$. From the remaining $2^{n-1} - 1$ clauses, we remove $\bar{x}$, and obtain $2^{n-1} - 1$ clauses over $n - 1$ variables, for which IH applies. Since every subset of $2^n - 1$ clauses is satisfiable, so is every smaller subset, for every n.   QED (6.9)

This follows, in fact, from a more general observation. Given a subset $T \subseteq C(V)$ of V-clauses, its models are determined exactly by the clauses in $C(V) \setminus T$. For an n-clause C, let $val(C)$ denote the valuation assigning $\mathbf{0}$ to all positive literals and $\mathbf{1}$ to all negative literals in C.

Fact 6.10 Let $T \subseteq C(V) : Mod(T) = \{val(C) : C \in C(V) \setminus T\}$.

This holds because each $C \in C(V) \setminus T$ differs from each $M \in T$ at least at one literal, say $l_{CM} \in M$ and $\bar{l}_{MC} \in C$ (where $l - \bar{l}$ denotes arbitrarily complementary pair of literals). Taking the complements of all literals in C will then make true $val(C) \models M \in T$, at least by the respective literal $\bar{l}_{CM} \in M$, on which the two differ. Since C contains such a literal for each clause from T, $val(C) \models T$.

Example 6.11
For $V = \{a, b, c\}$, let the theory T contain the 5 V-clauses in the first column.

		$C(V) \setminus T$								
		$a \vee \neg b \vee \neg c$			$\neg a \vee b \vee \neg c$			$\neg a \vee b \vee c$		
	$val(\_)$:	a	b	c	a	b	c	a	b	c
T :		0	1	1	1	0	1	1	0	0
1. $a \vee b \vee c$			1	1	1		1	1		
2. $a \vee b \vee \neg c$			1		1			1		1
3. $a \vee \neg b \vee c$				1	1	1	1	1	1	
4. $\neg a \vee \neg b \vee c$		1		1		1	1		1	
5. $\neg a \vee \neg b \vee \neg c$		1				1			1	1

The three clauses in $C(V) \setminus T$ give valuations in the second row. Each of them makes **1** the literals marked in the rows for the respective clauses of T.   □

Fact 6.10 finds important applications in the decision procedures for satisfiability and in counting the number of models of a theory. It is commonly applied as the so-called "semantic tree", representing the models of a theory. Order the atoms V of the alphabet in an arbitrary total ordering $v_1, v_2, \ldots, v_n$, and build a complete binary tree by staring with the empty root (level 0) and, at each level $i > 0$, adding two children, v_i and $\overline{v}_i$, to each node at the previous level $i - 1$. A complete branch in such a tree (i.e., each of the 2^n leaves) represents a possible assignment of the values **1** to each node v_i and **0** to each node $\overline{v}_i$ on the branch. According to Fact 6.10, each clause from $C(V)$ excludes one such branch, formed by negations of all literals in the clause. One says that the respective branch becomes "closed" and, on the drawing below, these closed branches, for the five clauses from T, are marked by ×:

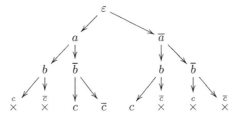

The literals on the open branches, terminating with the unmarked leaves, give the three models of T, as in Example 6.11.

In practice, the algorithms do not build the complete tree which quickly becomes prohibitively large as n grows. Instead, it is developed gradually, observing if there remain any open branches. Usually, a theory is given by clauses of various length, much shorter than the total number n of variables. Such a shorter clause excludes then *all* branches containing all its negated literals. E.g., if the theory is extended with the clause $\neg a \vee b$, all branches containing a and $\overline{b}$ become closed, as shown below. Once closed, a branch is never extended during the further construction of the tree.

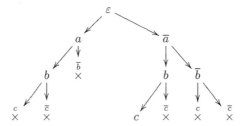

The resulting theory still has one model, represented by the open branch $\overline{a}, b, c$. Adding any clause closing this branch, i.e., $a \vee \neg b \vee \neg c$ or $\neg c$ or $\neg b$ or $\neg c \vee a$, etc., makes the theory inconsistent....................................[end optional]

6.3 Soundness

───────────────────────── a background story ─────────────────────────

The library offers its customers the possibility of ordering books on internet. From the main page one may ask the system to find the book one wishes to borrow. (We assume that appropriate search engine will always find the book one is looking for or else give a message that it could not be identified. In the sequel we are considering only the case when the book you asked for was found.)

The book (found by the system) may happen to be immediately available for loan. In this case, you may just reserve it and our story ends here. But the most frequent case is that the book is on loan or else must be borrowed from another library. In such a case, the system gives you the possibility to *order* it: you mark the book and the system will send you a message as soon as the book becomes available. (You need no message as long as the book is not available and the system need not inform you about that.) Simplicity of this scenario notwithstanding, this is actually our whole story.

There are two distinct assumptions which make us rely on the system when we order a book. The first is that when you get the message that the book is available *it really is*. The system will not play fool with you saying "Hi, the book is here" while it is still on loan with another user. We trust that what the system says ("The book is here") is true. This property is what we call "soundness" of the system – it never provides us with false information.

But there is also another aspect making up our trust in the system. Suppose that the book actually becomes available, but you do not get the appropriate message. The system is still sound – it does not give you any wrong information – but only because it does not give you any information whatsoever. It keeps silent although it should have said that the book is there and you can borrow it. The other aspect of our trust is that whenever there is a fact to be reported ('the book became available'), the system will do it – it is "complete".

Just like a system may be sound without being complete (keep silent even though the book arrived to the library), it may be complete without being sound. If it constantly informed you that the ordered book was available, it would eventually (when the book became available) report the true fact. However, in the meantime, it would provide you with a series of incorrect information – it would be unsound.

Thus, soundness of a system means that whatever it says is correct: it says "The book is here" *only if* it is here. Completeness means that everything that is correct will be said by the system: it says "The book is here" *if* (always when) the book is here. In the latter case, we should pay attention to the phrase "everything that is correct". It makes sense because our setting is very limited. We have one command 'order the book ...', and one possible response of the system: the message that the book became available. "Everything that is correct" means here simply that the book you ordered actually is available. It is only this limited context (i.e., limited and well-defined amount of true facts) which makes the notion of completeness meaningful.

In connection with axiomatic systems one often resorts to another analogy. The axioms and the deduction rules together define the scope of the system's knowledge about the world. If all aspects of this knowledge (all the theorems) are true about the world, the system is sound. This idea has enough intuitive content to be grasped with reference to vague notions of 'knowledge', 'the world', etc. and our illustration with the system saying "The book is here" only when it actually is, merely makes it more specific.

Completeness, on the other hand, would mean that everything that is true about the world (and expressible in the actual language), is also reflected in the system's knowledge (theorems). Here it becomes less clear what the intuitive content of 'completeness' might be. What can one possibly mean by "everything that is true"? In our library example, the user and the system use only very limited language allowing the user to 'order the book ...' and the system to state that it is available. Thus, the possible meaning of "everything" is limited to the book being available or not. One should keep this difference between 'real world' and 'availability of a book' in mind because the notion of completeness is as unnatural in the context of natural language and real world, as it is adequate in the context of bounded, sharply delineated worlds of formal semantics. The limited expressiveness of a formal language plays here crucial role of limiting the discourse to a well-defined set of expressible facts.

The library system should be both sound *and* complete to be useful. For axiomatic systems, the minimal requirement is that they are sound – completeness is a desirable feature which, typically, is much harder to obtain. There are axiomatic systems which are sound but *inherently incomplete but we will not study such systems.* [1] _____

Definition 5.11 introduced, among other concepts, the validity relation $\models A$, stating that A is satisfied by all structures. On the other hand, we studied the syntactic notion of a proof in a given axiomatic system $\mathcal{C}$, which we wrote as $\vdash_{\mathcal{C}} A$. We also saw a generalization of the provability predicate $\vdash_{\mathcal{H}}$ in Hilbert's system to the relation $\Gamma \vdash_{\mathcal{N}} A$, where Γ is a theory – a set of formulae. We now define the *semantic* relation $\Gamma \models A$ of "A being a (tauto)logical consequence of Γ".

Definition 6.12 For $\Gamma \subseteq \mathsf{WFF}_{\mathsf{PL}}$, $A \in \mathsf{WFF}_{\mathsf{PL}}$ and a valuation V, we write:
- $V \models \Gamma$ iff $V \models G$ for all $G \in \Gamma$ . . . . – V is a model of Γ
- $Mod(\Gamma) = \{V : V \models \Gamma\}$ . . . . – all models of Γ
- $\models \Gamma$ iff   for all $V : V \models \Gamma$ . . . . – Γ is valid
- $\Gamma \models A$ iff   for all $V \in Mod(\Gamma) : V \models A$ . . . . . . – A is a logical
  i.e., for all $V : V \models \Gamma \Rightarrow V \models A$ consequence of Γ.

The analogy between the symbols $\models$ and $\vdash$ is not accidental. The former refers to a semantic notion, while the later to a syntactic one and, ideally, these two notions should be equivalent in some sense. The following table gives the picture of the intended equivalences:

$$
\begin{array}{ccc}
\multicolumn{3}{c}{syntactic \ \ vs. \ \ semantic} \\
\hline
\vdash_{\mathcal{H}} A & \Leftrightarrow & \models A \\
\Gamma \vdash_{\mathcal{N}} A & \Leftrightarrow & \Gamma \models A
\end{array}
\tag{6.13}
$$

The implication $\Gamma \vdash_{\mathcal{C}} A \Rightarrow \Gamma \models A$ is called *soundness* of the proof system $\mathcal{C}$: whatever can be proven in $\mathcal{C}$ from the assumptions Γ, is true in every structure satisfying Γ. This is usually easy to establish as we will see shortly. The problematic implication is the other one – *completeness* – stating that any formula which is true in all models of Γ is provable in $\mathcal{C}$ from the assumptions Γ. ($\Gamma = \varnothing$ is a special case: the theorems $\vdash_{\mathcal{C}} A$ are tautologies and the formulae $\models A$ are those satisfied by all possible structures, since every structure V satisfies the empty set of assumptions.)

Remark 6.14 [Soundness and Completeness]
Another way of viewing these two implications is as follows. Given an axiomatic system $\mathcal{C}$ and a theory Γ, the relation $\Gamma \vdash_{\mathcal{C}} \_$ defines the set of formulae – the theorems – $Th_{\mathcal{C}}(\Gamma) = \{A : \Gamma \vdash_{\mathcal{C}} A\}$. On the other hand, the definition of $\Gamma \models \_$ gives a (possibly different) set of formulae, $\Gamma^* = \{B : \Gamma \models B\}$ – the set of (tauto)logical consequences of Γ. Soundness

---

[1]Thanks to Eivind Kolflaath for the library analogy.

of $\mathcal{C}$, i.e., the implication $\Gamma \vdash_{\mathcal{C}} A \Rightarrow \Gamma \models A$ means that each formula provable from Γ is Γ's (tauto)logical consequence, i.e., that $Th_{\mathcal{C}}(\Gamma) \subseteq \Gamma^*$. Completeness means that every (tauto)logical consequence of Γ is also provable and amounts to the opposite inclusion $\Gamma^* \subseteq Th_{\mathcal{C}}(\Gamma)$. $\qquad\Box$

For proving soundness of a system consisting, like $\mathcal{H}$ or $\mathcal{N}$, of axioms and proof rules, one has to show that the axioms are valid and the rules preserve validity. (Typically, a stronger form is shown, namely, preservation of truth: whenever a structure satisfies the assumptions of the rule, then it satisfies also the conclusion.) These two facts establish, by induction on the length of the proof $\Gamma \vdash_{\mathcal{N}} A$, that all theorems of the system are valid.

Theorem 6.15 [Soundness] For every set $\Gamma \subseteq \mathsf{WFF}_{\mathsf{PL}}$ and $A \in \mathsf{WFF}_{\mathsf{PL}}$:
$\Gamma \vdash_{\mathcal{N}} A \Rightarrow \Gamma \models A$.

Proof. We show validity of all axioms and that MP preserves truth (and hence, validity and logical consequence) in any structure.

A1–A3 :: In Exercise 5.2 we have seen that all axioms of $\mathcal{H}$ are valid, i.e., satisfied by any structure. In particular, the axioms are satisfied by all models of Γ, for any Γ.

A0 :: The axiom schema $A0$ allows us to conclude $\Gamma \vdash_{\mathcal{N}} B$ for any $B \in \Gamma$. This is obviously sound: any model V of Γ must satisfy all the formulae of Γ and, in particular, B.

MP :: Suppose $\Gamma \models A$ and $\Gamma \models A \to B$. Then, for an arbitrary $V \in Mod(\Gamma)$ we have $\widehat{V}(A) = \mathbf{1}$ and $\widehat{V}(A \to B) = \mathbf{1}$. Consulting the boolean table for $\to$: the first assumption reduces the possibilities for V to the two rows for which $\widehat{V}(A) = \mathbf{1}$, and then, the second assumption to the only possible row in which also $\widehat{V}(A \to B) = \mathbf{1}$. In this row $\widehat{V}(B) = \mathbf{1}$, so $V \models B$. Since V was an arbitrary model of Γ, it follows that $\Gamma \models B$.

QED (6.15)

Soundness of $\mathcal{H}$ follows by easy simplifications of this proof. The following corollary gives another formulation of soundness.

Corollary 6.16 Every satisfiable theory is consistent.

Proof. We show the equivalent statement that every inconsistent theory is unsatisfiable. Indeed, if $\Gamma \vdash_{\mathcal{N}} \bot$ then $\Gamma \models \bot$ by Theorem 6.15, hence Γ is not satisfiable (as there is no V with $V \models \bot$). QED (6.16)

Remark 6.17 [Equivalence of two soundness notions]
Soundness is often expressed as Corollary 6.16, since the two are equivalent:

   6.15. $\Gamma \vdash_{\mathcal{N}} A \Rightarrow \Gamma \models A$, and
   6.16. (exists $V : V \models \Gamma) \Rightarrow \Gamma \not\vdash_{\mathcal{N}} \bot$.

The implication 6.15 $\Rightarrow$ 6.16 is given in the proof of Corollary 6.16. For the opposite: if $\Gamma \vdash_{\mathcal{N}} A$ then $\Gamma \cup \{\neg A\}$ is inconsistent (Exercise 4.7) and hence (by 6.16) unsatisfiable, i.e., for any $V : V \models \Gamma \Rightarrow V \not\models \neg A$. But if $V \not\models \neg A$ then $V \models A$, and so, since V was arbitrary, $\Gamma \models A$. $\square$

6.4 COMPLETENESS

The proof of completeness involves several lemmata which we now proceed to establish. Just as there are two equivalent ways of expressing soundness, there are two equivalent ways of expressing completeness. One (corresponding to Corollary 6.16) says that every consistent theory is satisfiable and the other that every valid formula is provable.

Lemma 6.18 The two formulations of completeness are equivalent:

(1) $\Gamma \not\vdash_{\mathcal{N}} \bot \Rightarrow Mod(\Gamma) \neq \varnothing$
(2) $\Gamma \models A \Rightarrow \Gamma \vdash_{\mathcal{N}} A$.

> **Proof.** (1) $\Rightarrow$ (2). Assume (1) and $\Gamma \models A$, i.e., for any $V : V \models \Gamma \Rightarrow V \models A$. Then $\Gamma \cup \{\neg A\}$ has no model and, by (1), $\Gamma, \neg A \vdash_{\mathcal{N}} \bot$. By Deduction Theorem $\Gamma \vdash_{\mathcal{N}} \neg A \to \bot$, and so $\Gamma \vdash_{\mathcal{N}} A$ by Exercise 4.1.(6) and Lemma 4.11.(1).
> (2) $\Rightarrow$ (1). If $\Gamma \not\vdash_{\mathcal{N}} \bot$ then, taking (any instance of) $\bot$ for A, (2) gives that $\Gamma \not\models \bot$, i.e., for some valuation $V \in Mod(\Gamma)$ and $V \not\models \bot$. But this implies that $Mod(\Gamma) \neq \varnothing$, so (1) holds.   QED (6.18)

We prove the first of the above formulations: we take an arbitrary Γ and, assuming that it is consistent, i.e., $\Gamma \not\vdash_{\mathcal{N}} \bot$, we show that $Mod(\Gamma) \neq \varnothing$ by constructing a particular structure which we prove to be a model of Γ. This proof is not the simplest possible for PL. However, we choose to do it this way because it illustrates the general strategy used later in the completeness proof for FOL. Our proof uses the notion of a *maximal consistent* theory:

Definition 6.19 A theory $\Gamma \subset \mathsf{WFF}_{\mathsf{PL}}^{\Sigma}$ is *maximal consistent* iff it is consistent and, for any formula $A \in \mathsf{WFF}_{\mathsf{PL}}^{\Sigma}$, $\Gamma \vdash_{\mathcal{N}} A$ or $\Gamma \vdash_{\mathcal{N}} \neg A$.

From Exercise 4.7 we know that if Γ is consistent then for any A at most one, $\Gamma \vdash_{\mathcal{N}} A$ or $\Gamma \vdash_{\mathcal{N}} \neg A$, is the case, i.e.:

$$\Gamma \nvdash_{\mathcal{N}} A \text{ or } \Gamma \nvdash_{\mathcal{N}} \neg A \quad \text{or equivalently} \quad \Gamma \vdash_{\mathcal{N}} A \Rightarrow \Gamma \nvdash_{\mathcal{N}} \neg A. \qquad (6.20)$$

Consistent Γ cannot prove too much – if it proves something (A) then there is something else (namely $\neg A$) which it does not prove.

Maximality is a kind of the opposite – Γ cannot prove too little: if Γ does *not* prove something ($\neg A$) then there must be something else it proves (namely A):

$$\Gamma \vdash_{\mathcal{N}} A \text{ or } \Gamma \vdash_{\mathcal{N}} \neg A \quad \text{or equivalently} \quad \Gamma \vdash_{\mathcal{N}} A \Leftarrow \Gamma \nvdash_{\mathcal{N}} \neg A. \qquad (6.21)$$

If Γ is maximal consistent it satisfies both (6.20) and (6.21) and hence, for any formula A, *exactly* one of $\Gamma \vdash_{\mathcal{N}} A$ and $\Gamma \vdash_{\mathcal{N}} \neg A$ is the case.

For instance, given $\Sigma = \{a, b\}$, the theory $\Gamma = \{a \to b\}$ is consistent. But it is not maximal consistent because, for instance, $\Gamma \nvdash_{\mathcal{N}} a$ and $\Gamma \nvdash_{\mathcal{N}} \neg a$. (The same holds if we replace a by b.) The following, equivalent formulation of maximal consistency for PL, is easier to check.

Fact 6.22 A theory $\Gamma \subset \text{WFF}_{\text{PL}}^{\Sigma}$ is maximal consistent iff it is consistent and for all $a \in \Sigma : \Gamma \vdash_{\mathcal{N}} a$ or $\Gamma \vdash_{\mathcal{N}} \neg a$.

Proof. 'Only if' part, i.e. $\Rightarrow$, is trivial from Definition 6.19 which ensures that $\Gamma \vdash_{\mathcal{N}} A$ or $\Gamma \vdash_{\mathcal{N}} \neg A$ for all formulae, in particular all atomic ones. The opposite implication is shown by induction on the complexity of A.

A IS :

$a \in \Sigma$ :: This case is trivial, being exactly the assumption.

$\neg B$ :: By IH, we have that $\Gamma \vdash_{\mathcal{N}} B$ or $\Gamma \vdash_{\mathcal{N}} \neg B$. In the latter case, we are done ($\Gamma \vdash_{\mathcal{N}} A$), while in the former we obtain $\Gamma \vdash_{\mathcal{N}} \neg A$, i.e., $\Gamma \vdash_{\mathcal{N}} \neg\neg B$ from Lemma 4.11.

$C \to D$ :: By IH we have that either $\Gamma \vdash_{\mathcal{N}} D$ or $\Gamma \vdash_{\mathcal{N}} \neg D$. In the former case, we obtain $\Gamma \vdash_{\mathcal{N}} C \to D$ by Lemma 4.10. In the latter case, we have to consider two subcases – by IH either $\Gamma \vdash_{\mathcal{N}} C$ or $\Gamma \vdash_{\mathcal{N}} \neg C$. If $\Gamma \vdash_{\mathcal{N}} \neg C$ then, by Lemma 4.10, $\Gamma \vdash_{\mathcal{N}} \neg D \to \neg C$. Applying MP to this and axiom A3, we obtain $\Gamma \vdash_{\mathcal{N}} C \to D$. So, finally, assume $\Gamma \vdash_{\mathcal{N}} C$ (and $\Gamma \vdash_{\mathcal{N}} \neg D$). But then $\Gamma \vdash_{\mathcal{N}} \neg(C \to D)$ by Exercise 4.1.(4).   QED (6.22)

The maximality of a maximal consistent theory makes it easier to construct a model for it. We prove first this special case of the completeness theorem:

Lemma 6.23 Every maximal consistent theory is satisfiable.

Proof. Let Γ be any maximal consistent theory, and let Σ be the set of propositional variables. We define the valuation $V : \Sigma \to \{\mathbf{1}, \mathbf{0}\}$ by the equivalence

$$V(a) = \mathbf{1} \text{ iff } \Gamma \vdash_N a$$

for every $a \in \Sigma$. (Hence also $V(a) = \mathbf{0}$ iff $\Gamma \nvdash_N a$.) We show that V is a model of Γ, i.e., for any B : if $B \in \Gamma$ then $\widehat{V}(B) = \mathbf{1}$. In fact, we prove a stronger claim that for every formula B,

$$\widehat{V}(B) = \mathbf{1} \text{ iff } \Gamma \vdash_N B.$$

The proof goes by induction on (the complexity of) B:

a :: Immediate from the definition of V.

$\neg C$ :: $\widehat{V}(\neg C) = \mathbf{1}$ iff $\widehat{V}(C) = \mathbf{0}$. By IH, the latter holds iff $\Gamma \nvdash_N C$, i.e., iff $\Gamma \vdash_N \neg C$.

$C \to D$ :: We consider two cases:

- $\widehat{V}(C \to D) = \mathbf{1}$ implies that either $\widehat{V}(C) = \mathbf{0}$ or $\widehat{V}(D) = \mathbf{1}$. By the IH, this implies $\Gamma \nvdash_N C$ or $\Gamma \vdash_N D$, i.e., $\Gamma \vdash_N \neg C$ or $\Gamma \vdash_N D$. In the former case Exercise 4.1.(1), and in the latter Lemma 4.13.(2) gives that $\Gamma \vdash_N C \to D$.
- $\widehat{V}(C \to D) = \mathbf{0}$ implies $\widehat{V}(C) = \mathbf{1}$ and $\widehat{V}(D) = \mathbf{0}$, which by the IH imply $\Gamma \vdash_N C$ and $\Gamma \nvdash_N D$, i.e., $\Gamma \vdash_N C$ and $\Gamma \vdash_N \neg D$, which by Exercise 4.1.(4) and two applications of MP imply $\Gamma \vdash_N \neg (C \to D)$. Since Γ is consistent, this means that $\Gamma \nvdash_N C \to D$.   QED (6.23)

Next we use this result to show that *every* consistent theory is satisfiable. What we need, is a result stating that every consistent theory is a subset of some maximal consistent theory.

Lemma 6.24 Every consistent theory has a maximal consistent extension.

Proof. Let Γ be a consistent theory, and let $\{a_0, \ldots, a_n\}$ be the set of propositional variables used in Γ. [The case when $n = \omega$ (is countably infinite) is treated in the small font within the square brackets.]   Let

- $\Gamma_0 = \Gamma$
- $\Gamma_{i+1} = \begin{cases} \Gamma_i, a_i & \text{if this is consistent} \\ \Gamma_i, \neg a_i & \text{otherwise} \end{cases}$
- $\widehat{\Gamma} = \Gamma_{n+1}$ $[\, = \bigcup_{i < \omega} \Gamma_i, \text{ if } n = \omega] \,$.

We show by induction on i that for every i, Γ_i is consistent.

BASIS :: $\Gamma_0 = \Gamma$ is consistent by assumption.

IND. :: Suppose Γ_j is consistent. If Γ_{j+1} is inconsistent, then from the definition of Γ_{j+1} we know that both Γ_j, a_j and $\Gamma_j, \neg a_j$ are inconsistent, hence by Deduction Theorem $a_j \to \bot$ and $\neg a_j \to \bot$ are both provable from Γ_j. Exercise 4.1.(5)-(6) tells us that in this case both $\neg a_j$ and $\neg\neg a_j$ are provable from Γ_j, contradicting consistency of Γ_j (Exercise 4.7).

In particular, $\Gamma_n = \widehat{\Gamma}$ is consistent. [For the infinite case, we use the above proof, the fact that any finite subtheory of $\widehat{\Gamma}$ must be contained in a Γ_i for some i, and then the compactness Theorem 4.28.]

To finish the proof, we have to show that $\widehat{\Gamma}$ is not only consistent but also maximal, i.e., that for every A, either $\widehat{\Gamma} \vdash_{\mathcal{N}} A$ or $\widehat{\Gamma} \vdash_{\mathcal{N}} \neg A$. But this was shown in Fact 6.22, and so the proof is complete.   QED (6.24)

The completeness theorem is now an immediate consequence:

Theorem 6.25   Every consistent theory is satisfiable.

Proof. Let Γ be consistent. By Lemma 6.24 it can be extended to a maximal consistent theory $\widehat{\Gamma}$ which, by Lemma 6.23, is satisfiable, i.e., has a model V. Since $\Gamma \subseteq \widehat{\Gamma}$, this model satisfies also Γ.   QED (6.25)

Corollary 6.26   Let $\Gamma \subseteq \mathsf{WFF_{PL}}$ and $A \in \mathsf{WFF_{PL}}$. Then
  (1) $\Gamma \models A$ implies $\Gamma \vdash_{\mathcal{N}} A$,   and   (2) $\models A$ implies $\vdash_{\mathcal{H}} A$.

Proof. In view of Lemma 4.29, (2) follows from (1). But (1) follows from Theorem 6.25 by Lemma 6.18.   QED (6.26)

The soundness and completeness results are gathered in

Corollary 6.27   For any $\Gamma \subseteq \mathsf{WFF_{PL}}$, $A \in \mathsf{WFF_{PL}} : \Gamma \models A \Leftrightarrow \Gamma \vdash_{\mathcal{N}} A$.

Point (3) of the following corollary, giving a new, semantic characterization of consistency, follows from Remark 6.17 and Lemma 6.18. But it can be also seen as a consequence of soundness and completeness (if these were established without this remark and lemma).

Corollary 6.28 Γ is consistent iff any of the equivalent conditions holds:
  (1) $\Gamma \nvdash_{\mathcal{N}} \bot$, or
  (2) for every formula $A : \Gamma \nvdash_{\mathcal{N}} A$ or $\Gamma \nvdash_{\mathcal{N}} \neg A$, or
  (3) $Mod(\Gamma) \neq \varnothing$.

6.5 SOME APPLICATIONS

Having a sound and complete axiomatic system allows us to switch freely
between the syntactic (concerning provability) and semantic (concerning
validity) arguments – depending on which one is easier in a given context.

1. Is a formula valid?
Validity of a PL formula is, typically, easiest to verify by making the appropriate boolean table. But we have also proved several formulae. For
instance, asked whether $(A \to (B \to C)) \to ((A \to B) \to (A \to C))$ is
valid, we have a direct answer – it is axiom $A2$ of $\mathcal{H}$ and thus, **by soundness** of $\mathcal{H}$, we can immediately conclude that the formula is valid.

2. Is a formula provable?
In Theorem 4.31 we gave an argument showing decidability of membership
in $\vdash_{\mathcal{N}}$. In a bit roundabout way, we transformed $\mathcal{N}$ expressions into corresponding $\mathcal{G}$ expressions, and used $\mathcal{G}$ to decide their derivability (which, we
said, was equivalent to derivability in $\mathcal{N}$).

  Corollary 6.27 gives us another, semantic, way of deciding membership
in $\vdash_{\mathcal{N}}$. It says that $\mathcal{N}$-derivable formulae are exactly the ones which are
valid. Thus, to see if $G = A_1, \ldots, A_n \vdash_{\mathcal{N}} B$ is derivable in $\mathcal{N}$ it suffices
to see if $A_1, \ldots, A_n \models B$. Since G is derivable iff $G' = \vdash_{\mathcal{N}} A_1 \to (A_2 \to
\ldots (A_n \to B) \ldots)$ is, Lemma 4.30, the problem can be decided by checking
if G' is valid. But this is trivial! Just make the boolean table for G', fill
out all the rows and see if the last column contains only 1. If it does, G'
is valid and so, **by completeness**, derivable. If it does not (contains some
0), G' is not valid and, **by soundness**, is not derivable.

Example 6.29
Is $\vdash_{\mathcal{N}} A \to (B \to (B \to A))$? We make the boolean table:

A	B	$B{\to}A$	$B{\to}(B{\to}A)$	$A{\to}(B{\to}(B{\to}A))$
1	1	1	1	1
1	0	1	1	1
0	1	0	0	1
0	0	1	1	1

The table tells us that $\models A \to (B \to (B \to A))$ and thus, **by completeness of $\mathcal{N}$**, we conclude that the formula is derivable in $\mathcal{N}$.

  Now, is $\vdash_{\mathcal{N}} B \to (B \to A)$? The table is like the one above without the
last column. The formula is not valid (third row gives 0), $\not\models B \to (B \to A)$

and, **by soundness of** $\mathcal{N}$, we conclude that it is not derivable in $\mathcal{N}$. $\quad\square$

Notice that to *decide* provability by such a reference to semantics we need both properties – completeness guarantees that whatever is valid is provable, while soundness that whatever is not valid is not provable.

This application of soundness/completeness is not typical because, usually, axiomatic systems are designed exactly to facilitate answering the more complicated questions about validity of formulae. In PL, however, the semantics is so simple and decidable that it is easier to work with it directly than using respective axiomatic systems (except, perhaps, for $\mathcal{G}$).

3. Is a rule admissible?

For instance, is the rule $\dfrac{\vdash_{\mathcal{N}} A \to B \;\; ; \;\; \vdash_{\mathcal{N}} \neg B}{\vdash_{\mathcal{N}} \neg A}$ admissible in $\mathcal{N}$?

First, we have to verify if the rule itself is sound. So let V be an arbitrary structure (valuation) such that $V \models A \to B$ and $V \models \neg B$. From the latter we have that $V(B) = \mathbf{0}$ and so, using the definition of $\to$, we obtain that since $V(A \to B) = \mathbf{1}$, we must have $V(A) = \mathbf{0}$. This means that $V \models \neg A$. Since V was arbitrary, we conclude that the rule is sound.

Now comes the application of soundness/completeness of $\mathcal{N}$. If $\vdash_{\mathcal{N}} A \to B$ and $\vdash_{\mathcal{N}} \neg B$ then, **by soundness of** $\mathcal{N}$, we also have $\models A \to B$ and $\models \neg B$. Then, by soundness of the rule itself, $\models \neg A$. And finally, **by completeness of** $\mathcal{N}$, this implies $\vdash_{\mathcal{N}} \neg A$. Thus, the rule is, indeed, admissible in $\mathcal{N}$, even though we have not shown how exactly the actual proof of $\neg A$ would be constructed. This form of an argument can be applied to show that *any* *sound* rule will be admissible in a sound and complete axiomatic system.

On the other hand, if a rule is not sound, the soundness of the axiomatic system immediately implies that the rule will not be admissible in it. For instance, the rule $\dfrac{\vdash_{\mathcal{N}} A \to B \;\; ; \;\; \vdash_{\mathcal{N}} B}{\vdash_{\mathcal{N}} A}$ is not sound (verify it – find a valuation making both premises true and the conclusion false). **By soundness of** $\mathcal{N}$, we may conclude that it isn't admissible there.

EXERCISES 6.

EXERCISE 6.1 Translate the following arguments into formulae of PL and show either semantically (by setting up appropriate boolean tables) or syntactically (giving a proof in any of the reasoning systems we have seen, assuming its soundness and completeness) – that they are tautologies:

(1) If I am in Paris then I am in France and if I am in Rome then I am in Italy. Hence, if I am in Paris then I am in Italy or if I am in Rome then I am in France.

(2) If I press the gas pedal and turn the key, the car will start. Hence, either if I press the gas pedal the car will start or if I turn the key the car will start.

The confusion arises from the fact that in the daily language there is no clear distinction between $\to$ and $\Rightarrow$. The immediate understanding of the implications in the conclusions of these arguments will interpret them as $\Rightarrow$ rather than $\to$. This interpretation makes them sound absurd.

EXERCISE 6.2 Let F, G be two boolean functions given by
$$F(x,y) = 1 \text{ iff } x = 1, \qquad G(x,y,z) = 1 \text{ iff } y = 1 \text{ or } z = 0.$$
Write each of these functions as formulae in CNF and DNF.

EXERCISE 6.3 Find DNF and CNF formulae inducing the same boolean functions as the formulae:

(1) $(\neg a \wedge b) \to c$
(2) $(a \to b) \wedge (b \to c) \wedge (a \vee \neg c)$
(3) $(\neg a \vee b) \vee (a \vee (\neg a \wedge \neg b))$

[Recall first Exercise 5.14. You may follow the construction from the proof of Theorem 6.2. But you may just use the laws which you have learned so far to perform purely syntactic manipulations. Which of these two ways is simpler?]

EXERCISE 6.4 Define the binary $\uparrow$ (Sheffer's stroke), and its dual as follows:

$x \uparrow y$	1	0	$: y$		$x \downarrow y$	1	0	$: y$
$x:$ 1	1	1			$x:$ 1	0	0	
0	1	0			0	0	1	

Show that $\{\uparrow\}$ is expressively complete, and that so is $\{\downarrow\}$.

(Hint: Express some expressively complete set using the connective.)

EXERCISE 6.5 Define inductively the language $\mathcal{P}$ (for a given alphabet Σ):

$\bot \in \mathcal{P}$ and $\top \in \mathcal{P}$ – two propositional constants

$\Sigma \subset \mathcal{P}$ – the parameter alphabet

If $A, B, C \in \mathcal{P}$ then also $ite(A, B, C) \in \mathcal{P}$.

The semantics is defined extending any valuation $V : \Sigma \to \{1, 0\}$, by:

$\widehat{V}(\bot) = 0$ and $\widehat{V}(\top) = 1$

$\widehat{V}(a) = V(a)$ for $a \in \Sigma$

$$\widehat{V}(ite(A, B, C)) = \begin{cases} \widehat{V}(B) \text{ if } \widehat{V}(A) = 1 \\ \widehat{V}(C) \text{ if } \widehat{V}(A) = 0 \end{cases}$$

Show that the set $\{\top, \bot, ite(\_, \_, \_)\}$ is expressively complete.

EXERCISE 6.6 Show that $\{\vee, \wedge\}$ is not an expressively complete set.

EXERCISE 6.7 Let $\Sigma = \{a, b, c\}$ and consider two sets of formulae $\Delta = \{a \wedge (a \to b)\}$ and $\Gamma = \{a, a \to b, \neg c\}$. Give an example of

(1) an $A \in \mathsf{WFF}_{\mathsf{PL}}^{\Sigma}$ such that $\Delta \not\models A$ and $\Gamma \models A$
(2) a model (valuation) V such that $V \models \Delta$ and $V \not\models \Gamma$.

EXERCISE 6.8 Let $\Sigma = \{a, b, c\}$. Which of the following sets of formulae are maximal consistent?

(1) $\{a, b, \neg c\}$ (2) $\{\neg b \to a, \neg a \vee c\}$
(3) $\{\neg a \vee b, b \to c, \neg c\}$ (4) $\{\neg a \vee b, b \to c, \neg c, a\}$

EXERCISE 6.9 Show that Γ is maximal consistent (in $\mathcal{N}$) iff it has exactly one model.

EXERCISE 6.10 Apply de Morgan's laws to show directly (without using Corollaries 6.7-6.8) that

(1) if A is in CNF then $\neg A$ is equivalent to a formula in DNF
(2) if A is in DNF then $\neg A$ is equivalent to a formula in CNF.

EXERCISE 6.11 The following rules, called (respectively) the *constructive* and *destructive* dilemma, are often handy in constructing proofs

(CD) $\dfrac{A \vee B \; ; \; A \to C \; ; \; B \to D}{C \vee D}$ (DD) $\dfrac{\neg C \vee \neg D \; ; \; A \to C \; ; \; B \to D}{\neg A \vee \neg B}$

Show that they are sound, i.e., in any structure V which makes the assumptions of (each) rule true, the (respective) conclusion is true as well. Use this to show that these rules are admissible in $\mathcal{N}$. (If the syntax $X \vee Y$ seems confusing, it can be written as $\neg X \to Y$.)

EXERCISE 6.12 Recall compactness theorem 4.28. Using soundness and completeness theorem for $\mathcal{N}$, show its semantic counterpart, namely, that for every (propositional theory) Γ:

      Γ has a model iff every finite subtheory $\Delta \subseteq \Gamma$ has a model.

EXERCISE 6.13 Show the following claims (i.e., unprovability of the general schemata for arbitrary instances of A, B):

(1) $\not\vdash_{\mathcal{N}} \neg(B \to B)$ (2) $A \to B \not\vdash_{\mathcal{N}} B \to A$ (3) $A \to B, \neg A \not\vdash_{\mathcal{N}} \neg B$
Provide instances of (2) and (3) which *are* provable.

EXERCISE 6.14 [Completeness of Gentzen's system for PL]
The interpretation of sequents is defined as follows. A structure (valuation) V *satisfies* the sequent $\Gamma \vdash_{G} \Delta$, written $V \models \Gamma \vdash_{G} \Delta$, iff either there is a formula $\gamma \in \Gamma$ such that $V \not\models \gamma$, or there is a formula $\delta \in \Delta$ such that

$V \models \delta$, i.e., iff $V \models \bigwedge \Gamma \to \bigvee \Delta$. Writing the sequent explicitly:

$$V \models \gamma_1, \ldots, \gamma_g \vdash_{\mathcal{G}} \delta_1, \ldots, \delta_d \quad \Longleftrightarrow \quad V \models \gamma_1 \wedge \ldots \wedge \gamma_g \to \delta_1 \vee \ldots \vee \delta_d.$$

$\Gamma \vdash_{\mathcal{G}} \Delta$ is valid iff it is satisfied by every valuation, i.e., iff $\bigwedge \Gamma \Rightarrow \bigvee \Delta$.
We have remarked that the same formulae are provable whether in the axioms we require only atoms or allow general formulae. Here we take the version with *only* atomic formulae in the axioms (i.e., axioms are sequents $\Gamma \vdash_{\mathcal{G}} \Delta$ with $\Gamma \cap \Delta \neq \varnothing$ and where all formulae in $\Gamma \cup \Delta$ are atomic (i.e., propositional variables)). Also, we consider only the basic system with the rules for the connectives $\neg$ and $\to$.

(1) (a) Say (in *only one* sentence!) why the axioms of $\vdash_{\mathcal{G}}$ are valid, i.e., why every valuation V satisfies every axiom.

(b) Given an irreducible sequent S (containing only atomic formulae) which is *not* an axiom, describe a valuation V making $V \not\models S$ (a so called "counter-model" for S).

(2) Verify that the rules are *invertible*, that is, are sound in the direction "bottom up": for any valuation V, if V satisfies the conclusion of the rule, then V satisfies *all* the premises.

(3) Suppose that $\Gamma \not\vdash_{\mathcal{G}} \Delta$. Then, at least one of the branches of the proof (built bottom-up from this sequent) ends with a non-axiomatic sequent, S (containing only atoms). Hence, as shown in point 1.(b), S has a counter-model.

(a) On the basis of (one of) the above points, explain (in *only one* sentence) why this counter-model V for S will also be a counter-model for $\Gamma \vdash_{\mathcal{G}} \Delta$, i.e., why it will be the case that $V \not\models \bigwedge \Gamma \to \bigvee \Delta$.

(b) You have actually proved completeness of the system $\vdash_{\mathcal{G}}$. Explain (in *one* sentence or formula) why this is the case.

Soundness and above completeness result for Gentezen's system, together with the corresponding results for $\mathcal{N}$, establish now formally the equivalence (4.32), p. 137. Use these results to

(4) show the equivalence (4.32), for arbitrary finite Γ and formula B;

(5) show the admissibility of the (cut) rule in Gentzen's system (a more general version of (4.33) from p. 137), namely: $\dfrac{\Gamma \vdash_{\mathcal{G}} \Delta, A \ ; \ A, \Gamma \vdash_{\mathcal{G}} \Delta}{\Gamma \vdash_{\mathcal{G}} \Delta}$.

EXERCISE 6.15 [Łukasiewicz's 3-valued logic Ł3]

To the propositional language (with $\neg$ and $\to$) from Definition 4.4, we give a *3-valued semantics*, interpreting formulae over the set $\mathbb{L} = \{1, \frac{1}{2}, 0\}$. (The value $\frac{1}{2}$ represents "undetermined".) An assignment $V : \Sigma \to \mathbb{L}$

extends to all formulae, $\widehat{V}^3 : \mathsf{WFF}^\Sigma \to \mathbb{L}$, according to the two tables below to the left (coinciding with the ones from Definition 5.2 on the boolean values $\{1, 0\}$).

x	$\neg x$
1	0
$\frac{1}{2}$	$\frac{1}{2}$
0	1

$x \to y$	1	$\frac{1}{2}$	0	$: y$
$x:$ 1	1	$\frac{1}{2}$	0	
$\frac{1}{2}$	1	1	$\frac{1}{2}$	
0	1	1	1	

$V \models_3 A$ iff $\widehat{V}^3(A) = 1$

$\models_3 A$ iff for all $V : V \models_3 A$

$\mathsf{L} = \{A \in \mathsf{WFF}^\Sigma : \models_3 A\}$

Satisfaction $\models_3$ and tautologies L are defined as expected to the right, using only $1 \in \mathbb{L}$ as the designated element (marking "true" formulae). We are concerned only with the Ł3-tautologies L and, on the other hand, the classical ones, $Taut = \{A \in \mathsf{WFF}^\Sigma : \models A\}$.

(1) Show that $\mathsf{L} \subseteq Taut$, i.e., for every $A \in \mathsf{WFF}^\Sigma$, if for all $V_3 : V_3 \models_3 A$ then also for all $V : V \models A$, where V_3 ranges over 3-valued assignments to $\mathbb{L}$ while V over assignments to $\mathbb{B}$.

(Hint: Try a contrapositive argument, using the fact that the 3-valued tables reduce to the usual, boolean ones when applied only to the values $\{1, 0\}$.)

(2) Check which of the following schemata are Ł3-tautologies:
  (a) $\models_3 A \to (B \to A)$
  (b) $\models_3 (A \to (B \to C)) \to ((A \to B) \to (A \to C))$
  (c) $\models_3 (\neg B \to \neg A) \to (A \to B)$.

(3) Show that the Modus Ponens rule of Hilbert's system is Ł3-sound:
$$\frac{\models_3 A \;;\; \models_3 A \to B}{\models_3 B}.$$

(4) Using (2), explain if we can conclude that $\mathsf{L} \supseteq Taut$.

EXERCISE 6.16 [Independence of axiom A1]
The propositional formulae from Definition 4.4 are now interpreted in the set $\mathbb{T} = \{0, 1, 2\}$ and valuation $V : \Sigma \to \mathbb{T}$ is extended to all formulae $\widehat{V} : \mathsf{WFF}^\Sigma \to \mathbb{T}$ using the following tables:

x	$\neg x$
0	0
1	0
2	1

$x \to y$	0	1	2	$: y$
$x:$ 0	2	2	1	
1	2	1	2	
2	1	1	1	

The designated element is 1, i.e., V satisfies A, $V \models_T A$, iff $\widehat{V}(A) = 1$ and A is a $\mathbb{T}$-tautology, $\models_T A$, iff for every $V : V \models_T A$. Show that

(1) $\models_T (A \to (B \to C)) \to ((A \to B) \to (A \to C))$,
(2) $\models_T (\neg B \to \neg A) \to (A \to B)$,

(3) the Modus Ponens rule is sound with respect to this semantics,

(4) $\not\models_T A \to (B \to A)$.

(5) Explain how we can now conclude that the axiom A1 of the Hilbert's system is *independent* from the remaining axioms, i.e., is *not* derivable from them using the only inference rule MP.

Chapter 7

DIAGNOSING PARADOXES

———————————— a background story ——————————
"All Creteans are liars" said Epimenides from Creta some time
around 600 B.C. He did not seem to intend any irony but became
recorded as the first who formulated the liar paradox, which is today
expressed as "This sentence is false". The paradox amounts to violating the intuitive conviction that every declarative statement has a
specific truth-value. Motives for this tacit generalization abound in
daily statements like "Dogs bark", "It is sunny", etc. It appears valid
throughout all our discourse – until we encounter the liar or some of
his kin. For if the liar sentence is false, then what it claims is false,
but it claims its own falshood, so then it is true. And if it is true,
then what it claims is true, but it claims that it is false. There are
many statements which, entering logical or mathematical discourse,
challenged the accepted views and forced their adjustments. In Section
1.4, p.62, we saw Russell's paradox which demonstrated unsoundness
of general comprehension and led to its restriction. Semantic paradoxes, concerning the notions of truth and falsehood, have played a
similar role, but there is still no agreement on how to handle them.
Perhaps, the concept of truth is so fundamental, that its philosophical
scope extends beyond the abilities of mathematical formalization.

This chapter does not study the metaconcepts of truth and falsehood, but merely applies propositional logic for diagnosing paradoxical character of discourses. The liar paradox contains perhaps the
essence, but not the full range, of the phenomenon which, in every
situation, poses the question whether we actually encounter a paradox
at all. Its formalization will yield some general criteria for diagnosing
paradoxical character of many discourses. ———————————

7.1 SEMANTIC PARADOXES

To represent natural discourse which may, as it often does, refer to the truth of its own statements, we assume that every statement is uniquely identifed: in propositional logic, it has a unique name, which is equivalent to whatever it claims. If a is the statement "The next statement is false", i.e., $a \leftrightarrow$ "The next statement is false", while $b \leftrightarrow$ "The Moon is made of cheese", then the statement "The next statement is false and the Moon is made of cheese" is given a new name, say c, and captured by the equivalence $c \leftrightarrow a \wedge b$.[1] The liar is then represented simply as $l \leftrightarrow \neg l$.

Example 7.1
Consider the following two chains of statements – discourses:

a.	The next two statements are false.	a'.	The next statement is false.
		b'.	The next two statements are false.
b.	The last statement is false.		
c.	The next statement is false.	c'.	The next statement is false.
d.	The first statement was false.	d'.	The first statement is false.

At first, it is probably completely unclear whether any of these discourses makes sense. Represented as indicated, we obtain:

$$\Delta : a \leftrightarrow \neg b \wedge \neg c \qquad\qquad\qquad \Delta' : a' \leftrightarrow \neg b'$$
$$b \leftrightarrow \neg d \qquad\qquad\qquad\qquad b' \leftrightarrow \neg c' \wedge \neg d'$$
$$c \leftrightarrow \neg d \qquad\qquad\qquad\qquad c' \leftrightarrow \neg d'$$
$$d \leftrightarrow \neg a \qquad\qquad\qquad\qquad d' \leftrightarrow \neg a'$$

For Δ', we verify that the valuation $V(a') = V(c') = \mathbf{1}$ and $V(b') = V(d') = \mathbf{0}$ provides a model satisfying all formulae. Intuitively, Δ' is not paradoxical because we can assign truth-values to all involved statements in a consistent way, i.e., so that the statement (its name on the left of $\leftrightarrow$) is equivalent to whatever it states (its "content" on the right of $\leftrightarrow$).

To analyze Δ, we can use the rule of substituting provably equivalent formulae, Corollary 4.25. Substituting $\neg d$ for b and c, we obtain $a \leftrightarrow d$ and $d \leftrightarrow \neg a$, and then $a \leftrightarrow \neg a$. So Δ is inconsistent and, by soundness of our reasoning rule, it has no model, i.e., there is no way of assigning truth-values to all a, b, c, d so that the equivalences hold. Δ is a paradox exactly because any attempt ends up with a contradiction: assigning $\mathbf{0}$ to a forces us to assign it $\mathbf{1}$, while starting with $\mathbf{1}$ forces assigning it $\mathbf{0}$. Although much more complicated, this is exactly the problem with the liar $l \leftrightarrow \neg l$.□

---

[1] $A \leftrightarrow B$ abbreviates the conjunction $(A \rightarrow B) \wedge (B \rightarrow A)$.

Paradox is traditionally thought of as a single statement, but examples like this suggest that it is a feature of a whole discourse (single statement being only the simplest, special case.) We therefore define *a (propositional) paradox as an inconsistent discourse*, while a *discourse* as a theory in the specific form GNF, which was used above and is defined now.[2]

Definition 7.2 [GNF] A *basic* formula, over alphabet Σ, is an equivalence $x \leftrightarrow \bigwedge\{\neg x_i : i \in I\}$, where I is some index set, $x \in \Sigma$ and all $x_i \in \Sigma$. A theory is in GNF if it consists of basic formulae, where each $x \in \Sigma$ occurs exactly once on the left of such an equivalence.

Δ and Δ' above are both in GNF; $\{a \leftrightarrow \neg b \wedge \neg c, a \leftrightarrow \neg d\}$ is not since a occurs on the left of two basic formulae; $x \leftrightarrow \neg y \wedge z$ is not in GNF because z is not negated and neither y nor z occur on the left of $\leftrightarrow$.

Remark 7.3
We treat conjunction as a connective constructing a formula from a *set* of formulae, $\bigwedge S$, for $S \subseteq \mathsf{WFF}_{\mathsf{PL}}^{\Sigma}$. The conjunction we have been using so far can be seen as a special case for S with two elements. A valuation V of Σ is extended to all formulae as in Definition 5.5, with the additional clause: $\widehat{V}(\bigwedge S) = 1$ iff for all $s : s \in S$ implies $\widehat{V}(s) = \mathbf{1}$.

Consequently, conjunction of the empty set, $\bigwedge \varnothing$, evaluates to $\mathbf{1}$ (Exercise 5.16), since the antecedent of the implication is false for every s. GNF allows the form $x \leftrightarrow \bigwedge \varnothing$ (with the empty index set I), which corresponds to the inclusion of x among the non-logical axioms of the discourse.   □

7.2   FINITARY PARADOXES ARE CIRCULAR

———————————— a background story ————————————
A paradox, like the liar, may be expected to involve some form of circularity. It is, however, unclear what kind of circularity this may be and the tradition settled for the notion of self-reference. The paradox of the liar arises from its reference to its own truth-value. In 1936, this intuition obtained a specific form in the proof of Alfred Tarski's undefinability theorem which, informally, states that in a theory over a language which can express negation and refer to its own statements, it is impossible to define an adequate truth predicate (applying

---

[2]Paradox is not just an inconsistency but an unexpected inconsistency, which appears as a surpirse, leading from accepted premises, via apparently uncontroversial reasoning, to a false conclusion. Such features can hardly be captured by any formal definition.

exactly to the true statements of the language). The proof shows that
if such a predicate could be defined, the theory would entail the liar
statement, becoming thus inconsistent.

Many attempts to design "safe" languages, where no paradoxes
could appear, tried simply to ban self-reference. But this is certainly
too much, for there are forms of self-reference which are completely
innocent. "This statement is not false" (or "...is true") – the truth-
teller – is a typical example, which does not lead to any inconsistency.
It seems that self-reference must have a kind of "negative" character,
but it is hard to capture this intuition in precise, logical terms.

Tarski's theorem (or rather its proof) shows that, in the presence of
negation, self-reference may lead to a paradox, but not that absence of
self-reference ensures also absence of paradox. We will now represent
discourses in GNF as directed graphs, obtaining precise notions of
vicious circularity and its necessity for finitary paradoxes. _____

For a discourse Δ over alphabet Σ, we form a digraph[3] $\mathsf{G}(\Delta) = \langle \Sigma, \mathsf{N} \rangle$,
with the vertex set Σ and an edge from every variable occurring on the left
of $\leftrightarrow$ in a $B \in \Delta$ to all the variables occurring on the right of $\leftrightarrow$ in B.

Example 7.4

For the discourses from Example 7.1, we obtain the graphs:

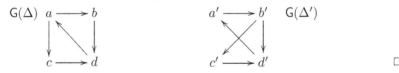

Conversely, given a directed graph $G = \langle G, \mathsf{N} \rangle$, we can form a discourse
$\mathsf{D}(G)$ in GNF consisting of the equivalences $x \leftrightarrow \bigwedge_{y \in \mathsf{N}(x)} \neg y$, for every
$x \in G$. Obviously, G and D are inverses of each other, i.e., for any discourse
$\Delta : \mathsf{D}(\mathsf{G}(\Delta)) = \Delta$ and for any graph $G : \mathsf{G}(\mathsf{D}(G)) = G$.

Vertices of the graph represent statements, so we try to assign truth-
values to them, observing the following rule:

$$V(x) = \mathbf{1} \Longleftrightarrow \big(\text{for all } y : y \in \mathsf{N}(x) \rightarrow V(y) = \mathbf{0}\big). \qquad (7.5)$$

Hence, $V(x) = \mathbf{0}$ whenever x has an edge to some y with $V(y) = \mathbf{1}$. If
V assignes truth-values to G so that (7.5) is satisfied, we say that it is
a *solution* for G, $V \in sol(G)$. For instance, in $\mathsf{G}(\Delta')$ above, assigning

---

[3]A directed graph (digraph) is a pair $\langle G, \mathsf{N} \rangle$ where G is a set of vertices and $\mathsf{N} \subseteq G \times G$
is the directed-edge relation, i.e., $\mathsf{N}(x, y)$ means that there is an edge from x to y. N is
also treated as a set-valued function: $\mathsf{N}(x) = \{y \in G : \mathsf{N}(x, y)\}$.

$a' = c' = 1$, requires that $b' = d' = 0$ and this assignment satisfies (7.5), i.e., solves the graph. Try to find $sol(\mathsf{G}(\Delta))$.

The rule reflects only the semantics of $\leftrightarrow$, used in the basic formulae of the graph's discourse. Verification of the following fact is an easy exercise.

Fact 7.6 For every graph $G : V \in sol(G)$ iff $V \models \mathsf{D}(G)$.
For every theory Δ in GNF : $V \models \Delta$ iff $V \in sol(\mathsf{G}(\Delta))$.

For this reason, we will no longer distinguish clearly between graphs and discourses, treating them as equivalent presentations.

Probably the simplest graph with no solution is $\bullet \overset{\frown}{\bigcirc}$. Its discourse is the liar $\bullet \leftrightarrow \neg\bullet$. On the other hand, the two-cycle $x \gtrless y$ admits two solutions, $x = 1, y = 0$ and $x = 0, y = 1$, showing that such a circularity is not problematic. x claims (via y) that x is not lying, or else it claims the falsity of y and y falsity of x – any one, but only one, may be right.

The graphical representation captures well the idea of circularity. A *cycle* in a graph is a standard notion – it is a path $x_1 x_2 ... x_n$, i.e., for all $1 \leq i < n : x_{i+1} \in \mathsf{N}(x_i)$, where also $x_1 \in \mathsf{N}(x_n)$. A cycle is odd/even when n is odd/even. A discourse (in GNF) without any self-reference is thus a graph without any cycles, so called DAG (directed acyclic graph).

Having now precise notion of discursive circularity, we can prove a simple fact that acyclic discourse is never paradoxical. But first, let us observe one consequence of (7.5). When $N(x) = \varnothing$, the implication on the right becomes vacuously true. Thus sinks (vertices of the graph with no outgoing edges, i.e., with $\mathsf{N}(x) = \varnothing$) must be assigned 1 by all solutions. We can think of them as indisputable facts, taken for granted in the discourse.

Fact 7.7 A finite acyclic discourse is not paradoxical.

Proof. We view the discourse as a graph, $G = \langle G_0, \mathsf{N} \rangle$ which, being a finite DAG, has some sinks $T_0 \subseteq G_0$. All T_0 must be assigned 1, and the rule (7.5) requires then all vertices with edges *to* T_0, $\mathsf{N}^-(T_0) = \{y \in G_0 : \mathsf{N}(y) \cap T_0 \neq \varnothing\}$, to be assigned 0. We set $F_0 = \mathsf{N}^-(T_0)$ and form a subgraph $G_1 = G_0 \setminus (T_0 \cup F_0)$ (with $\mathsf{N}_1 = \mathsf{N} \cap G_1 \times G_1$). The obtained subgraph is still a finite DAG, so we can repeat the same argument for it, and we do it inductively. Given G_i, we set

$T_i = sinks(G_i)$

$F_i = \mathsf{N}^-(T_i) = \{y \in G_i : \mathsf{N}(y) \cap T_i \neq \varnothing\}$

$G_{i+1} = G_i \setminus (T_i \cup F_i).$

Since G_0 is finite and every $T_i \neq \varnothing$, we reach eventually $G_k = \varnothing$. We then let $T = \bigcup T_i, F = \bigcup F_i$ and define $V = T \times \{\mathbf{1}\} \cup F \times \{\mathbf{0}\}$, which is a solution for G. It is well-defined, i.e., $F \cap T = \varnothing$, since for every $i : F_i \cap T_i = \varnothing$, while for $j \neq i : (F_j \cup T_j) \cap (F_i \cup T_i) = \varnothing$. V obviously satisfies (7.5) for all T_0 and F_0, so we assume IH that V is correct up to G_{i-1}. Then T_i become $sinks(G_i)$, having edges going only to F_{i-1}. Since all F_{i-1} are assigned $\mathbf{0}$, the assignment of $\mathbf{1}$ to all T_i satisfies (7.5). Then also assignment of $\mathbf{0}$ to $F_i = \mathsf{N}^-(T_i)$ becomes correct, so that the defined assignment remains correct for G_i. Being thus correct for all i, it is a solution for G.   QED (7.7)

Thus, a paradoxical discourse must contain some cycle. (Note that the claim is restricted to finite discourses!) The proof shows a fruitful application of graphical representation for obtaining logical results, but the result is still too weak. Excluding all cycles excludes all kinds of self-reference which, as we have seen, need not be dangerous. To yield a paradox, circularity must be vicious, and this turns out to depend on the way it is combined with negation. In purely graph-theoretic context, Richardson identified it as the odd cycle, proving in 1950's a much stronger result, namely:

Theorem 7.8 A finite graph with no *odd* cycle has a solution.

Intuitively, an odd cycle is a generalization of the liar which, passing through a series of intermediary stages, eventually negates itself, as exemplified by the following discourse:

The next statement is false.	$a \leftrightarrow \neg b$	
The next statement is false.	$b \leftrightarrow \neg c$	
The first statement is false.	$c \leftrightarrow \neg a$	

$$a \longrightarrow b$$

Trying to assign $\mathbf{1}$ to a, makes $b = \mathbf{0}$ and $c = \mathbf{1}$, which forces $a = \mathbf{0}$, and similarly trying $a = \mathbf{0}$ leads to $a = \mathbf{1}$.

Richardson's theorem says that such a cycle is necessary for a paradox or else, that no paradox arises from self-reference, as long as it does not express self-negation. The theorem can be seen as complementing the observation from the background story, that self-reference and negation – an odd cycle in graph representation – suffices for obtaining a paradox. (Of course, as seen already in $G(\Delta')$ from Example 7.4, odd cycle by itself does not mean that a paradox actually occurs. In the example above, it is enough to replace the last equivalence by $c \leftrightarrow \neg a \wedge \neg b$, i.e., to add to the graph the edge from c to b, to remove the paradox while retaining the odd cycle.)

7.3 ACYCLIC PARADOXES AND INFINITARY LOGIC

The previous section established necessity and sufficiency of odd cycles for the emergence of paradoxes in finite discourses. We emphasized the restriction to finite discourses which is, in fact, essential. In 1993, Stephen Yablo presented the discourse where infinitely many people stand in a row, everybody having infinitely many persons in front and saying "Everybody in front of me is lying."

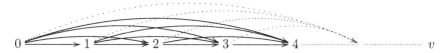

Represented as a graph, $Y = \langle \mathbb{N}, \{(i,j) : i < j\} \rangle$ has natural numbers as vertices and an edge from every number to all greater numbers. Assuming that some $i = \mathbf{1}$, then all $j > i$ must be $\mathbf{0}$. In particular, all $j > i + 1$ must be $\mathbf{0}$, but this implies that $i + 1 = \mathbf{1}$, contradicting $i = \mathbf{1}$. Hence, all $i = \mathbf{0}$ but then, for any i, all $j \in \mathsf{N}(i)$ are $\mathbf{0}$, so $i = \mathbf{1}$. The discourse is paradoxical, but no reasonable notion of circularity seems applicable to it.

One might think that the culprit is the infinity of the discourse but, as we will show in a moment, it is not sufficient for generating such acyclic paradoxes. It is necessary to change the logical language. The formula every i is stating, $i \leftrightarrow \bigwedge\{\neg j : j > i\}$, has the GNF form, but the essential difference from those considered earlier appears in the index set, which is now infinite. This is a crucial step, bringing us beyond propositional logic into its infinitary version. While in (the usual, finitary) propositional logic, we can form conjunctions over binary, or arbitrary finite, sets of propositions – the index set I in the definition 7.2 of basic formula must be finite, so in the infinitary logic it may be infinite.[4] Yablo's paradox is not expressible in finitary propositional logic, which we have studied until now.

Richardson's theorem is actually stronger than Theorem 7.8, showing solvability not only of finite graphs without odd cycle, but of *finitary* such graphs, namely, graphs which may be infinite but where $\mathsf{N}(x)$ is finite for every x. Having at our disposal logical means and representation of graphs as discourses, this becomes an easy corollary of Theorem 7.8.

Theorem 7.9 A *finitary* graph with no *odd* cycle has a solution.

---

[4]Varying admissible cardinality of sets which can enter under conjunction leads to different variants of infinitary logic, but we will not address such distinctions.

Proof. For such a graph G, $\mathsf{D}(G)$ is a finitary propositional theory Δ. For each finite subtheory $\Phi \subseteq \Delta$, its finite graph $\mathsf{G}(\Phi)$ has no odd cycle and is solvable by Theorem 7.8. Its solution is a model of Φ, by Fact 7.6. As every finite subtheory of Δ has a model so, by compactness, Exercise 6.12, Δ has a model which is a solution of G by Fact 7.6.

QED (7.9)

For instance, every acyclic discourse, where everybody accuses of lying at most a finite number of persons, other than oneself, is not paradoxical.

Example 7.10

In an infinite line of people (like in Yablo's paradox), everybody is saying: "The next 5 persons are lying". The graph $\langle \mathbb{N}, \{(x,y) : x < y \leq x + 5\}\rangle$ has no (odd) cycle and, for every vertex $x \in \mathbb{N}$, $\mathsf{N}(x)$ is finite (has only 5 following vertices) – by the above theorem, no paradox arises.   □

The results about finite and finitary logic/graphs do not extend in any obvious, direct way to infinitary cases. One of the main problems with infinitary logic is that the powerfull compactness property fails. It is not enough to consider all finite subtheories in order to determine consistency of an infinitary theory.

Example 7.11 [Failure of compactness]

(a) One of the simplest examples is the theory over countable alphabet Σ, containing literal $\neg x$ for every $x \in \Sigma$, and the formula $\neg \bigwedge\{\neg x : x \in \Sigma\}$. Using de Morgan law, which remains valid in infinitary logic, the latter can be written equivalently using infinitary disjunction, $\bigvee\{x : x \in \Sigma\}$. The theory is obviously inconsistent, since making all $\neg x$ true, i.e., setting $V(x) = \mathbf{0}$, makes $\widehat{V}(\bigvee\{x : x \in \Sigma\}) = \mathbf{0}$. But every finite subtheory is consistent. It is determined by a finite subset $F \subset \Sigma$, and contains only $\{\neg x : x \in F\}$, possibly, together with the infinite disjunction. So, assigning $V(x) = \mathbf{0}$ for all $x \in F$, leaves always some $y \in \Sigma \setminus F$ which can be assigned $V(y) = \mathbf{1}$ so that the infinite disjunction is also $\mathbf{1}$.

(b) Every finite subtheory of Y is consistent. Such a finite subtheory Φ is determined by a finite subset $F \subset \mathbb{N}$, so that $\Phi = \{i \leftrightarrow \bigwedge_{j>i} \neg j : i \in F\}$. We arrange F in descending order, i.e., as $F = \{i_1 > i_2 > ... > i_f\}$. Then all $j \in \mathbb{N} \setminus F \supseteq \{j > i_1\}$ can be assigned arbitrary values, forcing a unique value for the greatest $i_1 \in F : i_1 = \bigwedge_{j>i_1} \neg j$. We can now repeat the same for the next greatest element $i_2 \in F$, since now all $j > i_2$, including i_1, have obtained a value or can be assigned an arbitrary one. Repeating this for successive $i_n \in F$, until i_f, gives an assignment satisfying Φ. But, although every finite subtheory of Y has a model, the theory itself has none.   □

Yablo's paradox demonstrates that an analogue of Theorem 7.9 for infinitary discourses does not hold, while failure of compactness explains why we can not repeat its proof, which relies on compactness of finitary logic. Identification of sufficient conditions ensuring (non)paradoxical character of infinitary discourses, in a way reminiscent of Theorem 7.9, remains an open problem.

7.4 GNF AND SAT

GNF is motivated by the propositional representation and analysis of natural discourses, which may address truth of their statements. It allows the straightforward translation into graphs, and GNF stands for Graph Normal Form. For it is, in fact, a new normal form – for every propositional formula F there is a theory $GNF(F)$ in GNF such that F is consistent iff $GNF(F)$ is. Consequently, deciding if a discourse invovles a paradox is the same problem as deciding if an arbitrary theory is inconsistent. GNF, giving an equisatisfiable theory, differs from DNF and CNF which give logically equivalent formulae. Also unlike DNF/CNF, GNF requires new variables and several formulae in GNF to represent a single formula.

Theorem 7.12 [GNF] For every propositional formula F, there is a theory $GNF(F)$ in GNF, such that F has a model iff T has one.[5]

Proof. By Corollary 6.8, we can assume F to be in CNF, i.e., F is a conjunction of clauses $C_1 \wedge \ldots \wedge C_c$, each clause C_i being a disjunction of literals $l_1 \vee \ldots \vee l_{ni}$. First, for each propositional variable x_k occurring in F, we introduce a fresh one x'_k, and two equivalences $x_k \leftrightarrow \neg x'_k$ and $x'_k \leftrightarrow \neg x_k$. The resulting theory, call it Θ, is trivially satisfiable. Each clause $C_i, 1 \leq i \leq c$, consists of some positive and/or some negative literals (possibly $pi = 0$ or $ni = pi$):

$$C_i = x_1 \vee \ldots \vee x_{pi} \vee \neg x_{pi+1} \vee \ldots \vee \neg x_{ni}.$$

If there are any negative literals, we replace each such $\neg x_z$, for $pi+1 \leq z \leq ni$, by the positive x'_z. In the presence of Θ, the resulting clause C'_i is equivalent to C_i, i.e., $\Theta \models C_i \leftrightarrow C'_i$. We have also the logical

---

[5]Thanks to Kjetil Golid for simplifying the original proof.

equivalence $\models C'_i \leftrightarrow N_i$, where N_i is obtained from C'_i by de Morgan:

$$N_i = \neg(\neg x_1 \wedge \ldots \wedge \neg x_{pi} \wedge \neg x'_{pi+1} \wedge \ldots \wedge \neg x'_{ni}).$$

Finally, we introduce a fresh variable, c_i and form E_i as $c_i \leftrightarrow \neg N_i \wedge \neg c_i$:

$$c_i \leftrightarrow \neg x_1 \wedge \ldots \wedge \neg x_{pi} \wedge \neg x'_{pi+1} \wedge \ldots \wedge \neg x'_{ni} \wedge \neg c_i.$$

The theory $GNF(F)$ contains, in addition to Θ, all such equivalences E_i for each clause $C_i, 1 \leq i \leq c$, from the original formula F.

Each E_i is satisfiable only by making $c_i = \mathbf{0}$, which is possible only if one of $x_1 \ldots x_{pi}, x'_{pi+1} \ldots x'_{ni}$ becomes $\mathbf{1}$. Given that Θ are satisfied, this means that we satsify C_i. Since this holds for each C_i, we actually have that every model of $GNF(F)$, restricted to the original alphabet, gives a model of F, while every model of F can be trivially extended to a model of $GNF(F)$.   QED (7.12)

As a consequence, diagnosing paradox is the same as diagnosing inconsistency (the SAT problem). First, checking if a discourse is paradoxical is a special case of checking consistency, namely, of a GNF theory. But since every formula, and hence also every theory, can be represented as an equisatisfiable discourse, so every inconsistent theory gives rise to a paradox. This special case is thus fully general. In order to decide satisfiability of an arbitrary propositional theory, one can transform it into GNF and check if the resulting theory is satisfiable. The two problems are equivalent and so deciding if a paradox lurks in the complexity of a discourse is not easier than checking if an arbitrary propositional theory is consistent.

EXERCISES 7.

EXERCISE 7.1 Represent the following discourses in GNF and decide whether they are paradoxical. Draw the graphs, representing false claims by vertices with arrows to sinks. A statement claiming truth of s can be represented by sink s, but also as v claiming falsity of the negation of s, i.e., $v \rightarrow s' \rightarrow s$, where $\mathsf{N}(s') = \{s\}$.

(1) This and the next statment is false. The previous statement is false.
(2) This statment is false but the next one is not false. The previous statement is false.
(3) This statment is false but the next one is not. The previous statement is not false.

(4) Julius Caesar died of cancer and this statement is false.

(5) Julius Caesar was assassinated and this statement is false.

(6) The next statement is false. The next statement is false and Julius Caesar died of cancer. The first statement is false.

(7) a. The next statement is false.

    b. The next statement is false.

    c. The next and the last statement (f) are false.

    d. The next statement is false.

    e. The next statement is false.

    f. The first statement is false.

(8) Each of three people is saying: Everybody, including me, is lying.

(9) Each of three people is saying: Everybody, except me, is lying.

(10) There are two (countably) infinite parallel lines of people, $a_1 a_2 a_3 \dots$ and $b_1 b_2 b_3 \dots$, where

    (a) b_1 says: a_1 is lying.

    (b) For every $i > 0, b_{2i+1}$ says: b_{2i} and a_{2i+1} are both lying.

    (c) For every $i > 0, b_{2i}$ says: b_{2i-1} is lying.

    (d) For every $i \geq 0, a_{2i+1}$ says: a_{2i+2} is lying.

    (e) For every $i > 0, a_{2i}$ says: a_{2i+1} and b_{2i} are both lying.

Drawn as a graph, this discourse is as follows:

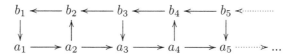

EXERCISE 7.2 Show the two claims from Fact 7.6.

EXERCISE 7.3 Graph-theory defines *kernel* of a directed graph as a subset of vertices $K \subseteq G$ such that $\mathsf{N}^-(K) = G \setminus K$, i.e., such that there are no edges between any two vertices in K, while every vertex outside of K has an edge to some vertex in K. Show that K is a kernel iff the assignment V_K defined by $V_K(x) = \mathbf{1} \Leftrightarrow x \in K$, is a solution of G, i.e., satisfies (7.5).

Part IV
FIRST ORDER LOGIC

Chapter 8

SYNTAX AND PROOF SYSTEMS OF FOL

──────────── a background story ────────────

Propositional language is very rudimentary and has limited expressive power. The only relation between various statements it can handle is that of identity and difference – the axiom schema $\vdash_{\mathcal{H}} A \to (B \to A)$ sees merely that the first and last statements must be identical while the middle one is arbitrary and may be different from A. Consider, however, the following argument:

	A:	Every man is mortal;
and	B:	Socrates is a man;
hence	C:	Socrates is mortal.

Since all the involved statements are different, the representation in PL will amount to $A \land B \to C$ which, obviously, is not a valid formula. The validity of this argument rests on the fact that the minor premise B makes Socrates, so to speak, an instance of the subject of the major premise A – whatever applies to every man, applies to each particular man.

In this particular case, one might try to refine the representation of the involved statements a bit. Say that we use the following propositional variables: Man (for 'being a man'), Mo (for 'being mortal') and So (for 'being Socrates'). Then we obtain: $So \to Man$ and $Man \to Mo$, which does entail $So \to Mo$. We were lucky! - because the argument had a rather simple structure. If we, for instance, wanted to say that "Some men are thieves" in addition to "All men are mortal", we could hardly use the same Man in both cases. The following argument, too, is very simple, but it illustrates the need for talking not only about atomic statements, but also about involved entities and relations between them:

Every horse is an animal;
hence   Every head of a horse is a head of an animal.

Its validity relies not so much on the form, let alone identity and difference, of the involved statements as on the relations between the involved entities, in particular, that of 'being a head of...' Since each horse is an animal, whatever applies to animals (or their heads) applies to horses as well. It is hard to imagine how such an argument might possibly be represented (as a valid statement) in PL.

Intuitively, the semantic "view of the world" underlying FOL can be summarised as follows.

(1) We have a universe of discourse U comprising all entities of (current) interest.

(2) We may have particular means of picking some entities.

  (a) For instance, a name, 'Socrates' can be used to designate a particular individual: Such names correspond to constants, or functions with 0 arguments.

  (b) An individual may also be picked by saying 'the father of Socrates'. Here, 'the father of ...' (just like 'the head of ...') is a function taking 1 argument (preferably a man, but generally an arbitrary entity from U, since the considered functions are total) and pointing at another individual. Functions may have arbitrary arities, e.g., 'the children of x and y' is a function of 2 arguments returning a set of children, 'the solutions of $x^2 - y^2 = 0$' is a function of 2 arguments returning a set of numbers.

  (c) To facilitate flexible means of expression, we may use variables, x, y, etc. to stand for arbitrary entities in more complex expressions.

(3) The entities from U can be classified and related using various predicates and relations:

  (a) We may identify subsets of U by means of (unary) predicates: the predicate $M(y)$ – true about those y's which are men, e.g. $M(Socrates)$, but not about inanimate things – identifies a subset of those elements of U about which it is true, i.e., $\{y \in U : M(y)\}$; the predicate $Mo(y)$ – true about the mortal beings and false about all other entities – identifies the subset of mortal beings; $H(y)$ – the subset of horses, $A(y)$ – the animals, etc.

(b) The entities may stand in various relations to each other: 'x is older than y' is a 2-argument relation which will hold between some individuals but not between others; similarly for 'x is the head of y', $Hd(x,y)$, etc.

(4) $Hd(x,y)$, stating that 'x is the head of y', is very different from the function $hd(y)$ returning, for every y, its head. The latter merely picks new individual objects. The former is a predicate – it states some fact. Predicates and relations are the means for stating the "atomic facts" about the world. (For this reason, first order logic is also called "predicate logic".) These can be then combined using the connectives, as in PL, 'and', 'or', 'not', etc. In addition, we may also state facts about indefinite, unnamed entities, for instance, 'for-every $x : M(x) \rightarrow Mo(x)$', 'for-no $x : H(x) \wedge M(x)$', etc.

Agreeing on such an interpretation of the introduced symbols, the opening arguments would be written :

$$
\begin{array}{lll}
\text{A:} & \text{for-every } y: & M(y) \rightarrow Mo(y) \\
\text{B:} & & M(Socrates) \\
\hline
\text{C:} & & Mo(Socrates)
\end{array}
$$

$$
\begin{array}{ll}
\text{for-every } y: & H(y) \rightarrow A(y) \\
\hline
\text{for-every } x: & (\text{there-is } y : H(y) \wedge Hd(x,y)) \rightarrow \\
& (\text{there-is } y : A(y) \wedge Hd(x,y))
\end{array}
$$

This illustrates the intention of the language of FOL, which we now begin to study, and its semantics which will be our object in the following chapters. _____

8.1 SYNTAX OF FOL

In propositional logic, the non-logical (i.e., relative to the actual application) part of the language was only the set Σ of propositional variables. In first order logic this part is much richer which also means that, in each particular application, the user can – and has to – make more detailed choices. Nevertheless, this non-logical part of the language has well defined components which will still make it possible to treat FOL in a uniform way, almost independently of such contextual choices.

Definition 8.1 The alphabet of FOL consist of two disjoint sets Σ and Φ:

Σ : the non-logical alphabet contains non-logical symbols:

- *individual constants*: $\mathcal{I} = \{a, b, c, \ldots\}$
- *individual variables*: $\mathcal{V} = \{x, y, z, \ldots\}$
- *function symbols*: $\mathcal{F} = \{f, g, h, \ldots\}$ each taking a fixed finite number of arguments, called its *arity*.
- *relation symbols*: $\mathcal{R} = \{P, Q, R, \ldots\}$, each with a fixed finite arity.

Φ : contains the logical symbols:

- the connectives of PL : $\neg, \rightarrow$
- *quantifier*: $\exists$

We make no assumption about the sizes of $\mathcal{I}, \mathcal{F}, \mathcal{R}$; any one of them may be infinite or empty, though typically each is finite. $\mathcal{V}$, on the other hand, is always a countably infinite set.

We also use some auxiliary symbols like parentheses and commas. Relation symbols are also called predicate symbols or just predicates. Individual variables and constants are usually called just variables and constants.

Example 8.2

Suppose we want to talk about stacks – standard data structures. We might start by setting up the following alphabet Σ_{Stack} (to indicate arities, we use the symbol U for the whole universe):

(1) $\mathcal{I} = \{empty\}$ – the only constant for representing empty stack;
(2) $\mathcal{V} = \{x, y, s, u, v, \ldots\}$ – we seldom lists these explicitly; just mark that something is a variable whenever it is used;
(3) $\mathcal{F} = \{top : U \rightarrow U, \ pop : U \rightarrow U, \ push : U^2 \rightarrow U\}$;
(4) $\mathcal{R} = \{St \subseteq U, \ El \subseteq U, \ \equiv \ \subseteq U^2\}$ – for identifying *St*acks, *El*ements, and for expressing equality. $\qquad\qquad\Box$

Unlike PL, the language of FOL is designed so that we may write not only formulae but also terms. The former, as before, will denote some boolean values. Terms are meant to refer to "individuals" or some "objects of the world". Formulae are built using propositional connectives, as in PL, from simpler expressions – atomic formulae – which, however, are not merely propositional variables but have some internal structure involving terms. In addition, we have a new formula-building operation of quantification over individual variables.

Definition 8.3 [**Terms**] The set of *terms* over Σ, $\mathcal{T}_\Sigma$, is defined inductively:

(1) all constants are terms, $\mathcal{I} \subseteq \mathcal{T}_\Sigma$.
(2) all variables are terms, $\mathcal{V} \subseteq \mathcal{T}_\Sigma$.

(3) if $f \in \mathcal{F}$ is an n-ary function symbol and $t_1, \ldots, t_n$ are in $\mathcal{T}_\Sigma$, then $f(t_1, \ldots, t_n) \in \mathcal{T}_\Sigma$.

Terms not containing any variables are called *ground terms,* and the set of ground terms is denoted $\mathcal{GT}_\Sigma$.

Example 8.4

Consider the alphabet of stacks from Example 8.2. The only ground terms are the constant *empty* and applications of functions to it, e.g., *pop(empty)*, *top(empty)*, *pop(pop(empty))*. Also other terms, like *push(empty, empty)*, *push(empty, pop(empty))*, are well-formed ground terms, even if they do not necessarily correspond to our intentions.

The non-ground terms will be of the same kind but will involve variables. For instance, when x, s are variables, the following are non-ground terms: *pop(s)*, *top(pop(s))*, *push(empty, x)*, *pop(push(x, empty))*, etc. □

Definition 8.5 [Formulae] The well-formed formulae of predicate logic over a given Σ, $\mathrm{WFF}_{\mathrm{FOL}}^\Sigma$, are defined inductively:

(1) If $P \in \mathcal{R}$ is an n-ary relation symbol and $t_1, \ldots, t_n$ are terms, then $P(t_1, \ldots, t_n) \in \mathrm{WFF}_{\mathrm{FOL}}^\Sigma$.
(2) If $A \in \mathrm{WFF}_{\mathrm{FOL}}^\Sigma$ and x is a variable, then $\exists x A$ is in $\mathrm{WFF}_{\mathrm{FOL}}^\Sigma$.
(3) If $A, B \in \mathrm{WFF}_{\mathrm{FOL}}^\Sigma$ then $\neg A, (A \rightarrow B) \in \mathrm{WFF}_{\mathrm{FOL}}^\Sigma$.

Formulae from point (1) are called *atomic,* those from (2) are *quantified.* Thus the FOL language "refines" the PL language in that propositional connectives connect not just propositional variables but more detailed atomic or quantified formulae.

Remark.

As in the case of PL, these definitions are parameterized by Σ, yielding a new instance of the FOL language for each particular choice of Σ. Nevertheless we will often speak about "the FOL language," taking Σ as an implicit, and arbitrary though usually non-empty, parameter. □

Example 8.6

Continuing the example of stacks, atomic formulae will be, for instance: *El(empty)*, *St(x)*, *empty* $\equiv$ *pop(empty)*, *push(s, x)* $\equiv$ *pop(pop(empty))*, etc. The non-atomic ones contain boolean combinations of atoms, e.g., *El(x)* $\wedge$ *St(s)* $\rightarrow$ *pop(push(s, x))* $\equiv$ *s*. □

The symbol $\exists$ is called the *existential quantifier* – $\exists x A$ reads as "there exists an x such that A". For convenience, we define the following abbreviation:

Definition 8.7 We define $\forall x A \overset{\text{def}}{=} \neg \exists x \neg A$.

$\forall$ is the *universal quantifier* and $\forall x A$ is read "for all x, A holds". Writing $Q x A$, we will mean any one of the two quantifiers, i.e., $\forall x A$ or $\exists x A$.

We will also use the abbreviations $\vee$ and $\wedge$ for propositional connectives. Sometimes (when it is "safe") the arguments to $\vee$ and $\wedge$ will be written without the surrounding parentheses. Similarly for $\rightarrow$.

8.2 SCOPE OF QUANTIFIERS

Quantification introduces several important issues into the syntax of FOL. This section explains the concept of the scope of quantification, the resulting distinction between the free and bound occurrences of variables, and the operation of substitution which has to respect this distinction.

Definition 8.8 In a quantified formula $Q x B$, B is the *scope* of $Q x$.

Parentheses or colons are used to disambiguate the scope: the notation $Q x(B)$, $Q x : B$, or $(Q x : B)$ indicates that B is the scope of $Q x$.

Example 8.9
The scopes of the various quantifiers are underlined:

	in formula	the scope		
(1)	$\forall x \exists y R(x,y)$	of $\exists y$	is	$R(x,y)$
		of $\forall x$	is	$\exists y R(x,y)$
(2)	$\forall x(R(x,y) \wedge R(x,x))$	of $\forall x$	is	$R(x,y) \wedge R(x,x)$
(3)	$\forall x R(x,y) \wedge R(x,x)$	of $\forall x$	is	$R(x,y)$
(4)	$\forall x R(x,x) \rightarrow \exists y Q(x,y)$	of $\forall x$	is	$R(x,x)$
		of $\exists y$	is	$Q(x,y)$
(5)	$\forall x(R(x,x) \rightarrow \exists y Q(x,y))$	of $\exists y$	is	$Q(x,y)$
		of $\forall x$	is	$R(x,x) \rightarrow \exists y Q(x,y)$
(6)	$\forall x(R(x,x) \rightarrow \exists x Q(x,x))$	of $\exists x$	is	$Q(x,x)$
		of $\forall x$	is	$R(x,x) \rightarrow \exists x Q(x,x)$. $\square$

Definition 8.10 For any formula A we say that

- an *occurrence* of a variable x which is not within the scope of any quantifier $Q x$ in A is *free* in A.

- An occurrence of a variable x which is not free in A is said to be *bound*. Moreover, it is bound *by* the innermost (i.e., closest to the left) quantifier of the form Qx in the scope of which it occurs.
- A is *closed* (a *sentence*) if it has no free occurrences of any variable.
- A is *open* if it is not closed.

For any A, $\mathcal{V}(A)$ is the set of variables with free occurrences in A. Thus A is closed iff $\mathcal{V}(A) = \varnothing$.

Example 8.11
In Example 8.9, y is *free* in (2) and (3). In (3) the occurrences of x in $R(x,x)$ are free too, but the occurrence of x in $R(x,y)$ is *bound*. Similarly, in (4) the occurrences of x in $R(x,x)$ are *bound*, but the one in $Q(x,y)$ is *free*. In (5) all occurrences of x are bound by the frontmost $\forall x$. In (6), however, the occurrences of x in $R(x,x)$ are bound by the frontmost $\forall x$, but the ones in $Q(x,x)$ are bound by the $\exists x$. Thus (2), (3) and (4) are open formulae while the others are closed. $\square$

Remark 8.12 [An analogy to programming]
As we will see in next chapter, the difference between bound and free variables is that the names of the former do not make any difference while of the latter do influence the interpretation of the formulae. As a convenient analogy, one may think about free variables in a formula A as global variables in an imperative program A. The bound variables correspond to the local variables and quantifier to a block with declaration of local variables. Consider program P1 on the left:

```
P1: begin                      P2:    begin
      int x,y;                           int x,y;
      x:=5; y:=10;                       x:=5; y:=10;
      begin                              begin
        int x,z;                           int w,z;
        x:=0; x:=x+3;                       w:=0; w:=w+3;
        z:=20; y:=30;                       z:=20; y:=30;
      end;                               end;
    end;                           end;
```

The global variable x is redeclared in the inner block. This can be said to make the global x "invisible" within this block. y is another global variable, while z is a local variable in the block. At the exit, we will have $x = 5$ and $y = 30$ since these global variables are not affected by the assignment to the local ones within the block. Also, z will not be available after the exit from the inner block.

A formula with a similar scoping effect would be:

$$A(x,y) \land \text{Q}x\text{Q}zB(x,z,y) \text{ or alternatively}$$
$$\text{Q}x\text{Q}y \left(A(x,y) \land \text{Q}x\text{Q}zB(x,z,y) \right)$$

(8.13)

where we ignore the meaning of the predicates A, B but concentrate only on the "visibility of the variables", i.e., the scope of the quantifiers. Variable y is free (in the first formula, and within the scope of the same quantifier in the other) and thus its occurrence in $B(\dots y)$ corresponds to the same entity as its occurrence in $A(\dots y)$. On the other hand, x in $A(x \dots)$ – in the outermost block – is one thing, while x in $B(x \dots)$ – in the inner block – a completely different one, since the latter is in the scope of the innermost quantifier $\text{Q}x$.

Program P2 on the right is equivalent to P1, the only difference being the renaming of the local x to w. In fact, the formula (8.13) will be equivalent to the one where the bound x has been renamed, e.g., to the following one

$$A(x,y) \land \text{Q}\mathbf{w}\text{Q}zB(\mathbf{w},z,y) \text{ or alternatively}$$
$$\text{Q}x\text{Q}y \left(A(x,y) \land \text{Q}\mathbf{w}\text{Q}zB(\mathbf{w},z,y) \right)$$

Renaming of free variables will not be allowed in the same way.   □

At this point, the meaning of the distinction free vs. bound may seem unclear. It is therefore best, for the time being, to accept Definition 8.10 at its face value. As illustrated by Example 8.9, it is easy to distinguish free and bounded occurrences by a simple syntactic check.

8.2.1 SOME EXAMPLES

Before we begin a closer discussion of the syntax of FOL, we give a few examples of vocabularies (alphabets) which can be used for describing some known structures.

Example 8.14 [Stacks]
Using the alphabet of stacks from Example 8.2, we may now set up the following (non-logical) axioms Γ_{Stack} for the theory of stacks ($x, s \in \mathcal{V}$):

(1) $St(empty)$
(2) $\forall x, s : El(x) \land St(s) \rightarrow St(push(s,x))$
(3) $\forall s : St(s) \rightarrow St(pop(s))$
(4) $\forall s : St(s) \rightarrow El(top(s))$.

These axioms describe merely the profiles of the functions and relate the extensions of the predicates. According to (1) *empty* is a stack, while (2) says that if x is an element and s is a stack then also the result of $push(s, x)$ is stack. (Usually, one uses some abbreviated notation to capture this information. In typed programming languages, for instance, it is taken care of by the typing system.) Further (non-logical) axioms, determining more specific properties of stacks, could then be:

(5) $pop(empty) \equiv empty$
(6) $\forall x, s : El(x) \wedge St(s) \rightarrow pop(push(s, x)) \equiv s$
(7) $\forall x, s : El(x) \wedge St(s) \rightarrow top(push(s, x)) \equiv x$.

Notice that we have not given any axioms for $\equiv$, intended to be the identity relation. Identity can not be axiomatized in FOL, but the following, standard axioms (8)-(10) ensure the interpretation as an equivalence relation, while (11) and (12) as a *congruence*, i.e., that $\equiv$-equivalent elements are indistinguishable under applications of functions or relations.

(8) $\forall x : x \equiv x$
(9) $\forall x, y : x \equiv y \rightarrow y \equiv x$
(10) $\forall x, y, z : x \equiv y \wedge y \equiv z \rightarrow x \equiv z$

and two axiom schemata:
(11) for every n-ary $f \in \mathcal{F}$:
$$\forall x_1, x_1' \ldots x_n, x_n' : x_1 \equiv x_1' \wedge \ldots \wedge x_n \equiv x_n' \rightarrow f(x_1 \ldots x_n) \equiv f(x_1' \ldots x_n')$$
(12) for every n-ary $R \in \mathcal{R}$:
$$\forall x_1, x_1' \ldots x_n, x_n' : x_1 \equiv x_1' \wedge \ldots \wedge x_n \equiv x_n' \wedge R(x_1 \ldots x_n) \rightarrow R(x_1' \ldots x_n').$$
$\square$

Example 8.15 [**Queues**]
We use the same alphabet as for stacks, although we intend different meaning to some symbols. Thus St is now to be interpreted as the set of (FIFO) queues, *pop* is to be interpreted as *tail*, *push* as *add* (at the end) and *top* as the head, the frontmost element of the queue. We only need to replace axioms (6)-(7) with the following:

(6a) $\forall x, s : El(x) \wedge St(s) \wedge s \equiv empty \quad \rightarrow \quad pop(push(s, x)) \equiv s$
(6b) $\forall x, s : El(x) \wedge St(s) \wedge \neg(s \equiv empty) \quad \rightarrow$
$$pop(push(s, x)) \equiv push(pop(s), x)$$
(7a) $\forall x, s : El(x) \wedge St(s) \wedge s \equiv empty \quad \rightarrow \quad top(push(s, x)) \equiv x$
(7b) $\forall x, s : El(x) \wedge St(s) \wedge \neg(s \equiv empty) \quad \rightarrow \quad top(push(s, x)) \equiv top(s)$

(6a) can be simplified to $\forall x : El(x) \rightarrow pop(push(empty, x)) \equiv empty$, when $\equiv$ *is* equality; similarly (7a). $\square$

Example 8.16 [Graphs]

All axioms in the above examples were universal formulae (with only $\forall$-quantifier in front). We now axiomatize graphs, which requires more specific formulae.

A graph is a structure with two sets: V – vertices, and E – edges. Each edge has a unique source and target vertex. Thus, we take as our non-logical alphabet Σ_{Graph}:

- $\mathcal{F} = \{sr, tr\}$ – for source and target functions
- $\mathcal{R} = \{V, E, \equiv\}$ – V, E unary predicates for the set of vertices and edges, binary $\equiv$ for the identity relation (axioms (8)-(12) from Example 8.14).

The axiom

(1) $\forall e : E(e) \rightarrow (\ V(sr(e)) \wedge V(tr(e))\ )$

determines the profile of the functions sr and tr. This is all that must be said to get arbitrary graphs. Typically, we think of a graph with at most one edge between two vertices. For that case, we need to add the axiom:

(2) $\forall e_1, e_2 : (\ sr(e_1) \equiv sr(e_2) \wedge tr(e_1) \equiv tr(e_2)\ ) \rightarrow e_1 \equiv e_2$.

The graphs, so far, are directed. An undirected graph can be seen as a directed graph where for any edge from x to y, there is also an opposite edge from y to x:

(3) $\forall x, y : (\exists e : sr(e) \equiv x \wedge tr(e) \equiv y) \rightarrow (\exists e : tr(e) \equiv x \wedge sr(e) \equiv y)$.

The intention of the above formula could be captured by a simpler one:

$\forall e_1 \exists e_2 : sr(e_1) \equiv tr(e_2) \wedge tr(e_1) \equiv sr(e_2)$.

Finally, we may make a special kind of *transitive* graphs: if there is an edge from x to y and from y to z, then there is also an edge from x to z, and this applies for every pair of edges:

(4) $\forall e_1, e_2 : \big(tr(e_1) \equiv sr(e_2) \rightarrow \exists e : (sr(e) \equiv sr(e_1) \wedge tr(e) \equiv tr(e_2)))\big)$. □

Example 8.17 [Simple graphs]

If we want to consider only simple graphs, i.e., graphs satisfying the axiom (2) from the previous example, we can choose a much more convenient vocabulary Σ_{SG}: our universe is the set of all possible vertices, we need no function symbols, and we use one binary relation symbol: E – the edge relation. Then axioms (1) and (2) become redundant. If we want to consider undirected graphs, we make E symmetric:

(1) $\forall x, y : E(x, y) \rightarrow E(y, x)$.

If we want to consider transitive graphs, we make E transitive:

(2) $\forall x, y, z : E(x, y) \wedge E(y, z) \rightarrow E(x, z)$.

Graphs *without* self-loops (i.e., where no edge leads from a vertex to itself) are ones where E is irreflexive:

(3) $\forall x : \neg E(x, x)$.

Using the representation from the previous example, this axiom would be: $\forall e : E(e) \rightarrow \neg(sr(e) \equiv tr(e))$. □

8.2.2 SUBSTITUTION

In a given term t or formula A, we may substitute a term for the *free* occurrences of any variable. This can be done simultaneously for several variables. A *substitution* is a function from some set of variables X to terms, $\sigma : X \rightarrow \mathcal{T}_\Sigma$, and the result of applying it to a term/formula, $\overline{\sigma}(t)/\overline{\sigma}(A)$, is defined inductively.

Definition 8.18 Application of a substitution σ to terms, $\overline{\sigma} : \mathcal{T}_\Sigma \rightarrow \mathcal{T}_\Sigma$, is defined inductively on the structure of terms:

$$x \in \mathcal{V} :: \overline{\sigma}(x) \stackrel{\text{def}}{=} \begin{cases} \sigma(x) & \text{if } x \in dom(\sigma) \\ x & \text{if } x \notin dom(\sigma) \end{cases}$$
$$f(t_1, \ldots, t_k) :: \overline{\sigma}(f(t_1, \ldots, t_k)) \stackrel{\text{def}}{=} f(\overline{\sigma}(t_1), \ldots, \overline{\sigma}(t_k)).$$

This determines the result $\overline{\sigma}(t)$ of substitution σ into any term t. Building on this, the application of substitution σ to *formulae*, $\overline{\sigma} : \text{WFF}^\Sigma \rightarrow \text{WFF}^\Sigma$, is defined by induction on the complexity of formulae:

$$\text{ATOMIC} :: \overline{\sigma}(P(t_1, \ldots, t_k)) \stackrel{\text{def}}{=} P(\overline{\sigma}(t_1), \ldots, \overline{\sigma}(t_k))$$
$$\neg B :: \overline{\sigma}(\neg B) \stackrel{\text{def}}{=} \neg \overline{\sigma}(B)$$
$$B \rightarrow C :: \overline{\sigma}(B \rightarrow C) \stackrel{\text{def}}{=} \overline{\sigma}(B) \rightarrow \overline{\sigma}(C)$$
$$\exists x A :: \overline{\sigma}(\exists x A) \stackrel{\text{def}}{=} \exists x \, \overline{\sigma}'(A) \text{ where } \sigma' = \sigma \setminus (\{x\} \times \mathcal{T}_\Sigma).$$

In the last case, the modification of σ to σ' removes x from the domain of the substitution, ensuring that no substitution for x occurs inside A.

Typically, the considered $dom(\sigma)$ has only one variable, $\sigma = \{\langle x, s \rangle\}$, and we write the result of such a substitution $\overline{\sigma}(t)/\overline{\sigma}(A)$ as t_s^x/A_s^x.

Example 8.19

In Example 8.9 formulae (1), (5) and (6) had no free variables, so the application of any substitution will leave these formulae unchanged. For formulae (2), (3) and (4) from that example, we obtain:

$$(2) \quad (\forall x(R(x,y) \wedge R(x,x)))_t^x \; = \; \forall x(R(x,y) \wedge R(x,x))$$
$$(2') \quad (\forall x(R(x,y) \wedge R(x,x)))_s^y \; = \; \forall x(R(x,s) \wedge R(x,x))$$
$$(3) \quad \quad (\forall x R(x,y) \wedge R(x,x))_t^x \; = \; \forall x R(x,y) \wedge R(t,t)$$
$$(4) \quad (\forall x R(x,x) \to \exists y Q(x,y))_t^x \; = \; \forall x R(x,x) \to \exists y Q(t,y). \qquad \square$$

The following example shows that some caution is needed when for a variable we substitute a term that itself contains variables (or even *is* a variable). If such variables are "captured" by quantifiers already present, there may be unexpected results:

Example 8.20

As mentioned in Remark 8.12, renaming bound variables results in equivalent formulae, e.g., the formulae $\forall y R(x,y)$ and $\forall z R(x,z)$ mean the same thing. However, performing the same substitution on the two produces formulae with (as we'll see later) very different interpretations:

$$(1) \; (\forall y R(x,y))_{f(z)}^x = \forall y R(f(z),y) \quad e.g. \; (\exists y(x < y))_{z+1}^x = \exists y(z+1 < y)$$
$$(2) \; (\forall z R(x,z))_{f(z)}^x = \forall z R(f(z),z) \quad e.g. \; (\exists z(x < z))_{z+1}^x = \exists z(z+1 < z)$$

The variable z is free throughout the first example, but in (2) gets captured by $\forall z/\exists z$ after substitution. $\qquad \square$

To guard against such unintended behaviour, we introduce the next definition. It makes z *substitutable for* x in $\forall y R(x,y)$ but not in $\forall z R(x,z)$.

Definition 8.21 Let A be a formula, x a variable and s a term. The property "s is substitutable for x in A" is defined by induction on A as follows:

Atomic :: If A is atomic, then s is substitutable for x in A.

$\neg B$ :: s is substitutable for x in $\neg B$ iff s is substitutable for x in B.

$B \to C$ :: s is substitutable for x in $B \to C$ iff s is substitutable for x in both B and C.

$\exists x A$ :: s is substitutable for x in $\exists x A$.
       (Since no substitution in fact takes place.)

$\exists y A$ :: If $y \neq x$ then s is substitutable for x in $\exists y A$ iff either x does not occur free in A, or both s does not contain the variable y, and s is substitutable for x in A.

Briefly put, s is substitutable for x in A (or the substitution A_s^x is *legal*) iff there are no free occurrences of x in A inside the scope of any quantifier that binds any variable occurring in s.

We may occasionally talk more generally about *replacing* an arbitrary term by another term, rather than just *substituting* a term for a variable.

8.3 THE AXIOMATIC SYSTEM $\mathcal{N}$

The proof system $\mathcal{N}$ for FOL uses the predicate $\vdash_{\mathcal{N}} \subseteq \mathcal{P}(\mathsf{WFF_{FOL}}) \times \mathsf{WFF_{FOL}}$ and is an extension of the system $\mathcal{N}$ for PL.

Definition 8.22 The $\mathcal{N}$ system for FOL consists of:

AXIOMS :: A0: $\Gamma \vdash_{\mathcal{N}} B$, for all $B \in \Gamma$;

A1: $\Gamma \vdash_{\mathcal{N}} A \to (B \to A)$;

A2: $\Gamma \vdash_{\mathcal{N}} (A \to (B \to C)) \to ((A \to B) \to (A \to C))$;

A3: $\Gamma \vdash_{\mathcal{N}} (\neg B \to \neg A) \to (A \to B)$;

A4: $\Gamma \vdash_{\mathcal{N}} A_t^x \to \exists x A$ if A_t^x is legal;

RULES :: MP: $\dfrac{\Gamma \vdash_{\mathcal{N}} A \ ; \ \Gamma \vdash_{\mathcal{N}} A \to B}{\vdash_{\mathcal{N}} B}$;

$\exists$I: $\dfrac{\Gamma \vdash_{\mathcal{N}} A \to B}{\Gamma \vdash_{\mathcal{N}} \exists x A \to B}$ if $x \notin \mathcal{V}(B)$.

Since the system inherits all the axiom schemata and rules from Definition 4.12, every theorem in that earlier system has a counterpart in this system. For instance, Lemmata 4.8 and 4.16 can be taken over directly. The only difference is that the uppercase letters – in the above rules, axioms and the following results – stand now for arbitrary FOL-formulae. For instance, $\exists x A \to (\exists x \neg \exists y D \to \exists x A)$ is an instance of A1. Thus any derivations we performed using propositional variables in PL, can be now performed in the same way, provided that the FOL-formulae involved are syntactically identical whenever required (like in the above instance of A1).

Admissible rules are a different matter, but in most cases these also carry over to the extended system. Thus the next lemma corresponds exactly to Lemmata 4.10 and 4.13, and is proved in exactly the same way.

Lemma 8.23 The following rules are admissible in $\mathcal{N}$:

(1) $\dfrac{\Gamma \vdash_{\mathcal{N}} A \to B \ ; \ \Gamma \vdash_{\mathcal{N}} B \to C}{\Gamma \vdash_{\mathcal{N}} A \to C}$

(2) $\dfrac{\Gamma \vdash_{\mathcal{N}} B}{\Gamma \vdash_{\mathcal{N}} A \to B}$.

The next lemma illustrates the use of the new elements, i.e., A4 and the "∃ introduction" rule ∃I.

Lemma 8.24 Formula (1) is provable and rules (2)-(5) are admissible in $\mathcal{N}$:

(1) $\Gamma \vdash_{\mathcal{N}} \forall x A \rightarrow A$

(2) $\dfrac{\Gamma \vdash_{\mathcal{N}} A \rightarrow B}{\Gamma \vdash_{\mathcal{N}} \exists x A \rightarrow \exists x B}$

(3) $\forall$I : $\dfrac{\Gamma \vdash_{\mathcal{N}} B \rightarrow A}{\Gamma \vdash_{\mathcal{N}} B \rightarrow \forall x A}$ if $x \notin \mathcal{V}(B)$

(4) $\forall$G : $\dfrac{\Gamma \vdash_{\mathcal{N}} A}{\Gamma \vdash_{\mathcal{N}} \forall x A}$

(5) SB: $\dfrac{\Gamma \vdash_{\mathcal{N}} A}{\Gamma \vdash_{\mathcal{N}} A_t^x}$ if A_t^x is legal.

Proof. If something is provable using only A0–A3 and MP we write just PL to the right – these parts have been proven earlier or are left as exercises. (In view of the completeness theorem of PL, it is sufficient to convince oneself that it corresponds to a tautology.)

(1) $1 : \Gamma \vdash_{\mathcal{N}} \neg A \rightarrow \exists x \neg A$ A4
$2 : \Gamma \vdash_{\mathcal{N}} (\neg A \rightarrow \exists x \neg A) \rightarrow (\neg \exists x \neg A \rightarrow A)$ PL
$3 : \Gamma \vdash_{\mathcal{N}} \neg \exists x \neg A \rightarrow A$ $MP : 1, 2$

(2) $1 : \Gamma \vdash_{\mathcal{N}} A \rightarrow B$ *assumption*
$2 : \Gamma \vdash_{\mathcal{N}} B \rightarrow \exists x B$ A4
$3 : \Gamma \vdash_{\mathcal{N}} A \rightarrow \exists x B$ $L.8.23.(1) : 1, 2$
$4 : \Gamma \vdash_{\mathcal{N}} \exists x A \rightarrow \exists x B$ ∃I : 3

(3) $1 : \Gamma \vdash_{\mathcal{N}} B \rightarrow A$ *assumption*
$2 : \Gamma \vdash_{\mathcal{N}} \neg A \rightarrow \neg B$ $MP : 1, L.4.16$
$3 : \Gamma \vdash_{\mathcal{N}} (\exists x \neg A) \rightarrow \neg B$ ∃I : $2 + x$ not free in B
$4 : \Gamma \vdash_{\mathcal{N}} ((\exists x \neg A) \rightarrow \neg B) \rightarrow (B \rightarrow \neg \exists x \neg A)$ PL
$5 : \Gamma \vdash_{\mathcal{N}} B \rightarrow \neg \exists x \neg A$ $MP : 3, 4$

(4) $1 : \Gamma \vdash_{\mathcal{N}} A$ *assumption*
$2 : \Gamma \vdash_{\mathcal{N}} (\exists x \neg A) \rightarrow A$ $L.8.23.(2) : 1$
$3 : \Gamma \vdash_{\mathcal{N}} (\exists x \neg A) \rightarrow \neg \exists x \neg A$ $\forall$I : 2
$4 : \Gamma \vdash_{\mathcal{N}} ((\exists x \neg A) \rightarrow \neg \exists x \neg A) \rightarrow \neg \exists x \neg A$ PL
$5 : \Gamma \vdash_{\mathcal{N}} \neg \exists x \neg A$ $MP : 3, 4$

(5) $1 : \Gamma \vdash_{\mathcal{N}} A$ *assumption*
$2 : \Gamma \vdash_{\mathcal{N}} \neg \exists x \neg A$ $L.8.24.(4)$
$3 : \Gamma \vdash_{\mathcal{N}} \neg A_t^x \rightarrow \exists x \neg A$ A4
$4 : \Gamma \vdash_{\mathcal{N}} (\neg \exists x \neg A) \rightarrow A_t^x$ PL : 3
$5 : \Gamma \vdash_{\mathcal{N}} A_t^x$ $MP : 2, 4$ QED (8.24)

8.3.1 Deduction Theorem in FOL

Notice the difference between

(i) the provability $\Gamma \vdash_{\mathcal{N}} A \to B$ and (ii) the admissible rule $\dfrac{\Gamma \vdash_{\mathcal{N}} A}{\Gamma \vdash_{\mathcal{N}} B}$.

Point (1) of Lemma 8.24 enables us to conclude, by a single application of MP, that the rule inverse to the one from point (4), namely, $\dfrac{\Gamma \vdash_{\mathcal{N}} \forall x A}{\Gamma \vdash_{\mathcal{N}} A}$ is admissible. In fact, quite generally, (i) implies (ii), for having i) and the assumption of (ii), single application of MP yields the conclusion of (ii).

Now, in the $\mathcal{N}$ system for PL, the opposite implication holds, too. For assuming admissibility of (ii), we have $\Gamma, A \vdash_{\mathcal{N}} A$ by axiom A0, from which $\Gamma, A \vdash_{\mathcal{N}} B$ follows by (ii), and then $\Gamma \vdash_{\mathcal{N}} A \to B$ by Deduction Theorem.

In FOL, however, the implication from (ii) to (i) is not necessarily true, and this is related to the limited validity of Deduction Theorem. For instance, point (4) does not allow us to conclude that also $\Gamma \vdash_{\mathcal{N}} A \to \forall x A$. In fact, this is not the case, but we have to postpone a precise argument showing that until we have discussed semantics. At this point, let us only observe that if this formula were provable, then we would also have $\Gamma \vdash_{\mathcal{N}} \exists x A \to \forall x A$ by a single application of $\exists$I. But this looks unsound: "if there exists an x such that A, then for all x A". (Sure, in case the assumption of the rule (4) from Lemma 8.24 is satisfied we can obtain: $\Gamma \vdash_{\mathcal{N}} A$, $\Gamma \vdash_{\mathcal{N}} \forall x A$, and so, by Lemma 8.23.(2) $\Gamma \vdash_{\mathcal{N}} A \to \forall x A$. But this is only a very special case dependent on the assumption $\Gamma \vdash_{\mathcal{N}} A$.)

Example 8.25
Let us return to the example with horse-heads and animal-heads from the background story of this chapter. Using the alphabet with two predicates, H for 'being a horse' and A for 'being an animal', and a binary relation $Hd(x, y)$ for 'x being a head of y', the argument becomes:

$$\frac{\vdash_{\mathcal{N}} \forall y(H(y) \to A(y))}{\vdash_{\mathcal{N}} \forall x(\, \exists y(H(y) \land Hd(x,y)) \to \exists y(A(y) \land Hd(x,y)) \,)} \tag{8.26}$$

This (a bit peculiar) rule is admissible in $\mathcal{N}$:

1 : $\Gamma \vdash_{\mathcal{N}} \forall y(H(y) \to A(y))$ *assumption*

2 : $\Gamma \vdash_{\mathcal{N}} H(y) \to A(y)$ $MP : 1, L.8.24.(1)$

3 : $\Gamma \vdash_{\mathcal{N}} H(y) \land Hd(x,y) \to A(y) \land Hd(x,y)$ PL : 2.

4 : $\Gamma \vdash_{\mathcal{N}} \exists y(H(y) \land Hd(x,y)) \to \exists y(A(y) \land Hd(x,y))$ $L.8.24.(2) : 3.$

5 : $\Gamma \vdash_{\mathcal{N}} \forall x(\, \exists y(H(y) \land Hd(x,y)) \to \exists y(A(y) \land Hd(x,y)) \,)$ $\forall$G , $L.8.24.(4) : 4.$

This shows the admissibility of (8.26) for arbitrary Γ. In particular, if we take $\Gamma = \{\forall y(H(y) \to A(y))\}$, the first line (assumption) becomes an instance of the axiom $A0$ and the last line becomes an unconditional statement:

$$\forall y(H(y) \to A(y)) \vdash_{\mathcal{N}}$$
$$\forall x(\ \exists y(H(y) \land Hd(x,y)) \to \exists y(A(y) \land Hd(x,y))\ ). \tag{8.27}$$

As observed before this example, admissibility of a rule (ii), like (8.26), does not mean that the corresponding implication (i) is provable. We are *not* entitled to conclude from (8.27) that also the following holds:

$$\vdash_{\mathcal{N}} \forall y(H(y) \to A(y)) \to$$
$$\forall x(\ \exists y(H(y) \land Hd(x,y)) \to \exists y(A(y) \land Hd(x,y))\ ). \tag{8.28}$$

This could be obtained from (8.27) if we had Deduction Theorem for FOL– we now turn to this issue.   □

As a matter of fact, the unrestricted Deduction Theorem from PL is not an admissible proof rule of FOL. It will be easier to see why when we turn to the semantics. For now we just prove the weaker version that does hold. Note the restriction on A.

Theorem 8.29 [**Deduction Theorem for** FOL] If $\Gamma, A \vdash_{\mathcal{N}} B$, and A is *closed* then $\Gamma \vdash_{\mathcal{N}} A \to B$.

Proof. By induction on the length of the proof $\Gamma, A \vdash_{\mathcal{N}} B$. The cases of axioms and MP are treated exactly as in the proof of the theorem for PL, 4.14 (using Lemmata 4.8 and 8.23.(2)). We have to verify the induction step for the last case of the proof using $\exists$I, i.e.:

$$\frac{\Gamma, A \vdash_{\mathcal{N}} C \to D}{\Gamma, A \vdash_{\mathcal{N}} \exists x C \to D} \quad x \text{ not free in } D.$$

By IH, we have the first line of the following proof:

1 : $\Gamma \vdash_{\mathcal{N}} A \to (C \to D)$
2 : $\Gamma \vdash_{\mathcal{N}} C \to (A \to D)$ PL : $C.4.18$
3 : $\Gamma \vdash_{\mathcal{N}} \exists x C \to (A \to D)$ $\exists$I : $2 + A$ closed and x not free in D
4 : $\Gamma \vdash_{\mathcal{N}} A \to (\exists x C \to D)$ PL : $C.4.18$.   QED (8.29)

Revisiting (8.27) from Example 8.25, we see that the assumption of Deduction Theorem is satisfied: $\forall y(H(y) \to A(y))$ is closed. Thus, in this particular case, we actually may conclude that also (8.28) holds. However, due to the restriction in Deduction Theorem, such a transition will not be possible in general.

Just like in the case of PL, MP is a kind of dual to this theorem and we have the corollary corresponding to 4.17, with the same proof.

Corollary 8.30 If A is closed then: $\Gamma, A \vdash_N B$ iff $\Gamma \vdash_N A \to B$.

The assumption that A is closed is needed because of the deduction theorem, i.e., only for the implication $\Rightarrow$. The opposite $\Leftarrow$ does not require A to be closed and is valid for any A.

8.4   GENTZEN'S SYSTEM FOR FOL

Recall that the system $\mathcal{G}$ for PL works with sequents $\Gamma \vdash_{\mathcal{G}} \Delta$, where both Γ, Δ are (finite) sets of formulae. The axioms and rules for propositional connectives give a sound and complete Gentzen system for PL – with all the connectives. Since the set $\{\neg, \to\}$ is expressively complete we restricted earlier our attention to these connectives but the complete set of rules is easier to use in the presence of other connectives. It is easy to see that treating, for instance, $\vee$ and $\wedge$ as abbreviations, the corresponding rules $(\vee\vdash)$, $(\vdash\vee)$, $(\wedge\vdash)$ and $(\vdash\wedge)$ are derivable from the other rules (cf. Subsection 4.8.2). Gentzen's system for FOL is obtained by adding four quantifier rules.

$Ax\ \Gamma \vdash_{\mathcal{G}} \Delta$ where $\Gamma \cap \Delta \neq \varnothing$

$$(\vdash\vee)\ \frac{\Gamma \vdash_{\mathcal{G}} A, B, \Delta}{\Gamma \vdash_{\mathcal{G}} A \vee B, \Delta} \qquad\qquad (\vee\vdash)\ \frac{\Gamma, A \vdash_{\mathcal{G}} \Delta \ ; \ \Gamma, B \vdash_{\mathcal{G}} \Delta}{\Gamma, A \vee B \vdash_{\mathcal{G}} \Delta}$$

$$(\vdash\wedge)\ \frac{\Gamma \vdash_{\mathcal{G}} A, \Delta \ ; \ \Gamma \vdash_{\mathcal{G}} B, \Delta}{\Gamma \vdash_{\mathcal{G}} A \wedge B, \Delta} \qquad\qquad (\wedge\vdash)\ \frac{\Gamma, A, B \vdash_{\mathcal{G}} \Delta}{\Gamma, A \wedge B \vdash_{\mathcal{G}} \Delta}$$

$$(\vdash\neg)\ \frac{\Gamma, B \vdash_{\mathcal{G}} \Delta}{\Gamma \vdash_{\mathcal{G}} \neg B, \Delta} \qquad\qquad (\neg\vdash)\ \frac{\Gamma \vdash_{\mathcal{G}} B, \Delta}{\Gamma, \neg B \vdash_{\mathcal{G}} \Delta}$$

$$(\vdash\to)\ \frac{\Gamma, A \vdash_{\mathcal{G}} B, \Delta}{\Gamma \vdash_{\mathcal{G}} A \to B, \Delta} \qquad\qquad (\to\vdash)\ \frac{\Gamma \vdash_{\mathcal{G}} \Delta, A \ ; \ \Gamma, B \vdash_{\mathcal{G}} \Delta}{\Gamma, A \to B \vdash_{\mathcal{G}} \Delta}$$

$$(\vdash\exists)\ \frac{\Gamma \vdash_{\mathcal{G}} \Delta, \exists x A, A_t^x}{\Gamma \vdash_{\mathcal{G}} \Delta, \exists x A}\ A_t^x\ \text{legal} \qquad (\exists\vdash)\ \frac{\Gamma, A_{x'}^x \vdash_{\mathcal{G}} \Delta}{\Gamma, \exists x A \vdash_{\mathcal{G}} \Delta}\ x'\ \text{fresh}$$

$$(\vdash\forall)\ \frac{\Gamma \vdash_{\mathcal{G}} A_{x'}^x, \Delta}{\Gamma, \vdash_{\mathcal{G}} \forall x A, \Delta}\ x'\ \text{fresh} \qquad (\forall\vdash)\ \frac{A_t^x, \forall x A, \Gamma \vdash_{\mathcal{G}} \Delta}{\forall x A, \Gamma \vdash_{\mathcal{G}} \Delta}\ A_t^x\ \text{legal}$$

The requirement on a variable x to be fresh' means that it must be a new variable not occurring in the sequent. (One may require only that it does not occur freely in the sequent, but we will usually mean that it does not occur at all.) This, in particular, means that its substitution for x is legal.

Notice the peculiar repetition of $\exists x A / \forall x A$ in rules $(\vdash\exists)$ and $(\forall\vdash)$. The need for it, related to the treatment of quantifiers, is illustrated in Exercise

8.13. Here, we mention its two consequences. Semantically, it makes the rules trivially invertible (as the propositional rules were shown in Exercise 6.14). In terms of building a bottom-up proof, it enables us – or a machine! – to choose all the time new witnesses until, eventually and hopefully, an instance of the axiom is reached. For instance, the axiom A4 of $\mathcal{N}$ can be proved as follows:

> 3. $A_t^x \vdash_{\mathcal{G}} \exists x A, A_t^x$ Ax
> 2. $A_t^x \vdash_{\mathcal{G}} \exists x A$ $(\vdash \exists)$ A_t^x legal
> 1. $\vdash_{\mathcal{G}} A_t^x \to \exists x A$ $(\vdash \to)$ (A_t^x legal by assumption).

We see easily in line 2 that only the substitution A_t^x will match the left side of $\vdash_{\mathcal{N}}$. But a machine for automated theorem proving may be forced to try all terms in some predefined order. If it removed $\exists x A$, replacing it by a "wrong" A_s^x, it would have to stop. Retaining the formula allows it to continue past "unlucky substitutions", like A_s^x in line 3 below:

> 4. $A_t^x \vdash_{\mathcal{G}} \exists x A, A_s^x, A_t^x$ Ax
> 3. $A_t^x \vdash_{\mathcal{G}} \exists x A, A_s^x$ $(\vdash \exists)$ A_t^x legal
> 2. $A_t^x \vdash_{\mathcal{G}} \exists x A$ $(\vdash \exists)$ A_s^x legal
> 1. $\vdash_{\mathcal{G}} A_t^x \to \exists x A$ $(\vdash \to)$ (A_t^x legal by assumption).

This is related to the difficulties with treatment of the quantifiers (Exercise 8.13). There are typically infinitely many terms which might be substituted, so such a repetition, although helpful, does not solve the problem of *undecidability* of $\mathcal{G}$ for FOL. If new terms can be generated indefinitely, there may be no way of deciding that the proof will never terminate by looking only at its initial, finite part. As we will see in Theorem 10.38, validity in FOL is generally undecidable, so this is no fault of the system $\mathcal{G}$.

Note also that the rule $(\vdash \to)$ is an *unrestricted* Deduction Theorem (unlike that in Theorem 8.29). We will discuss this in the following chapter.

Example 8.31

Below, we prove the formula from (8.28) in $\mathcal{G}$. As in PL, we use $\mathcal{G}$ for constructing the proof bottom-up.

All $x, y, z, w \in \mathcal{V}$. In applications repeating a formula, $(\vdash \exists)$ and $(\forall \vdash)$, we have dropped these repetitions to make the proof more readable. Of course, we have to choose the substituted terms so that the proof goes through, in particular, so that the respective substitutions are legal. The F, resp. G in the marking 'F/G_y^w *legal*', refer to the formula where the substitution actually takes place – F to $A(y) \wedge H(z, y)$, and G to $H(y) \to A(y)$. The highest three lines use abbreviations: $H = H(w)$, $A = A(w)$, $Hd = Hd(z, w)$:

$$(\to\vdash)\ \dfrac{\dfrac{H,Hd\vdash_{\mathcal{G}} A,H \qquad H,Hd,A\vdash_{\mathcal{G}} A}{H,Hd,H\to A\vdash_{\mathcal{G}} A} \qquad H,Hd,H\to A\vdash_{\mathcal{G}} Hd}{}$$

$(\vdash\wedge)\ \dfrac{}{H,Hd,H\to A\vdash_{\mathcal{G}} A\wedge Hd}$

abbrv.:

$(\forall\vdash)\ \dfrac{H(w),Hd(z,w),H(w)\to A(w)\vdash_{\mathcal{G}} A(w)\wedge Hd(z,w)}{H(w),Hd(z,w),\forall y(H(y)\to A(y))\vdash_{\mathcal{G}} A(w)\wedge Hd(z,w)}\ G_w^y\ \text{legal}$

$(\wedge\vdash)\ \dfrac{}{H(w)\wedge Hd(z,w),\forall y(H(y)\to A(y))\vdash_{\mathcal{G}} A(w)\wedge Hd(z,w)}$

$(\vdash\exists)\ \dfrac{}{H(w)\wedge Hd(z,w),\forall y(H(y)\to A(y))\vdash_{\mathcal{G}} \exists y(A(y)\wedge Hd(z,y))}\ F_w^y\ \text{legal}$

$(\exists\vdash)\ \dfrac{}{\exists y(H(y)\wedge Hd(z,y)),\forall y(H(y)\to A(y))\vdash_{\mathcal{G}} \exists y(A(y)\wedge Hd(z,y))}\ w\ \text{fresh}$

$(\vdash\to)\ \dfrac{}{\forall y(H(y)\to A(y))\vdash_{\mathcal{G}} \exists y(H(y)\wedge Hd(z,y))\to \exists y(A(y)\wedge Hd(z,y))}$

$(\vdash\forall)\ \dfrac{}{\forall y(H(y)\to A(y))\vdash_{\mathcal{G}} \forall x(\exists y(H(y)\wedge Hd(x,y))\to \exists y(A(y)\wedge Hd(x,y)))}\ z\ \text{fresh}$

$(\vdash\to)\ \dfrac{}{\vdash_{\mathcal{G}} \forall y(H(y)\to A(y))\to \forall x(\exists y(H(y)\wedge Hd(x,y))\to \exists y(A(y)\wedge Hd(x,y)))}$

Note that the rules requiring introduction of fresh variables in the assumption are applied before (seen bottom-up) the rules performing arbitrary legal substitutions. **This standard should be observed whenever constructing (bottom-up) proofs in $\mathcal{G}$** – one always tries first to use rules $(\vdash\forall)$ or $(\exists\vdash)$, introducing fresh variables, and only later $(\vdash\exists)$ or $(\forall\vdash)$, which require merely legal substitutions. The choice of terms introduced by the latter can be then adjusted to other terms occurring already in the sequent, easing the way towards axioms.   □

EXERCISES 8.

EXERCISE 8.1 Define inductively the function $\mathcal{V}:\mathsf{WFF}_{\mathsf{FOL}}^{\Sigma}\to\mathcal{P}(\mathcal{V})$ returning the set of variables occurring freely in a formula.

EXERCISE 8.2 Formulate the following claims about sets in an appropriate FOL-language:

(1) $A\subseteq B$ (2) $(A\cup B)\subseteq(A\cap B)$
(3) $A\cap B\cap C\neq\varnothing\to(A\cap B\neq\varnothing\wedge A\cap C\neq\varnothing)$

EXERCISE 8.3 Prove the following:

(1) $\vdash_{\mathcal{N}} \exists y\exists x A\to \exists x\exists y A$
   Hint: Complete the following proof by filling out appropriate things for '?':
   1. $\vdash_{\mathcal{N}} A\to\exists y A$ A4
   2. ?   A4
   3. $\vdash_{\mathcal{N}} A\to\exists x\exists y A$ L.8.23.(1) : 1, 2
   4. $\vdash_{\mathcal{N}} \exists x A\to\exists x\exists y A$? : 3 (x not free in $\exists x\exists y A$)
   5. $\vdash_{\mathcal{N}} \exists y\exists x A\to\exists x\exists y A$? : 4 (?)

(2) $\vdash_{\mathcal{N}} \exists y A_y^x\to \exists x A$ – if y is substitutable for x in A, and not free in A
   (Hint: Two steps only! First an instance of A4, and then $\exists$I.)

(3) $\vdash_{\mathcal{N}} \exists x \forall y A \;\rightarrow\; \forall y \exists x A$ (Hint: Lemma 8.24.(1), then (2) and (3).)

(4) $\vdash_{\mathcal{N}} \forall x A \rightarrow \exists x A$

    (Hint: Lemma 8.24.(1), A4, and then Lemma 8.23.(1).)

EXERCISE 8.4 Is the following proof correct? If not, what is wrong?

1 : $\vdash_{\mathcal{N}} \forall x (P(x) \vee R(x)) \rightarrow (P(x) \vee R(x))$ $L.8.24.(1)$

2 : $\vdash_{\mathcal{N}} \forall x (P(x) \vee R(x)) \rightarrow (\forall x P(x) \vee R(x))$ $\forall I$

3 : $\vdash_{\mathcal{N}} \forall x (P(x) \vee R(x)) \rightarrow (\forall x P(x) \vee \forall x R(x))$ $\forall I$

EXERCISE 8.5 Show admissibility of the rule $\dfrac{\Gamma \vdash_{\mathcal{N}} \forall x A}{\Gamma \vdash_{\mathcal{N}} A_t^x}$ for legal A_t^x.

EXERCISE 8.6 Re-work Example 8.25 for the other argument from the background story at the beginning of this chapter, i.e.:

(1) Design a FOL alphabet for expressing the argument:

> Every man is mortal;
>
> and   Socrates is a man;
> ———————————————
> hence Socrates is mortal.

(2) Express it as a rule and show its admissibility in $\mathcal{N}$.

(3) Write it now as a single formula (analogous to (8.28)) – decide which connective(s) to choose to join the two premises of the rule into the antecedent of the implication. Use the previous point and Theorem 8.29 to show that this formula is provable in $\mathcal{N}$.

(4) Prove now your formula from the previous point using $\mathcal{G}$.

EXERCISE 8.7 The formula $(A \rightarrow \exists x B) \rightarrow \exists x (A \rightarrow B)$ can be proven in $\mathcal{N}$ as follows:

1. $\vdash_{\mathcal{N}} B \rightarrow (A \rightarrow B)$ A1

2. $\vdash_{\mathcal{N}} \exists x B \rightarrow \exists x (A \rightarrow B)$ $L.8.24.(2)$

3. $\vdash_{\mathcal{N}} (A \rightarrow B) \rightarrow \exists x (A \rightarrow B)$ A4

4. $\vdash_{\mathcal{N}} \neg A \rightarrow \exists x (A \rightarrow B)$ PL : 3

5. $\vdash_{\mathcal{N}} (A \rightarrow \exists x B) \rightarrow \exists x (A \rightarrow B)$ PL : 2, 4

Verify that the lines 4. and 5. are valid transitions according to PL. (Hint: They correspond to provability (or validity!), for instance in $\mathcal{G}$, of:

$\vdash ((A \rightarrow B) \rightarrow Z) \rightarrow (\neg A \rightarrow Z)$ and $B \rightarrow Z, \neg A \rightarrow Z \vdash (A \rightarrow B) \rightarrow Z$.)

EXERCISE 8.8 Assuming that x has no free occurrences in A, complete the following proofs:

(1) $\vdash_{\mathcal{N}} \exists x (A \rightarrow B) \rightarrow (A \rightarrow \exists x B)$

    1. $\vdash_{\mathcal{N}} B \rightarrow \exists x B$ A4

    2. $\vdash_{\mathcal{N}} A \rightarrow (B \rightarrow \exists x B)$ (?)

    3. $\vdash_{\mathcal{N}} (A \rightarrow (B \rightarrow \exists x B)) \rightarrow ((A \rightarrow B) \rightarrow (A \rightarrow \exists x B))$ A2

    $\ldots$

(2) $\vdash_{\mathcal{N}} \forall x(B \to A) \to (\exists x B \to A)$
    1. $\vdash_{\mathcal{N}} \forall x(B \to A) \to (B \to A)$ (?)
    2. $\vdash_{\mathcal{N}} B \to (\forall x(B \to A) \to A)$ PL(?)
    ...

(3) $\vdash_{\mathcal{N}} \exists x(B \to A) \to (\forall x B \to A)$
    1. $\vdash_{\mathcal{N}} (\forall x B \to B) \to ((B \to A) \to (\forall x B \to A))$ (?)
    2. $\vdash_{\mathcal{N}} \forall x B \to B$ (?)
    ...

EXERCISE 8.9 Show that:

(1) $\vdash_{\mathcal{N}} \forall x(A \to B) \to \neg \exists x(A \wedge \neg B)$
(2) $\vdash_{\mathcal{N}} \neg \exists x(A \wedge \neg B) \to \forall x(A \to B)$

EXERCISE 8.10 Show that the rule $\dfrac{\vdash A \to B}{\vdash \forall x A \to \forall x B}$ is admissible in $\mathcal{N}$ without any side-conditions on x.

(Hint: Lemma 8.24.(2), and two applications of some relevant result from PL.)

EXERCISE 8.11 Prove now all the formulae from Exercises 8.3 and 8.8 using Gentzen's system.

EXERCISE 8.12 Using Gentzen's system

(1) Show provability:
    $\vdash_{\mathcal{G}} \forall x(A \to B) \to (\forall x A \to \forall x B)$.
(2) Now try to construct a proof for the opposite implication, i.e.,
    $\vdash_{\mathcal{G}} (\forall x A \to \forall x B) \to \forall x(A \to B)$.
    Can you tell why this proof will not "succeed"?

EXERCISE 8.13 Prove $\exists x(\exists y P(x, y) \to \forall z \exists x P(z, x))$ in Gentzen's system.
(Hint: Retaining the formula from the conclusion in the premise of rules $(\forall \vdash)$, $(\vdash \exists)$ may be necessary.)

Chapter 9

SEMANTICS OF FOL

9.1 THE BASIC DEFINITIONS

Definition 9.1 A FOL *structure* M for an alphabet Σ consists of

(1) a non-empty set $\underline{M}$ – the *interpretation domain*
(2) for each constant symbol $a \in \mathcal{I}$, an element $[\![a]\!]^M \in \underline{M}$
(3) for each function symbol $f \in \mathcal{F}$ with arity n, a function $[\![f]\!]^M : \underline{M}^n \to \underline{M}$
(4) for each relation symbol $P \in \mathcal{R}$ with arity n, a subset $[\![P]\!]^M \subseteq \underline{M}^n$.

Thus, a structure is simply a set where all constant, function and relation symbols have been interpreted arbitrarily. The only restriction concerns the arities of the function and relation symbols which have to be respected by their interpretation.

Notice that according to Definition 9.1, variables do not receive any fixed interpretation. Thus, it does not provide sufficient means for assigning meaning to *all syntactic expressions* of the language. The meaning of variables, and expressions involving variables, will depend on the choice of the assignment.

Definition 9.2 [**Interpretation of terms**] Any function $v : \mathcal{V} \to \underline{M}$ is called a *variable assignment* or just an assignment. Given a structure M and an assignment v, the interpretation of terms is defined inductively:

(1) for $x \in \mathcal{V} : [\![x]\!]_v^M = v(x)$
(2) for $a \in \mathcal{I} : [\![a]\!]_v^M = [\![a]\!]^M$
(3) for n-ary $f \in \mathcal{F}$ and $t_1, \ldots, t_n \in \mathcal{T}_\Sigma$:
    $[\![f(t_1, \ldots, t_n)]\!]_v^M = [\![f]\!]^M([\![t_1]\!]_v^M, \ldots, [\![t_n]\!]_v^M)$.

According to points (2) and (3), interpretation of constants and function symbols does not depend on variable assignment. Consequently, interpretation of an arbitrary ground term $t \in \mathcal{GT}$ is fixed in a given structure M, i.e., for any assignment $v : [\![t]\!]_v^M = [\![t]\!]^M$.

Example 9.3

Let Σ contain the constant symbol $\odot$, one unary function symbol s and a binary function symbol $\oplus$. Here are some examples of Σ-structures:

(1) A is the natural numbers $\mathbb{N}$ with $[\![\odot]\!]^A = 0$, $[\![s]\!]^A = +1$ and $[\![\oplus]\!]^A = +$.
   Here, we understand $ss\odot \oplus sss\odot$ as 5, i.e., $[\![ss\odot \oplus sss\odot]\!]^A = 2 + 3 = 5$.
   For an assignment v with $v(x) = 2$ and $v(y) = 3$, we get $[\![x \oplus y]\!]^A_v = [\![x]\!]^A_v [\![\oplus]\!]^A [\![y]\!]^A_v = 2 + 3 = 5$.
   For an assignment w with $w(x) = 4$ and $w(y) = 7$, we get $[\![x \oplus y]\!]^A_w = [\![x]\!]^A_w [\![\oplus]\!]^A [\![y]\!]^A_w = 4 + 7 = 11$.

(2) B is the natural numbers $\mathbb{N}$ with $[\![\odot]\!]^B = 0$, $[\![s]\!]^B = +1$ and $[\![\oplus]\!]^B = *$.
   Here we have : $[\![ss\odot \oplus sss\odot]\!]^B = 2 * 3 = 6$.

(3) C is the integers $\mathbb{Z}$ with $[\![\odot]\!]^C = 1$, $[\![s]\!]^C = +2$ and $[\![\oplus]\!]^C = -$.
   Here we have : $[\![ss\odot \oplus sss\odot]\!]^C = 5 - 7 = -2$.
   What will be the values of $x \oplus y$ under the assignments from (1)?

(4) Given a non-empty set (e.g., $S = \{a, b, c\}$), we let the domain of D be S^* (the finite strings over S), with $[\![\odot]\!]^D = \epsilon$ (the empty string), $[\![s]\!]^D(\epsilon) = \epsilon$ and $[\![s]\!]^D(wx) = w$ (where x is the last element in the string wx), and $[\![\oplus]\!]^D(p, t) = pt$ (i.e., concatenation of strings). $\qquad\square$

As we can see, the requirements on something being a Σ-structure are very weak – a non-empty set with *arbitrary* interpretation of constant, function and relation symbols respecting merely arity. Consequently, there is a huge number of structures for any alphabet – in fact, so huge that it is not even a set but a class. We will not, however, be concerned with this distinction. If necessary, we will denote the collection of *all* Σ-structures by $\mathsf{Str}(\Sigma)$.

A Σ-structure M, together with an assignment, induces the interpretation of all $\mathsf{WFF}^\Sigma_{\mathsf{FOL}}$. As in the case of PL, such an interpretation is a function $[\![\text{-}]\!]^M_v : \mathsf{WFF}^\Sigma_{\mathsf{FOL}} \to \{1, 0\}$.

Definition 9.4 [Interpretation of formulae] M determines a boolean value for every formula relative to every variable assignment v, according to the following rules:

(1) If $P \in \mathcal{R}$ is n-ary and $t_1, \ldots, t_n$ are terms, then
   $$[\![P(t_1, \ldots, t_n)]\!]^M_v = 1 \quad \Leftrightarrow \quad \langle [\![t_1]\!]^M_v, \ldots, [\![t_n]\!]^M_v \rangle \in [\![P]\!]^M.$$

(2) Propositional connectives are combined as in PL. For $A, B \in \mathsf{WFF}^\Sigma_{\mathsf{FOL}}$:
   $$[\![\neg A]\!]^M_v = 1 \Leftrightarrow [\![A]\!]^M_v = 0$$
   $$[\![A \to B]\!]^M_v = 1 \Leftrightarrow [\![A]\!]^M_v \text{ implies } [\![B]\!]^M_v$$
   $$\Leftrightarrow [\![A]\!]^M_v = 0 \text{ or } [\![B]\!]^M_v = 1.$$

(3) Quantified formulae:
   $$[\![\exists x A]\!]^M_v = 1 \Leftrightarrow \text{ there is an } \underline{a} \in \underline{M} : [\![A]\!]^M_{v[x \mapsto \underline{a}]} = 1.$$

Recall from Remark 1.5 that the notation $v[x \mapsto a]$ denotes the function v modified at one point, namely so that now $v(x) = a$. Thus the definition says that the value assigned to the bound variable x by valuation v is inessential when determining the boolean value $[\exists x A]_v$ – no matter what $v(x)$ is, it will be "modified" $v[x \mapsto \underline{a}]$ if an appropriate $\underline{a}$ can be found. We will observe consequences of this fact for the rest of the current subsection.

The following fact justifies the reading of $\forall x$ as "for all x".

Fact 9.5 For every structure M and assignment v:
$$[\forall x A]_v^M = \mathbf{1} \Leftrightarrow \text{ for all } \underline{a} \in \underline{M} : [A]_{v[x \mapsto \underline{a}]}^M = \mathbf{1}.$$

Proof. We only expand $\forall x$ according to Definition 8.7 and apply Definition 9.4.

$$[\forall x A]_v^M = \mathbf{1} \overset{8.7}{\Longleftrightarrow} [\neg \exists x \neg A]_v^M = \mathbf{1} \overset{9.4}{\Longleftrightarrow} [\exists x \neg A]_v^M = \mathbf{0}$$
$$\overset{9.4}{\Longleftrightarrow} [\neg A]_{v[x \mapsto \underline{a}]}^M = \mathbf{1} \text{ for no } \underline{a} \in \underline{M}$$
$$\overset{9.4}{\Longleftrightarrow} [A]_{v[x \mapsto \underline{a}]}^M = \mathbf{0} \text{ for no } \underline{a} \in \underline{M}$$
$$\overset{9.4}{\Longleftrightarrow} [A]_{v[x \mapsto \underline{a}]}^M = \mathbf{1} \text{ for all } \underline{a} \in \underline{M}. \qquad \text{QED (9.5)}$$

Again, notice that to evaluate $[\forall x A]_v$, it does not matter what $v(x)$ is – we will have to check *all* possible modifications $v[x \mapsto \underline{a}]$ anyway.

Point (3) of Definition 9.4 captures the crucial difference between free and bound variables – the truth of a formula depends on the *names* of the free variables but not of the bound ones. More precisely, a closed formula (sentence) is either true or false in a given structure – its interpretation according to the above definition will *not* depend on the assignment v. An open formula is neither true nor false – since we do not know what objects the free variables refer to. To determine the truth-value of an open formula, the above definition requires us to provide an assignment to its free variables. The resulting value will often depend on the chosen assignment.

Example 9.6
Consider the alphabet with unary and binary relation symbols P, R, and a structure M with $\underline{M} = \{\underline{a}, \underline{b}\}$ and $[P]^M = \{\underline{a}\}, [R]^M = \{\langle \underline{a}, \underline{b} \rangle\}$. We evaluate $[B]^M$ under two different assignments:
$$v(x) = \underline{a} : [P(x)]_v^M = \mathbf{1} \Leftrightarrow [x]_v^M \in [P]^M \Leftrightarrow v(x) = \underline{a} \in \{\underline{a}\} - \text{holds};$$
$$w(x) = \underline{b} : [P(x)]_w^M = \mathbf{1} \Leftrightarrow [x]_w^M \in [P]^M \Leftrightarrow w(x) = \underline{b} \in \{\underline{a}\} - \text{fails}.$$
Thus $[P(x)]_v^M = \mathbf{1}$ while $[P(x)]_w^M = \mathbf{0}$. The following illustrates the same point, but on a more complex formula $A = \exists x R(x, y)$. We check whether $[A]_v^M = \mathbf{1}$, resp. $[A]_w^M = \mathbf{1}$ for the two assignments v, w:

$$v = \{x \mapsto \underline{a}, y \mapsto \underline{a}\}$$

$$[\![\exists x R(x,y)]\!]_v^M = 1 \Leftrightarrow [\![R(x,y)]\!]_{v[x\mapsto\underline{x}]}^M = 1 \text{ for some } \underline{x} \in \underline{M}$$

$$\Leftrightarrow \langle v[x \mapsto \underline{x}](x), v[x \mapsto \underline{x}](y)\rangle \in [\![R]\!]^M \text{ for some } \underline{x} \in \underline{M}$$

$$\Leftrightarrow \langle v[x \mapsto \underline{x}](x), \underline{a}\rangle \in [\![R]\!]^M \text{ for some } \underline{x} \in \underline{M}$$
$$(\text{since } v[x \mapsto \underline{x}](y) = v(y) = \underline{a})$$

$$\Leftrightarrow \text{ at least one of the following holds}$$
$$(\text{since } v[x \mapsto \underline{x}](x) = \underline{x})$$

$v[x \mapsto \underline{a}]$	$v[x \mapsto \underline{b}]$
$\langle \underline{a}, \underline{a}\rangle \in [\![R]\!]^M$	$\langle \underline{b}, \underline{a}\rangle \in [\![R]\!]^M$
but	
$\langle \underline{a}, \underline{a}\rangle \notin [\![R]\!]^M$	$\langle \underline{b}, \underline{a}\rangle \notin [\![R]\!]^M$

$$\text{so} \quad [\![\exists x R(x,y)]\!]_v^M = 0$$

$$w = \{x \mapsto \underline{a}, y \mapsto \underline{b}\}$$

$$[\![\exists x R(x,y)]\!]_w^M = 1 \Leftrightarrow [\![R(x,y)]\!]_{w[x\mapsto\underline{x}]}^M = 1 \text{ for some } \underline{x} \in \underline{M}$$

$$\Leftrightarrow \langle w[x \mapsto \underline{x}](x), w[x \mapsto \underline{x}](y)\rangle \in [\![R]\!]^M \text{ for some } \underline{x} \in \underline{M}$$

$$\Leftrightarrow \langle w[x \mapsto \underline{x}](x), \underline{b}\rangle \in [\![R]\!]^M \text{ for some } \underline{x} \in \underline{M}$$
$$(\text{since } w[x \mapsto \underline{x}](y) = w(y) = \underline{b})$$

$$\Leftrightarrow \text{ at least one of the following holds}$$

$w[x \mapsto \underline{a}]$	$w[x \mapsto \underline{b}]$
$\langle \underline{a}, \underline{b}\rangle \in [\![R]\!]^M$	$\langle \underline{b}, \underline{b}\rangle \in [\![R]\!]^M$

$$\text{so} \quad [\![\exists x R(x,y)]\!]_w^M = 1 \text{ with } x \mapsto \underline{a} \text{ providing a witness}$$

Thus, $[\![A]\!]_v^M = 0$ while $[\![A]\!]_w^M = 1$. Notice that the values assigned to the bound variable x by v and w do not matter – one has to consider $v[x \mapsto \underline{x}]$, resp. $w[x \mapsto \underline{x}]$ for all possible cases of $\underline{x}$. What makes the difference was the fact that $v(y) = \underline{a}$ – for which no $\underline{x}$ could be found with $\langle \underline{x}, \underline{a}\rangle \in [\![R]\!]^M$, while $w(y) = \underline{b}$ – for which we found such an $\underline{x}$, namely $\langle \underline{a}, \underline{b}\rangle \in [\![R]\!]^M$.

Universal quantifier $\forall x$ has an entirely analogous effect – with the above two assignments, you may use Fact 9.5 directly to check that $[\![\forall x \neg R(x,y)]\!]_v^M = 1$, while $[\![\forall x \neg R(x,y)]\!]_w^M = 0$. $\qquad\square$

This influence of the *names* of free variables, but not of the bound ones, on the truth of formulae is expressed in the following lemma.

Lemma 9.7 Let M be a structure, A a formula and v and w two assignments such that for every $x \in \mathcal{V}(A) : v(x) = w(x)$. Then $[\![A]\!]_v^M = [\![A]\!]_w^M$.

Proof. Induction on A. For atomic A, the claim is obvious (for terms in A: $[\![t]\!]_v^M = [\![t]\!]_w^M$), and induction passes trivially through the connectives. So let A be a quantified formula $\exists y B$

$$[\![A]\!]_v^M = 1 \Leftrightarrow \text{ for some } \underline{a} \in \underline{M} : [\![B]\!]_{v[y \mapsto \underline{a}]}^M = 1$$
$$\overset{\text{IH}}{\Leftrightarrow} \text{ for some } \underline{a} \in \underline{M} : [\![B]\!]_{w[y \mapsto \underline{a}]}^M = 1$$
$$\Leftrightarrow [\![A]\!]_w^M = 1.$$

By IH, $[\![B]\!]_{v[y \mapsto \underline{a}]}^M = [\![B]\!]_{w[y \mapsto \underline{a}]}^M$, since $v(x) = w(x)$ for all free variables $x \in \mathcal{V}(A)$, and the modification $[y \mapsto \underline{a}]$ makes them agree on y as well, and hence on all the free variables of B. \hfill QED (9.7)

Remark.
The interpretation of constants, function and relation symbols does not depend on the assignment. Similarly, the interpretation of ground terms and ground formulae is independent of the assignments. For formulae even more is true – by Lemma 9.7, the boolean value of any *closed* formula A does not depend on the assignment (as $\mathcal{V}(A) = \varnothing$, for all assignments v, w and all $x \in \mathcal{V} : x \in \varnothing \to v(x) = w(x)$.) In Example 9.6, $[\![\exists x P(x)]\!]_v^M = 1$ and $[\![\exists y \exists x R(x, y)]\!]_v^M = 1$ for all v.

For this reason, we often drop v in $[\![A]\!]_v^M$ and write simply $[\![A]\!]^M$ in case A is closed, just as we may drop v in $[\![t]\!]^M$ if t is a ground term. \hfill □

Example 9.8
We design a structure M for the alphabet Σ_{Stack} from Example 8.2. (We identify elements of A with 1-letter strings in A^*, i.e., $A \subset A^*$.)

- the underlying set $\underline{M} = A^*$ consists of all finite strings over a non-empty set A; below $w, v \in \underline{M}$ are arbitrary, while $a \in A$:
- $[\![empty]\!]^M = \epsilon \in A^*$ – the empty string
- $[\![pop]\!]^M(wa) = w$ and $[\![pop]\!]^M(\epsilon) = \epsilon$
- $[\![top]\!]^M(wa) = a$ and $[\![top]\!]^M(\epsilon) = a_0$ for some $a_0 \in A$
- $[\![push]\!]^M(w, v) = wv$ – the string w with v appended at the end
- finally, we let $[\![El]\!]^M = A$, $[\![St]\!]^M = A^*$ and $[\![\equiv]\!]^M = \{\langle m, m \rangle : m \in \underline{M}\}$, i.e., the identity relation.

We have given an interpretation to all the symbols from Σ_{Stack}, so M is a Σ_{Stack}-structure. Notice that all functions are total.

We now check the value of all the axioms from Example 8.14 in M. We apply Definition 9.4, Fact 9.5 and Lemma 9.7 (its consequence from the remark before this example, allowing us to drop assignments whenever interpreting ground terms and sentences).

(1) $[\![St(empty)]\!]^M = [\![empty]\!]^M \in [\![St]\!]^M = \epsilon \in A^* = \mathbf{1}$.

(2) $[\![\forall x, s : El(x) \wedge St(s) \rightarrow St(push(s, x))]\!]^M = \mathbf{1}$

$\Leftrightarrow$ for all $c, d \in \underline{M} : [\![El(x) \wedge St(s) \rightarrow St(push(s, x))]\!]^M_{[x \mapsto c, s \mapsto d]} = \mathbf{1}$

$\Leftrightarrow$ for all $c, d \in \underline{M} : [\![El(x) \wedge St(s)]\!]^M_{[x \mapsto c, s \mapsto d]} = \mathbf{0}$
$\qquad\qquad\qquad\qquad$ or $[\![St(push(s, x))]\!]^M_{[x \mapsto c, s \mapsto d]} = \mathbf{1}$

$\Leftrightarrow$ for all $c, d \in \underline{M} : [\![El(x)]\!]^M_{[x \mapsto c, s \mapsto d]} = \mathbf{0}$ or $[\![St(s)]\!]^M_{[x \mapsto c, s \mapsto d]} = \mathbf{0}$
$\qquad\qquad\qquad\qquad$ or $[\![St(push(s, x))]\!]^M_{[x \mapsto c, s \mapsto d]} = \mathbf{1}$

$\Leftrightarrow$ for all $c, d \in \underline{M} : c \notin [\![El]\!]^M$ or $d \notin [\![St]\!]^M$
$\qquad\qquad\qquad\qquad$ or $[\![push(s, x)]\!]^M_{[x \mapsto c, s \mapsto d]} \in [\![St]\!]^M$

$\Leftrightarrow$ for all $c, d \in \underline{M} : c \notin A$ or $d \notin A^*$ or $dc \in A^*$

$\Leftrightarrow$ true, since whenever $c \in A$ and $d \in A^*$ then also $dc \in A^*$.

(3) We drop axioms (3) and (4) and go directly to (5) and (6).

(5) $[\![pop(empty) \equiv empty]\!]^M = \mathbf{1}$

$\Leftrightarrow \langle [\![pop(empty)]\!]^M, [\![empty]\!]^M \rangle \in [\![\equiv]\!]^M$

$\Leftrightarrow \langle [\![pop]\!]^M(\epsilon), \epsilon \rangle \in [\![\equiv]\!]^M \Leftrightarrow \langle \epsilon, \epsilon \rangle \in [\![\equiv]\!]^M \Leftrightarrow \epsilon = \epsilon \Leftrightarrow$ true.

(6) $[\![\forall x, s : El(x) \wedge St(s) \rightarrow pop(push(s, x)) \equiv s]\!]^M = \mathbf{1} \Leftrightarrow$

for all $c, d \in \underline{M} : [\![El(x) \wedge St(s) \rightarrow pop(push(s, x)) \equiv s]\!]^M_{[x \mapsto c, s \mapsto d]} = \mathbf{1}$

$\Leftrightarrow$ for all $c, d \in \underline{M} : [\![El(x) \wedge St(s)]\!]^M_{[x \mapsto c, s \mapsto d]} = \mathbf{0}$
$\qquad\qquad\qquad\qquad$ or $[\![pop(push(s, x)) \equiv s]\!]^M_{[x \mapsto c, s \mapsto d]} = \mathbf{1}$

$\Leftrightarrow$ for all $c, d \in \underline{M} : c \notin [\![El]\!]^M$ or $d \notin [\![St]\!]^M$
$\qquad\qquad\qquad\qquad$ or $\langle [\![pop]\!]^M([\![push]\!]^M(d, c)), d \rangle \in [\![\equiv]\!]^M$

$\Leftrightarrow$ for all $c, d \in \underline{M} : c \notin A$ or $d \notin A^*$ or $[\![pop]\!]^M([\![push]\!]^M(d, c)) = d$

$\Leftrightarrow$ for all $c, d \in \underline{M} :$ if $c \in A$ and $d \in A^*$ then $[\![pop]\!]^M(dc) = d$

$\Leftrightarrow$ for all $c, d \in \underline{M} :$ if $c \in A$ and $d \in A^*$ then $d = d \Leftrightarrow$ true.

For the last transition it is essential that $c \in A$ – if c is a non-empty string, e.g. ab then $[\![pop]\!]^M(dab) = da \neq d$. Axiom (7) can be verified along the same lines. $\qquad\qquad\qquad\qquad\qquad\qquad\qquad\qquad\qquad$ □

Typically, infinitely many formulae evaluate to $\mathbf{1}$ in a structure. In this example, for instance, also the following formula evaluates to $\mathbf{1}$:

$$\forall w, x : St(x) \wedge \neg(x \equiv empty) \rightarrow pop(push(w, x)) \equiv push(w, pop(x)).$$

It goes counter our intuitive understanding of stacks – we do not push stacks on stacks, only elements. This is an accident caused by the fact that all functions must be total and, more importantly, by the specific definition of our structure M. What in programming languages are called "types" is expressed in FOL by additional predicates. In FOL, we do not have types of stacks, elements (integers, etc.), preventing one from applying the function *top* to the elements. We have predicates identifying parts of the domain. The axioms, as in the example above, specify what should hold when the arguments come from the right parts of the domain (have right types). But the requirement of totality on all functions forces them to yield some results also when applied outside such intended definition domains. Thus, the axioms from Example 8.14 do not describe *uniquely* the particular data type stacks we might have in mind. They only define some properties which such data type should satisfy.

The following result is crucial for proving the soundness of $A4$ (Exercise 9.10). A useful special case arises if t_1 or t_2 is x itself. In that case the lemma says that the identity $v(x) = [\![t]\!]_v^M$ implies the identity $[\![A]\!]_v^M = [\![A_t^x]\!]_v^M$, provided t is substitutable for x in A.

Lemma 9.9 If t_1, t_2 are both substitutable for x in A and $[\![t_1]\!]_v^M = [\![t_2]\!]_v^M$, then $[\![A_{t_1}^x]\!]_v^M = [\![A_{t_2}^x]\!]_v^M$.

Proof. By induction on the complexity of A.

BASIS :: It is easy to show (by induction on the complexity of term s) that $[\![t_1]\!]_v^M = [\![t_2]\!]_v^M$ implies $[\![s_{t_1}^x]\!]_v^M = [\![s_{t_2}^x]\!]_v^M$. Hence, it implies also $[\![R(s_1, \ldots, s_n)_{t_1}^x]\!]_v^M = [\![R(s_1, \ldots, s_n)_{t_2}^x]\!]_v^M$.

IND. :: The induction steps for $\neg$ and $\to$ are trivial, as is the case for $\exists x$ for the same variable x. As an example, for $A = \neg B$, we simply check: $[\![\neg B_{t_1}^x]\!]_v^M = 1 \Leftrightarrow [\![B_{t_1}^x]\!]_v^M = 0 \overset{\text{IH}}{\Leftrightarrow} [\![B_{t_2}^x]\!]_v^M = 0 \Leftrightarrow [\![\neg B_{t_2}^x]\!]_v^M = 1$.

Now let A be $\exists y B$ and $y \neq x$. As t_1 and t_2 are substitutable for x in A, so either (1) x does not occur free in B, in which case the proof is trivial, or (2a) y does not occur in t_1, t_2 and (2b) t_1, t_2 are both substitutable for x in B. Now $(\exists y B)_{t_i}^x = \exists y (B_{t_i}^x)$ and hence $[\![(\exists y B)_{t_i}^x]\!]_v^M = 1$ iff $[\![B_{t_i}^x]\!]_{v[y \mapsto a]}^M = 1$ for some $\underline{a}$. By (2a) we know that $[\![t_1]\!]_{v[y \mapsto a]}^M = [\![t_2]\!]_{v[y \mapsto a]}^M$ for all $\underline{a}$, so by (2b) and the IH we know that $[\![B_{t_1}^x]\!]_{v[y \mapsto a]}^M = 1$ for some $\underline{a}$ iff $[\![B_{t_2}^x]\!]_{v[y \mapsto a]}^M = 1$ for some $\underline{a}$. QED (9.9)

9.2 SEMANTIC PROPERTIES

The following definition introduces the central semantic concepts of FOL. They correspond to the concepts from PL, given in Definition 5.11, except for the relation $\models_v$, which depends on an assignment to free variables.

Definition 9.10 Let A, M and v range over formulae, structures, and variable-assignments into $\underline{M}$, respectively.

A is	iff	condition holds	notation:
true with respect to M and v	iff	$[\![A]\!]_v^M = 1$	$M \models_v A$
false with respect to M and v	iff	$[\![A]\!]_v^M = 0$	$M \not\models_v A$
satisfied in M	iff	for all $v : M \models_v A$	$M \models A$
not satisfied in M	iff	there is a $v : M \not\models_v A$	$M \not\models A$
valid	iff	for all $M : M \models A$	$\models A$
not valid	iff	there is an $M : M \not\models A$	$\not\models A$
satisfiable	iff	there is an $M : M \models A$	
unsatisfiable	iff	for all $M : M \not\models A$	

Remark [Terminology and notation]

Sometimes, instead of saying "A is satisfied (not satisfied) in M", one says that "A is true (false)", or even that it is "valid (not valid) in M".

Lemma 9.7 tells us that only a part of v is relevant in "$M \models_v A$", namely, the partial function that identifies the values for $x \in \mathcal{V}(A)$. If $\mathcal{V}(A) \subseteq \{x_1, \ldots, x_n\}$ and $\underline{a}_i = v(x_i)$, we may write just "$M \models_{\{x_1 \mapsto \underline{a}_1, \ldots, x_n \mapsto \underline{a}_n\}} A$." We may even drop the curly braces, since they only clutter up the expression. □

Example 9.11

In Example 9.8 we have shown that the structure M models each axiom ϕ of stacks, $M \models \phi$, from Example 8.14.

Recall now Example 8.17 of an alphabet Σ_{SG} for simple graphs containing only one binary relation symbol E. Any set $\underline{U}$ with a binary relation R on it is an Σ_{SG}-structure, i.e., we let $[\![E]\!]^U = R \subseteq \underline{U} \times \underline{U}$.

We now want to make it to satisfy axiom (1) from 8.17: $U \models \forall x, y : E(x, y) \to E(y, x)$. By Definition 9.10, this requires

$$\text{for all assignments } v : [\![\forall x, y : E(x, y) \to E(y, x)]\!]_v^U = 1. \qquad (9.12)$$

Since the formula is closed, we may ignore assignments – by Fact 9.5 we must show:

$$\text{for all } \underline{a}, \underline{b} \in \underline{U} : [\![E(x, y) \to E(y, x)]\!]_{[x \mapsto \underline{a}, y \mapsto \underline{b}]}^U = 1. \qquad (9.13)$$

By Definition 9.4.(2) this means:

$$\text{for all } \underline{a}, \underline{b} \in \underline{U} : [\![E(x, y)]\!]_{[x \mapsto \underline{a}, y \mapsto \underline{b}]}^U = 0 \text{ or } [\![E(y, x)]\!]_{[x \mapsto \underline{a}, y \mapsto \underline{b}]}^U = 1, \qquad (9.14)$$

i.e., by Definition 9.4.(1):

$$\text{for all } \underline{a}, \underline{b} \in \underline{U} : \langle \underline{a}, \underline{b} \rangle \notin [\![E]\!]^U \text{ or } \langle \underline{b}, \underline{a} \rangle \in [\![E]\!]^U, \qquad (9.15)$$

and since $[\![E]\!]^U = R$:

$$\text{for all } \underline{a}, \underline{b} \in \underline{U} : \langle \underline{a}, \underline{b} \rangle \notin R \text{ or } \langle \underline{b}, \underline{a} \rangle \in R. \qquad (9.16)$$

But this says just that for any pair of elements $\underline{a}, \underline{b} \in \underline{U}$, if $R(\underline{a}, \underline{b})$ then also $R(\underline{b}, \underline{a})$. Thus the axiom holds only in the structures where the relation (here R) interpreting the symbol E is symmetric and does not hold in those where it is not symmetric. Put a bit differently – and this is the way one uses (non-logical) axioms – the axiom *selects only* those Σ_{SG}-structures where the relation interpreting E is symmetric – it narrows the relevant structures to those satisfying the axiom.

Quite an analogous procedure would show that the other axioms from Example 8.17, would narrow the possible interpretations to those where R is transitive, resp. irreflexive. □

9.3 OPEN VS. CLOSED FORMULAE

Semantic properties of sentences bear some resemblance to the respective properties of formulae of PL. However, this resemblance does not apply with equal strength to open formulae. We start with a comparison to PL.

Remark 9.17 [Comparing with PL]
Comparing the table from Definition 9.10 with Definition 5.11, we see that the last three double rows correspond exactly to the definition for PL. The first double row is new because now, in addition to formulae and structures, we also have valuation of individual variables. As before, *contradictions* are the unsatisfiable formulae, and the structure M is a *model* of A if A is satisfied (valid) in $M : M \models A$. (Notice that if A is valid *in* an M, it is not valid (in general) but only satisfiable.)

An important difference from PL concerns the relation between $M \not\models A$ and $M \models \neg A$. In PL, we had that for any structure V and formula A

$$V \not\models A \Rightarrow V \models \neg A \qquad (9.18)$$

simply because any V induced unique boolean value for all formulae. The corresponding implication does not, in general, hold in FOL:

$$M \not\models A \nRightarrow M \models \neg A. \qquad (9.19)$$

In fact, we may have:

$$M \not\models A \text{ and } M \not\models \neg A. \tag{9.20}$$

(Of course, we will never get $M \models A$ and $M \models \neg A$.) To see (9.20), consider a formula $A = P(x)$ and a structure M with two elements $\{0, 1\}$ and $[\![P]\!]^M = \{1\}$. Then $M \not\models A \Leftrightarrow M \not\models_v A$ for some $v \Leftrightarrow [\![A]\!]^M_v = \mathbf{0}$ for some v, and this is indeed the case for $v(x) = 0$ since $0 \notin [\![P]\!]^M$. On the other hand, we also have $M \not\models \neg A$ since $[\![\neg A]\!]^M_w = \mathbf{0}$ for $w(x) = 1$.

It is the presence of free variables which causes the implication in (9.19) to fail because then the interpretation of a formula is not uniquely determined unless one specifies an assignment to the free variables. Given an assignment, we do have

$$M \not\models_v A \Rightarrow M \models_v \neg A. \tag{9.21}$$

Consequently, the implication (9.19) holds for sentences.   □

Summarizing this remark, we can set up the following table which captures some of the essential difference between free and bound variables in terms of relations between negation of satisfaction of a formula, $M \not\models F$, and satisfaction of its negation, $M \models \neg F$ – M is an arbitrary structure:

$$
\begin{array}{c|c}
F \text{ closed (or PL)} & F \text{ open} \\
\hline
M \not\models F \Leftrightarrow M \models \neg F & M \not\models F \not\Rightarrow M \models \neg F \\
 & M \not\models F \Leftarrow M \models \neg F
\end{array}
\tag{9.22}
$$

For closed formulae, we can say that "negation commutes with satisfaction" (or else "satisfaction of negation *is the same as* negation of satisfaction"), while this is not the case for open formulae.

The above remark leads to another important difference between FOL and PL, which we mentioned earlier. The respective Deduction Theorems 4.14 and 8.29 differ in that the latter has the additional restriction of closedness. The reason for this is the semantic complication introduced by the free variables. For PL we defined two notions $B \Rightarrow A$ and $B \models A$ which, as a matter of fact, coincided. We had there the following picture:

$$\forall V : V \models B \to A \overset{\text{D.5.11}}{\Longleftrightarrow} \models B \to A \qquad B \models A \overset{\text{D.6.12}}{\Longleftrightarrow} \forall V : \text{ if } V \models B$$
$$\quad \Updownarrow \text{ C.6.27} \qquad\qquad \text{C.6.27 } \Updownarrow \qquad\qquad \text{then } V \models A$$
$$\vdash_{\mathcal{N}} B \to A \overset{\text{C.4.17}}{\Longleftrightarrow} B \vdash_{\mathcal{N}} A$$

The vertical equivalences follow from the soundness and completeness theorems, while the horizontal one follows from Deduction Theorem 4.14. This chain of equivalences is a roundabout way to verify that for PL : $B \models A$ iff $B \Rightarrow A$. This picture isn't that simple in FOL. We preserve the two Definitions 5.11 and 6.12:

Definition 9.23 For arbitrary FOL-formulae A, B:

- $B \models A$ iff for every structure M : if $M \models B$ then $M \models A$.
- A is a logical consequence of B, written $B \Rightarrow A$, iff $\models B \rightarrow A$.

However, $M \models B$ means now that $M \models_v B$ holds *for all assignments v*. Thus the definitions read:

$B \models A$	$B \Rightarrow A$
for all M :	for all M :
if (for all $v : M \models_v B$) then (for all $u : M \models_u A$)	for all $v : M \models_v B \rightarrow A$

It is not obvious that the two are equivalent – in fact, they are not, if there are free variables involved (Exercise 9.14). If there are no free variables then, by Lemma 9.7, each formula has a fixed boolean value in any structure (and we can remove the quantification over v's and u's from the above table). The difference is expressed in the restriction of Deduction Theorem 8.29 requiring the formula B to be closed. (Assuming soundness and completeness, Chapter 11, this restriction leads to the same equivalences as above for PL.) We write '$A \Leftrightarrow B$' for '$A \Rightarrow B$ and $B \Rightarrow A$' which, by the above remarks (and Exercise 9.14) is *not* the same as '$A \models B$ and $B \models A$'. $A \Leftrightarrow B$ is read as "A and B are logically equivalent".

Fact 9.24 For any formula $A : \neg \forall x A \Leftrightarrow \exists x \neg A$.

Proof. By Definition 8.7, $\neg \forall x A = \neg \neg \exists x \neg A$, which is equivalent to the right-hand side (Example 5.12).   QED (9.24)

Fact 9.25 $\forall x A \Rightarrow A$.

Proof. We have to show $\models \forall x A \rightarrow A$, that is, for an arbitrary structure M and assignment $v : M \models_v \forall x A \rightarrow A$. Let M, v be arbitrary. If $[\![\forall x A]\!]_v^M = \mathbf{0}$ then we are done, so assume $[\![\forall x A]\!]_v^M = \mathbf{1}$.

$$
\begin{aligned}
[\![\forall x A]\!]_v^M = \mathbf{1} &\Leftrightarrow [\![\neg \exists x \neg A]\!]_v^M = \mathbf{1} && \text{Def. 8.7} \\
&\Leftrightarrow [\![\exists x \neg A]\!]_v^M = \mathbf{0} && \text{Def. 9.4.(2)} \\
&\Leftrightarrow \text{for no } \underline{a} \in \underline{M} : [\![\neg A]\!]_{v[x \mapsto \underline{a}]}^M = \mathbf{1} && \text{Def. 9.4.(3)} \\
&\Leftrightarrow \text{for all } \underline{a} \in \underline{M} : [\![\neg A]\!]_{v[x \mapsto \underline{a}]}^M = \mathbf{0} && \text{same as above} \\
&\Rightarrow [\![\neg A]\!]_{v[x \mapsto v(x)]}^M = \mathbf{0} && \text{special case } \underline{a} = v(x) \\
&\Leftrightarrow [\![\neg A]\!]_v^M = \mathbf{0} && \text{since } v = v[x \mapsto v(x)] \\
&\Leftrightarrow [\![A]\!]_v^M = \mathbf{1} && \text{Def. 9.4.(2)}
\end{aligned}
$$

Since M and v were arbitrary, the claim follows.   QED (9.25)

For a sentence A we have, obviously, equivalence of A and $\forall x A$, since the latter does not modify the meaning of the formula. For open formulae, however, the implication opposite to this from Fact 9.25 does not hold.

Fact 9.26   There is an A with a free variable $x : A \not\Rightarrow \forall x A$.

> **Proof.** Let A be $P(x)$, where P is a predicate symbol. To see that $\not\models$
> $P(x) \rightarrow \forall x P(x)$, consider a structure M with $\underline{M} = \{p, q\}, \llbracket P \rrbracket^M = \{p\}$
> and an assignment $v(x) = p$. We have that $M \not\models_v P(x) \rightarrow \forall x P(x)$,
> since $\llbracket P(x) \rrbracket_v^M = \mathbf{1}$ while $\llbracket \forall x P(x) \rrbracket_v^M = \llbracket \forall x P(x) \rrbracket^M = \mathbf{0}$.   QED (9.26)

In spite of this fact, we have another close relation between satisfaction of A and $\forall x A$. Given a (not necessarily closed) formula A, the *universal closure* of A, written $\forall(A)$, is the formula $\forall x_1 \ldots \forall x_n A$, where $\{x_1, \ldots, x_n\} = \mathcal{V}(A)$. The following fact shows that satisfaction of a – possibly open! – formula in a given structure is, in fact, equivalent to the satisfaction of its universal closure.

Fact 9.27   For any structure M and formula A, the following are equivalent:

$$(1)\quad M \models A \qquad\qquad (2)\quad M \models \forall(A)$$

Proof. Strictly speaking, we should proceed by induction on the number of free variables in A but this does not change anything essential in the proof. We treat them uniformly and mark as $\overline{x}$.

$$M \models A \overset{9.10}{\Longleftrightarrow} \text{ for all } v : M \models_v A$$
$$\Longleftrightarrow \text{ for all } v, \text{ for all } \overline{m} \in \underline{M} : M \models_{v[\overline{x} \mapsto \overline{m}]} A$$
$$\overset{9.5}{\Longleftrightarrow} \text{ for all } v : M \models_v \forall \overline{x} A$$
$$\overset{9.10}{\Longleftrightarrow} M \models \forall \overline{x} A \qquad\qquad\qquad \text{QED (9.27)}$$

9.3.1   DEDUCTION THEOREM IN $\mathcal{G}$ AND $\mathcal{N}$

Observe that Gentzen's rules $(\vdash\vee)$ and $(\wedge\vdash)$, Section 8.4, indicate the semantics of sequents. $A_1 \ldots A_n \vdash_{\mathcal{G}} B_1 \ldots B_m$ corresponds by these rules to $A_1 \wedge \ldots \wedge A_n \vdash_{\mathcal{G}} B_1 \vee \ldots \vee B_m$, and by rule $(\vdash\rightarrow)$ to $\vdash_{\mathcal{G}} (A_1 \wedge \ldots \wedge A_n) \rightarrow (B_1 \vee \ldots \vee B_m)$ which is a simple formula with the expected semantics corresponding to the semantics of the original sequent.

   Now $\mathcal{G}$ for FOL, unlike $\mathcal{N}$, is a *truly* natural deduction system. Rule $(\vdash\rightarrow)$ is the unrestricted Deduction Theorem built into $\mathcal{G}$. Recall that it was not so for $\mathcal{N}$ – Deduction Theorem 8.29 allowed us to use a restricted version of the rule: $\dfrac{\Gamma, A \vdash_{\mathcal{N}} B}{\Gamma \vdash_{\mathcal{N}} A \to B}$ only if A is closed! Without this restriction, the rule would be unsound, e.g.:

1. $A \vdash_{\mathcal{N}} A$ $A0$
2. $A \vdash_{\mathcal{N}} \forall x A$ $L.8.24.(4)$
3. $\vdash_{\mathcal{N}} A \to \forall x A$ $DT!$
4. $\vdash_{\mathcal{N}} \exists x A \to \forall x A$ $\exists I$

The conclusion of this proof is obviously invalid (verify this) and we could derive it only using a wrong application of DT in line 3.

In $\mathcal{G}$, such a proof cannot proceed beyond step 1. Rule $(\vdash \forall)$ requires replacement of x from $\forall x A$ by a *fresh* x', i.e., not occurring (freely) in the sequent. Attempting this proof in $\mathcal{G}$ would lead to the following:

4. $A(x') \vdash_{\mathcal{G}} A(x)$
3. $A(x') \vdash_{\mathcal{G}} \forall x A(x)$ $(\vdash \forall),\ x$ fresh $(x \neq x')$
2. $\exists x A(x) \vdash_{\mathcal{G}} \forall x A(x)$ $(\exists \vdash),\ x'$ fresh
1. $\vdash_{\mathcal{G}} \exists x A(x) \to \forall x A(x)$ $(\vdash \to)$

But $A(x) \neq A(x')$ so line 4. is not an axiom. (If x does not occur in A, i.e., the quantification $\forall x A$ is somehow redundant, then this would be an axiom and everything would be fine.)

It is this, different than in $\mathcal{N}$, treatment of variables (built into the different quantifier rules) which enables $\mathcal{G}$ to use unrestricted Deduction Theorem. It is reflected at the semantic level in that the semantics of $\vdash_{\mathcal{G}}$ is different from $\vdash_{\mathcal{N}}$. According to Definition 9.23, $A \models B$ iff $\models \forall(A) \to \forall(B)$ and this is reflected in $\mathcal{N}$, e.g., in the fact that from $A \vdash_{\mathcal{N}} B$ we can deduce that $\forall(A) \vdash_{\mathcal{N}} \forall(B)$ from which $\vdash_{\mathcal{N}} \forall(A) \to \forall(B)$ follows now by Deduction Theorem 8.29.

The semantics of $A \vdash_{\mathcal{G}} B$ is different – such a sequent is interpreted as $A \Rightarrow B$, that is, $\models \forall(A \to B)$. The free variables occurring in both A and B are now interpreted in the same way across the sign $\vdash_{\mathcal{G}}$. Using Definition 9.23, one translates easily between sequents and formulae of $\mathcal{N}$. The first equivalence gives the general form from which the rest follows. The last line expresses perhaps most directly the difference in the treatment of free variables indicated above. (M is an arbitrary structure.)

$$M \models (A_1, \ldots, A_n \vdash_{\mathcal{G}} B_1, \ldots, B_m) \iff$$
$$M \models \forall((A_1 \wedge \ldots \wedge A_n) \to (B_1 \vee \ldots \vee B_m))$$

$$M \models (\varnothing \vdash_{\mathcal{G}} B_1, \ldots, B_m) \iff M \models \forall(B_1 \vee \ldots \vee B_m)$$
$$M \models (A_1, \ldots, A_n \vdash_{\mathcal{G}} \varnothing) \iff M \models \forall(\neg(A_1 \wedge \ldots \wedge A_n))$$
$$\models (\forall(A_1), \ldots, \forall(A_n) \vdash_{\mathcal{G}} B) \iff \{A_1 \ldots A_n\} \models B.$$

Exercises 9.

EXERCISE 9.1 Consider the formula $F = \forall x \exists y (R(x, y) \wedge \neg R(y, x))$.
(1) Specify a strucure M with $\underline{M} = \{a, b, c\}$ such that $M \models F$.
(2) Can R in (1) be interpreted as reflexive relation? Explain.
(3) Specify a structure N with $\underline{N} = \{a, b, c\}$ such that $N \not\models F$.

EXERCISE 9.2 Let P, Q be binary relation symbols and N be a structure with $\underline{N} = \mathbb{N}$, $[\![P]\!]^N = \{\langle a, b \rangle \mid a + b \geq 5\}$ and $[\![Q]\!]^N = \{\langle a, b \rangle \mid b = a + 2\}$. Answer the following questions for each of the following formulae:
$F_1 = \forall x P(x, y) \rightarrow \exists x Q(x, y)$ and
$F_2 = \forall x P(x, y) \rightarrow \forall x Q(x, y)$.
(1) Is it the case that $N \models_v F_i$, for the valuation $v(y) = 6$?
(2) Is it the case that $N \models_v F_i$, for the valuation $v(y) = 1$?
(3) Is it the case that $N \models F_i$?

EXERCISE 9.3 Translate each of the following sentences into a FOL language (choose the needed relation symbols yourself):

(1) *Everybody loves somebody.*
(2) *If everybody is loved (by somebody) then somebody loves everybody.*
(3) *If everybody loves somebody and John does not love anybody then John is nobody.*

At least two formulae are not valid. Which ones? Is the third one valid?

EXERCISE 9.4 Using the previous exercise, construct a structure where the implication $\forall y \exists x A \rightarrow \exists x \forall y A$, opposite to the one from Exercise 8.3.(3), does not hold.

EXERCISE 9.5 Design a FOL language for expressing the following two statements. Make it simplest possible, e.g., intending an interpretation among 'all men', there is no need for the predicate 'is a man'.

(1) *There is a man who shaves all men who do not shave themselves.*
(2) *There is no man who shaves all and only men who do not shave themselves.*

(1.a) Write statement (1) as a FOL sentence in your language.
(1.b) Is it satisfiable? Prove your answer.
(1.c) Is it valid? If the answer is 'No', give a counter-example, while if it is 'Yes', prove (1) using Gentzen's system.

(2.a) Write statement (2) as a FOL sentence in your language.
(2.b) Is it valid or not? If the answer is 'No', give a counter-example, while if it is 'Yes', prove (2) using Gentzen's system.

EXERCISE 9.6 Write the axioms for an equivalence relation (reflexivity, symmetry and transitivity) as three FOL sentences and let Θ denote the resulting set. Show that the axioms are independent, i.e., for each $T \in \Theta$ show that $\Theta \setminus \{T\} \not\models T$, by providing a model of $\Theta \setminus \{T\}$ which is not a model of T.

EXERCISE 9.7 Give an example contradicting the opposite implication to the one from Lemma 8.24.(1), i.e., show that $\not\models A \to \forall x A$.

EXERCISE 9.8 Show that
(1) $\exists x(A(x) \to B(x)) \Leftrightarrow (\forall x A(x) \to \exists x B(x))$.
(2) $\forall x(A(x) \to B(x)) \not\Leftrightarrow (\exists x A(x) \to \forall x B(x))$.

EXERCISE 9.9 Consider the two formulae:

$$F = \exists x(B(x) \wedge C(y)) \quad \text{and} \quad G = \exists y(B(y) \wedge C(y)).$$

For each of the following questions, either give a proof for the positive answer or a counter-example for the negative one. Does it hold that
(1) $G \Rightarrow F$?   (2) $F \Rightarrow G$?

EXERCISE 9.10 Verify the following claims:

(1) $\exists x A \Leftrightarrow \exists y A_y^x$, when y does not occur in A.
(2) $\models \forall x A \leftrightarrow \forall y A_y^x$, when y does not occur in A.
   ($\models A \leftrightarrow B$ denotes $\models A \to B$ and $\models B \to A$.)
(3) For both points above, give an example showing that the equivalence may fail when y occurs in A (even when the substitution A_y^x is legal).
(4) $A_t^x \Rightarrow \exists x A$, when t is substitutable for x in A.
   Show that the assumption about substitutability is necessary, i.e., give an example of a non-valid formula of the form $A_t^x \to \exists x A$, when t is *not* substitutable for x in A.

EXERCISE 9.11 Let Γ be arbitrary theory and A, B arbitrary formulae:

(1) Use Fact 9.27 to show the equivalence: $\Gamma, A \models B \Leftrightarrow \Gamma, \forall(A) \models B$.
(2) Show soundness of the following version of Deduction Theorem: if $\Gamma, A \models B$ then $\Gamma \models \forall(A) \to B$.

EXERCISE 9.12 Show that $M \models \neg A$ implies $M \not\models A$. Give a counter-example to the opposite implication, i.e. an M and A over appropriate alphabet such that $M \not\models A$ *and* $M \not\models \neg A$.   (Hint: A must have free variables.)

EXERCISE 9.13 Show that $(\forall x A \vee \forall x B) \Rightarrow \forall x(A \vee B)$. Give a counter-example demonstrating that the opposite implication (which you hopefully did not manage to prove in Exercise 8.4) does not hold.

EXERCISE 9.14 Recall Remark 9.17 and discussion after Definition 9.23 (as well as Exercise 8.12).

(1) Show that $\forall(A \to B) \Rightarrow (\forall(A) \to \forall(B))$.
   – Use this to show that $A \Rightarrow B$ implies $A \models B$.
(2) Give an argument (an example of A, B and structure) falsifying the opposite implication, i.e., showing $\not\models (\forall(A) \to \forall(B)) \to \forall(A \to B)$.
   – Use this to show that $A \models B$ does not imply $A \Rightarrow B$.
(3) Show that the unrestricted Deduction Theorem, $\dfrac{\Gamma, A \vdash_{\mathcal{N}} B}{\Gamma \vdash_{\mathcal{N}} A \to B}$, is not sound: with $\Gamma = \varnothing$, provide instances of A and B such that $A \models B$ but $\not\models A \to B$.

EXERCISE 9.15 Lemma 8.24.(5) showed admissibility of the substitution rule in $\mathcal{N}$, SB: $\dfrac{\Gamma \vdash_{\mathcal{N}} A}{\Gamma \vdash_{\mathcal{N}} A_t^x}$ if t is substitutable for x in A.

(1) Show now that this rule is sound, i.e., for any Γ and FOL-structure $M \models \Gamma$: if $M \models A$ then also $M \models A_t^x$ when the substitution is legal.
(2) Show that the requirement of legality of A_t^x is necessary: give an example of Γ and a formula A with $\Gamma \models A$, where substitution A_t^x is illegal, and a structure $M \models \Gamma$ which satisfies A but not A_t^x.

EXERCISE 9.16 Show that the rule of $\exists$-introduction is not sound, i.e., that $\Gamma \models A \to B$ does not necessarily imply $\Gamma \models (\exists x A) \to B$, when we drop the side-condition requiring $x \notin \mathcal{V}(B)$.

EXERCISE 9.17 Using the semantics of $\vdash_{\mathcal{G}}$ from Subsection 9.3.1, show that the four quantifier rules of Gentzen's system are sound and invertible.

(Hint: Show first a stronger claim, namely, that for each of these rules $\dfrac{\Gamma_i \vdash_{\mathcal{G}} \Delta_i}{\Gamma \vdash_{\mathcal{G}} \Delta}$ and for arbitrary structure M, it holds that

$$M \models \bigwedge \Gamma \to \bigvee \Delta \iff M \models \bigwedge \Gamma_i \to \bigvee \Delta_i.$$

Explain how this entails the desired conclusion.)

Chapter 10

MORE SEMANTICS

10.1 PRENEX NORMAL FORM

We have seen in Corollaries 6.7 and 6.8 that every PL formula can be written equivalently in DNF and CNF. A normal form which is particularly useful in the study of FOL is Prenex Normal Form.

Definition 10.1 [PNF] A formula A is in Prenex Normal Form iff it has the form $Q_1x_1 \ldots Q_nx_nB$, where Q_i are quantifiers and B contains no quantifiers.

The quantifier part $Q_1x_1 \ldots Q_nx_n$ is called the *prefix*, and the quantifier-free part B the *matrix* of A.

To show that each formula is equivalent to some formula in PNF we need the next lemma.

Lemma 10.2 Let A, B be formulae, $F[A]$ be a formula with some occurrence(s) of A, and $F[B]$ be the same formula with the occurrence(s) of A replaced by B. If $A \Leftrightarrow B$ then $F[A] \Leftrightarrow F[B]$.

Proof. Exercise 5.15 showed the version for PL. The proof is by induction on the complexity of $F[A]$, with a special case considered first:

$\underline{F[A] \text{ IS} :}$

    A :: This is a special case in which we have trivially $F[A] = A \Leftrightarrow B = F[B]$.

        In the rest of the proof, assume that we are not in the special case.

  ATOMIC :: If $F[A]$ is atomic then either we have the special case, or no replacement is made, i.e., $F[A] = F[B]$, since F has no subformula A.

  $\neg C[A]$ :: By IH $C[A] \Leftrightarrow C[B]$. So $\neg C[A] \Leftrightarrow \neg C[B]$.

$C[A] \to D[A]$ :: Again, IH gives $C[A] \Leftrightarrow C[B]$ and $D[A] \Leftrightarrow D[B]$, from which the conclusion follows easily (as in Exercise 5.15).

$\exists x C[A]$ :: By IH, $C[A] \Leftrightarrow C[B]$ which means that for all assignments to the free variables, including x, the two have the same value. Hence $\exists x C[A] \Leftrightarrow \exists x C[B]$.   QED (10.2)

The following lemma lists transformations of formulae allowing to construct PNF by purely syntactic manipulation. It also ensures that the result of such transformations is logically equivalent to the original formula.

Lemma 10.3 The *prenex operations* are given by the following equivalences:
(1) Quantifier movement along $\to$:

$$\text{if } x \text{ not free in } A: \quad A \to \forall x B \;\Leftrightarrow\; \forall x (A \to B)$$
$$\text{and} \quad A \to \exists x B \;\Leftrightarrow\; \exists x (A \to B)$$
$$\text{if } x \text{ not free in } B: \quad \forall x A \to B \;\Leftrightarrow\; \exists x (A \to B)$$
$$\text{and} \quad \exists x A \to B \;\Leftrightarrow\; \forall x (A \to B).$$

(2) Quantifier movement along $\neg$:

$$\neg \exists x A \;\Leftrightarrow\; \forall x \neg A$$
$$\neg \forall x A \;\Leftrightarrow\; \exists x \neg A.$$

(3) Renaming of bound variables:

$$\text{when } y \text{ does not occur in } A: \quad \exists x A \;\Leftrightarrow\; \exists y A_y^x$$
$$\text{and} \quad \forall x A \;\Leftrightarrow\; \forall y A_y^x.$$

Proof. (3) was proved in Exercise 9.10. We show the first and the third equivalences from (1); the rest is left for Exercise 10.3.

Let M be an arbitrary structure and v an arbitrary assignment to the free variables occurring in A and $\forall x B$. We have

$M \models_v A \to \forall x B$
1. $\Leftrightarrow$ $[\![A \to \forall x B]\!]_v^M = 1$
2. $\Leftrightarrow$ $[\![A]\!]_v^M = \mathbf{0}$ or $[\![\forall x B]\!]_v^M = \mathbf{1}$
3. $\Leftrightarrow$ $[\![A]\!]_v^M = \mathbf{0}$ or for all $\underline{a} \in \underline{M} : [\![B]\!]_{v[x \mapsto \underline{a}]}^M = \mathbf{1}$
4. $\Leftrightarrow$ for all $\underline{a} \in \underline{M}$ ($[\![A]\!]_v^M = \mathbf{0}$ or $[\![B]\!]_{v[x \mapsto \underline{a}]}^M = \mathbf{1}$)
5. $\Leftrightarrow$ for all $\underline{a} \in \underline{M}$ ($[\![A]\!]_{v[x \mapsto \underline{a}]}^M = \mathbf{0}$ or $[\![B]\!]_{v[x \mapsto \underline{a}]}^M = \mathbf{1}$)
6. $\Leftrightarrow$ $[\![\forall x (A \to B)]\!]_v^M = \mathbf{1}$
7. $\Leftrightarrow$ $M \models_v \forall x (A \to B)$

The equivalence between lines 4 and 5 follows from Lemma 9.7 because x is not free in A. Since M and v were arbitrary, we can conclude that, when x is not free in A then $(A \to \forall x B) \Rightarrow \forall x (A \to B)$.

The other equivalence of (1) is established as follows:

$M \models_v \forall x A \to B$

1. $\Leftrightarrow$ $[\![\forall x A \to B]\!]_v^M = 1$

2. $\Leftrightarrow$ $[\![\forall x A]\!]_v^M = 0$ or $[\![B]\!]_v^M = 1$

3. $\Leftrightarrow$ $[\![\neg \exists x \neg A]\!]_v^M = 0$ or $[\![B]\!]_v^M = 1$

4. $\Leftrightarrow$ $[\![\exists x \neg A]\!]_v^M = 1$ or $[\![B]\!]_v^M = 1$

5. $\Leftrightarrow$ (for some $\underline{a} \in \underline{M}$ $[\![\neg A]\!]_{v[x \mapsto \underline{a}]}^M = 1$) or $[\![B]\!]_v^M = 1$

6. $\Leftrightarrow$ (for some $\underline{a} \in \underline{M}$ $[\![A]\!]_{v[x \mapsto \underline{a}]}^M = 0$) or $[\![B]\!]_v^M = 1$

7. $\Leftrightarrow$ for some $\underline{a} \in \underline{M}$ ($[\![A]\!]_{v[x \mapsto \underline{a}]}^M = 0$ or $[\![B]\!]_{v[x \mapsto \underline{a}]}^M = 1$)

8. $\Leftrightarrow$ for some $\underline{a} \in \underline{M}$ $[\![A \to B]\!]_{v[x \mapsto \underline{a}]}^M = 1$

9. $\Leftrightarrow$ $[\![\exists x (A \to B)]\!]_v^M = 1$

10. $\Leftrightarrow$ $M \models_v \exists x (A \to B)$

Again, the crucial equivalence of lines 6 and 7 follows from Lemma 9.7 because x is not free in B.   QED (10.3)

Remark.

Notice the change of quantifier in the last two equivalences in point (1) of Lemma 10.3. The last one, $\exists x A \to B \Leftrightarrow \forall x (A \to B)$, assuming that $x \notin \mathcal{V}(B)$, can be illustrated as follows. Let $R(x)$ stand for 'x raises his voice' and T for 'there will be trouble' (which has no free variables). The sentence "If somebody raises his voice there will be trouble" can be represented as

$$\exists x R(x) \to T. \qquad (10.4)$$

The intention here is to say that *no matter who* raises his voice, the trouble will ensue. Intuitively, it is equivalent to say: "If *anyone* raises his voice, there will be trouble." and this sentence can be easier seen to correspond to $\forall x (R(x) \to T)$. The third equivalence from point (1) is not so natural. We would like to read $\forall x R(x) \to T$ as "If all raise their voices, there will be trouble." But it is equivalent to

$$\exists x (R(x) \to T), \qquad (10.5)$$

which might be saying something like "There exists somebody (a specific x) such that if x raises his voice, there will be trouble." Intuitively, the two say very different things. Often, one tends to ignore the different scope of the quantifier in (10.5) and in (10.4) and read both the same way. This is, again, only a remainder that one has to be careful with formalizing natural language expressions and should always keep in mind the important differences arising from the different scopes of quantifiers. $\qquad$ $\square$

Theorem 10.6 Every formula B is equivalent to a formula B_P in PNF.

Proof. By induction on the complexity of B. The IH gives us a PNF for the subformulae and Lemma 10.2 allows us to replace these subformulae by their PNF. The equivalences from Lemma 10.3 are applied from left to right.

ATOMIC :: Having no quantifiers, B is obviously in PNF.

$\neg A$ :: By IH, A has a PNF, and by Lemma 10.2, $B \Leftrightarrow \neg A_P$. Using (2) of Lemma 10.3, we can move $\neg$ inside changing all the quantifiers. The result will be B_P.

$\exists x A$ :: Replacing A with A_P gives a PNF $B_P = \exists x A_P$.

$A \rightarrow C$ :: By IH and Lemma 10.2, this is equivalent to $A_P \rightarrow C_P$. First, use (3) of Lemma 10.3 to rename all bound variables in A_P so that they are distinct from all the variables (bound or free) in C_P. Then do the same with C_P. Use Lemma 10.3.(1) to move the quantifiers outside the whole implication. (Because of the renaming, no bound variable will at any stage occur freely in the other formula.) The result is B_P. $\qquad$ QED (10.6)

Example 10.7

We obtain PNF using the prenex operations from Lemma 10.3:

$$\forall x \exists y A(x, y) \rightarrow \neg \exists x B(x) \quad \Leftrightarrow \quad \exists x (\exists y A(x, y) \rightarrow \neg \exists x B(x)) \quad (1)$$
$$\Leftrightarrow \quad \exists x \forall y (A(x, y) \rightarrow \neg \exists x B(x)) \quad (1)$$
$$\Leftrightarrow \quad \exists x \forall y (A(x, y) \rightarrow \forall x \neg B(x)) \quad (2)$$
$$\Leftrightarrow \quad \exists x \forall y (A(x, y) \rightarrow \forall z \neg B(z)) \quad (3)$$
$$\Leftrightarrow \quad \exists x \forall y \forall z (A(x, y) \rightarrow \neg B(z)) \quad (1)$$

Formulae with abbreviated connectives may be first rewritten to the form with $\neg$ and $\rightarrow$ only, before applying the prenex transformations:

$$\exists x A(x, y) \vee \forall y B(y) \quad \Leftrightarrow \quad \neg \exists x A(x, y) \rightarrow \forall y B(y)$$
$$\Leftrightarrow \quad \forall x \neg A(x, y) \rightarrow \forall y B(y) \quad (2)$$
$$\Leftrightarrow \quad \exists x (\neg A(x, y) \rightarrow \forall y B(y)) \quad (1)$$
$$\Leftrightarrow \quad \exists x (\neg A(x, y) \rightarrow \forall z B(z)) \quad (3)$$
$$\Leftrightarrow \quad \exists x \forall z (\neg A(x, y) \rightarrow B(z)) \quad (1)$$
$$\Leftrightarrow \quad \exists x \forall z (A(x, y) \vee B(z))$$

Alternatively, we may use prenex operations derivable from Lemma 10.3:

$$(\mathsf{Q}xA \vee B) \Leftrightarrow \mathsf{Q}x(A \vee B) \qquad \text{provided } x \notin \mathcal{V}(B)$$
$$(\mathsf{Q}xA \wedge B) \Leftrightarrow \mathsf{Q}x(A \wedge B) \qquad \text{provided } x \notin \mathcal{V}(B) \qquad \square$$

Notice that PNF is not unique, since the order in which we apply the prenex operations may be chosen arbitrarily.

Example 10.8
Let B be $(\forall x\ x > 0) \rightarrow (\exists y\ y = 1)$. We can apply the prenex operations to B in two ways:

$$(\forall x\ x > 0) \rightarrow (\exists y\ y = 1)$$
$$\Leftrightarrow$$

$\exists x\ (x > 0 \rightarrow (\exists y\ y = 1))$ $\Leftrightarrow \exists x\ \exists y\ (x > 0 \rightarrow y = 1)$	$\exists y\ ((\forall x\ x > 0) \rightarrow y = 1) \Leftrightarrow$ $\exists y\ \exists x\ (x > 0 \rightarrow y = 1)$

Obviously, since the order of the quantifiers of the same kind does not matter (Exercise 8.3.(1)), the two resulting formulae are equivalent. However, the quantifiers may also be of different kinds:

$$(\exists x\ x > 0) \rightarrow (\exists y\ y = 1)$$
$$\Leftrightarrow$$

$\forall x\ (x > 0 \rightarrow (\exists y\ y = 1))$ $\Leftrightarrow \forall x\ \exists y\ (x > 0 \rightarrow y = 1)$	$\exists y\ ((\exists x\ x > 0) \rightarrow y = 1) \Leftrightarrow$ $\exists y\ \forall x\ (x > 0 \rightarrow y = 1)$

Although it is not true *in general* that $\forall x \exists y A \Leftrightarrow \exists y \forall x A$, so prenex operations – preserving equivalence, due to renaming of bound variables and avoidance of name clashes with variables in other subformulae – ensure that the results (like the two formulae above) are equivalent. $\square$

10.1.1 PNF HIERARCHY

The existence of PNF allows us to "measure the complexity" of formulae. Comparing the prefixes, we would say that $A_1 = \exists x \forall y \exists z B$ is "more complex" than $A_2 = \exists x \exists y \exists z B$. Roughly, a PNF formula is the more complex, the more changes of quantifiers in its prefix.

Definition 10.9 A formula A is Δ_0 iff it has no quantifiers. It is:

- Σ_1 iff $A = \exists x_1 \dots \exists x_n B$, where B is Δ_0.
- Π_1 iff $A = \forall x_1 \dots \forall x_n B$, where B is Δ_0.
- Σ_{i+1} iff $A = \exists x_1 \dots \exists x_n B$, where B is Π_i.
- Π_{i+1} iff $A = \forall x_1 \dots \forall x_n B$, where B is Σ_i.

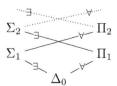

A formula has PNF index Σ_i/Π_i if its PNF is Σ_i/Π_i. Since PNF is not unique, neither is formula's index and, typically, saying that a formula has index Σ_i/Π_i, one means that this is the least such i.

Also, a formula may have both index Π_i and Σ_i, in which case it is said to have index Δ_i. In Example 10.8, the second formula is equivalent to both $\forall x \exists y B$ and to $\exists y \forall x B$, having thus indices Π_2 and Σ_2, i.e., Δ_2.

10.2 SUBSTRUCTURES: AN EXAMPLE FROM MODEL THEORY

Roughly, model theory studies the properties of model classes. A model class is not just an arbitrary collection K of FOL-structures – it is a collection of *models of some set* Γ of formulae, i.e., such that $\mathsf{K} = Mod(\Gamma)$. The syntactic form of the formulae in Γ may have a heavy influence on the properties of its model class. On the other hand, knowing some properties of a given class of structures, model theory may sometimes tell what syntactic forms of axioms are necessary/sufficient for axiomatizing this class. In general, there exist non-axiomatizable classes K, i.e., such that for no FOL-theory Γ, one can get $\mathsf{K} = Mod(\Gamma)$.

As an elementary example of the property of a class of structures we consider closure under substructures and superstructures.

Definition 10.10 Let M, N be Σ-structures for a FOL alphabet Σ: N is a *substructure* of M (M is a *superstructure/extension* of N), $N \sqsubseteq M$, iff:

- $\underline{N} \subseteq \underline{M}$
- For all $a \in \mathcal{I}$: $[\![a]\!]^N = [\![a]\!]^M$
- For all $f \in \mathcal{F}$, and $\underline{a}_1, \ldots, \underline{a}_n \in \underline{N}$:
  $[\![f]\!]^N(\underline{a}_1, \ldots, \underline{a}_n) = [\![f]\!]^M(\underline{a}_1, \ldots, \underline{a}_n) \in \underline{N}$
- For all $R \in \mathcal{R}$, and $\underline{a}_1, \ldots, \underline{a}_n \in \underline{N}$:
  $\langle \underline{a}_1, \ldots, \underline{a}_n \rangle \in [\![R]\!]^N \Leftrightarrow \langle \underline{a}_1, \ldots, \underline{a}_n \rangle \in [\![R]\!]^M$

We say that a class of structures K is closed under:

- substructures if whenever $M \in \mathsf{K}$ and $N \sqsubseteq M$, then also $N \in \mathsf{K}$, and
- superstructures if whenever $N \in \mathsf{K}$ and $N \sqsubseteq M$, then also $M \in \mathsf{K}$.

Thus $N \sqsubseteq M$ iff N has a more restricted interpretation domain than M, but all constants, function and relation symbols are interpreted identically within this restricted domain. Obviously, every structure is its own substructure, $M \sqsubseteq M$. If $N \sqsubseteq M$ and $N \neq M$, which means that $\underline{N}$ is a *proper* subset of $\underline{M}$, then we say that N is a proper substructure of M.

Example 10.11

Let Σ be given by $\mathcal{I} = \{c\}$ and $\mathcal{F}_2 = \{\odot\}$. The structure Z with $\underline{Z} = \mathbb{Z}$ being the integers, $[\![c]\!]^Z = 0$ and $[\![\odot]\!]^Z(x,y) = x + y$ is a Σ-structure. The structure N with $\underline{N} = \mathbb{N}$ being only the natural numbers with zero, $[\![c]\!]^N = 0$ and $[\![\odot]\!]^N(x,y) = x + y$ is obviously a substructure $N \sqsubseteq Z$.

Restricting furthermore the domain to the even numbers, i.e. taking P with $\underline{P}$ being the even numbers greater or equal zero, $[\![c]\!]^P = 0$ and $[\![\odot]\!]^P(x,y) = x + y$ yields again a substructure $P \sqsubseteq N$.

The class $\mathsf{K} = \{Z, N, P\}$ is not closed under substructures. One can easily find other Σ-substructures not belonging to K (for instance, all negative numbers with zero and addition is a substructure of Z).

Notice that, in general, to obtain a substructure it is *not* enough to select an arbitrary subset of the underlying set. If we restrict N to the set $\{0, 1, 2, 3\}$ it will not yield a substructure of N – because, for instance, $[\![\odot]\!]^N(1,3) = 4$ and this element is not in our set. Any structure, and hence a substructure in particular, must be "closed under all operations", i.e., applying any operation to elements of the (underlying set of the) structure must produce an element in the structure.

On the other hand, a subset of the underlying set may fail to be a substructure if the operations are interpreted in different way. Let M be like Z only that now we let $[\![\odot]\!]^M(x,y) = x - y$. Neither N nor P are substructures of M since, in general, for $x,y \in \underline{N}$ (or $\in \underline{P}$): $[\![\odot(x,y)]\!]^N = x + y \neq x - y = [\![\odot(x,y)]\!]^M$. Modifying N so that $[\![\odot]\!]^{N'}(x,y) = x - y$ does not yield a substructure of M either, because this does not define $[\![\odot]\!]^{N'}$ for $x > y$. No matter how we define this operation for such cases (for instance, to return 0), we won't obtain a substructure of M – the result will be different than in M. □

Remark 10.12

Given a FOL alphabet Σ, we may consider *all* Σ-structures, $\mathsf{Str}(\Sigma)$, which is obviously closed under Σ-substructures. With the substructure relation, $\langle \mathsf{Str}(\Sigma), \sqsubseteq \rangle$ forms a weak partial ordering (Definition 1.15), as the following properties of the relation $\sqsubseteq$ follow easily from Definition 10.10:

- $\sqsubseteq$ is obviously reflexive (any structure is its own substructure),
- transitive (substructure of a substructure of Y is a substructure of Y)
- antisymmetric (if X is a substructure of Y and Y of X then $X = Y$).

□

As a consequence of Theorem 10.6, any axiomatizable class K can be axiomatized by formulae in PNF. This fact has a model theoretic flavour, but model theory studies more specific relations between the classes of structures and the syntactic form of their axiomatizations. The following gives a simple example of a model theoretic result. Point (1) says that satisfiaction of an existential formula is preserved when passing to the superstructures – the model class of any theory containing only sentences with index Σ_1 is closed under superstructures.[1] Dually, (2) implies that model class of any theory containing only universal sentences is closed under substructures.

Theorem 10.13 Let A be a Σ_1 sentence and B a Π_1 sentence over some alphabet Σ, and $N \sqsubseteq M$ be two Σ-structures. If

(1) $N \models A$ then $M \models A$
(2) $M \models B$ then $N \models B$.

> **Proof.** (1) A is closed Σ_1, i.e., it is (equivalent to) $\exists x_1 \ldots \exists x_n A'$ where A' has no quantifiers nor variables other than $x_1, \ldots, x_n$. So, if $N \models A$ then there exist $\underline{a}_1, \ldots, \underline{a}_n \in \underline{N}$ such that $N \models_{x_1 \mapsto \underline{a}_1, \ldots, x_n \mapsto \underline{a}_n} A'$. Since $N \sqsubseteq M$, we have $\underline{N} \subseteq \underline{M}$ and the interpretation of all symbols is the same in M as in N. Hence $M \models_{x_1 \mapsto \underline{a}_1, \ldots, x_n \mapsto \underline{a}_n} A'$, i.e., $M \models A$.
> (2) This is a dual argument. Since $M \models \forall x_1 \ldots \forall x_n B'$, B' is true for all elements of $\underline{M}$ and $\underline{N} \subseteq \underline{M}$, so B' will be true for all element of this subset as well.   QED (10.13)

The theorem can be applied in at least two different ways which we illustrate in the following two examples. We consider only case (2), i.e., when the formulae of interest are Π_1 (universal).

Example 10.14 [**Constructing new structures for Π_1 axioms**]
Given a set of Π_1 axioms and any structure M satisfying them, the theorem implies that every substructure of M will also satisfy the axioms.

Let Σ contain only one binary relation symbol R. Recall Definition 1.15 – a strict partial ordering is axiomatized by two formulae

(1) $\forall x \forall y \forall z : R(x,y) \land R(y,z) \to R(x,z)$ – transitivity, and
(2) $\forall x : \neg R(x,x)$ – irreflexivity.

Let N be an arbitrary strict partial ordering, i.e., an arbitrary Σ-structure satisfying these axioms. For instance, let $N = \langle \mathbb{N}, < \rangle$ be the natural numbers with less-than relation. Since both axioms are Π_1, the theorem tells

---

[1]This is the easy part of Łoś-Tarski theorem, which states also that preservation in all superstructures implies that the formula has index Σ_1.

us that any substructure of N, i.e., any subset of $\underline{N}$ with the interpretation of R restricted as in Definition 10.10, will itself be a strict partial ordering. For instance, the subset of odd natural numbers, with the same less-than relation, is by the theorem a strict partial ordering. □

The following example illustrates another application of the theorem.

Example 10.15 [**Non-axiomatizability by Π_1 sentences**]
The theorem can be used to show that some class is not Π_1-axiomatizable.

Let Σ be as in the previous example. A (strict partial) ordering is *dense* if, in addition to the two axioms from the previous example, it also has the following property:

(3) $\forall x, y \; \exists z : R(x, y) \to R(x, z) \wedge R(z, y)$.

For instance, the closed interval of all real numbers $M = \langle [0, 1], < \rangle$ is a dense strict partial ordering. Now, remove all the numbers from the open interval $(0, 1)$ – this leaves us with just two elements $\{0, 1\}$ ordered $0 < 1$. This is a Σ substructure of M (Σ contains no ground terms, so any elements from the underlying set can be removed). But this is not a dense ordering! The class of dense orderings over Σ is not closed under substructures and so, by the theorem, is not axiomatizable using only formula with index Π_1. As shown by (3), Π_2 formulae suffice. □

10.3 "Syntactic" semantics

Model theory classifies primarily structures, and formulae are only possible means of doing that. As we have just seen, one can ask if a given class of structures can be axiomatized by, say, Π_1 formulae or else if it is closed under, say, superstructures. But one can also ask about the existence of finite models for a given theory, about the existence of infinite models, countable models etc. We now present one such special type of semantic structures, which are generated from the mere syntax of the theory. They are of particular importance both in logic (in proofs of completeness or consistency) and in computer science (establishing a direct connection between first order logic and mechanical computability). The rest of this chapter introduces the concept of such syntactically generated models and relates them to the Turing computable functions. The application in proving completeness will be shown in the following chapter.

10.3.1   REACHABLE AND TERM STRUCTURES

Definition 10.16   A Σ-structure T is *reachable* iff for each $\underline{a} \in \underline{T}$ there is a ground term $t \in \mathcal{GT}_\Sigma$ with $\underline{a} = [\![t]\!]^T$. A *reachable model* of a Σ-theory Γ is a Σ-reachable structure M such that $M \models \Gamma$.

Intuitively, a reachable structure for a Σ contains only elements which can be denoted (reached) by some ground term.

Example 10.17
Let Σ contain only three constants a, b, c. Define a Σ-structure M by:

- $\underline{M} = \mathbb{N}$, and
- $[\![a]\!]^M = 0$, $[\![b]\!]^M = 1$, $[\![c]\!]^M = 2$.

M contains a lot of "junk" elements (all natural numbers greater than 2) which are not required to be there in order for M to be a Σ-structure – M is not a *reachable* Σ-structure. Define N by

- $\underline{N} = \{0, 1, 2, 3, 4\}$, and
- $[\![a]\!]^N = 0$, $[\![b]\!]^N = 1$, $[\![c]\!]^N = 2$.

Obviously, $N \sqsubseteq M$. Still N contains unreachable elements 3, 4. Restricting $\underline{N}$ to $\{0, 1, 2\}$, we obtain a new substructure $T \sqsubseteq N$. T is the only reachable structure of the three. Yet another structure is given by:

- $\underline{S} = \{0, 1\}$, and
- $[\![a]\!]^S = 0$, $[\![b]\!]^S = 1$, $[\![c]\!]^S = 1$.

S is reachable too, but it is not a substructure of any previous one: although $\underline{S} \subset \underline{T}$, we have that $[\![c]\!]^S = 1 \neq 2 = [\![c]\!]^T$. $\square$

Fact 10.18   A reachable Σ-structure T has no proper substructure.

> **Proof.** The claim is that there is no Σ-structure M with $M \sqsubseteq T$ and $M \neq T$, i.e., such that $\underline{M} \subset \underline{T}$. Indeed, since each element $\underline{a} \in \underline{T}$ is the (unique) interpretation of some ground term $t \in \mathcal{GT}_\Sigma$, $\underline{a} = [\![t]\!]^T$, if we remove $\underline{a}$ from $\underline{T}$, t will have no interpretation in the resulting subset, which could coincide with its interpretation $[\![t]\!]^T$ in T.   QED (10.18)

The proposition shows also a particular property of reachable structures with respect to the partial ordering $\langle \mathsf{Str}(\Sigma), \sqsubseteq \rangle$ defined in Remark 10.12.

Corollary 10.19   A Σ-reachable structure is minimal in $\langle \mathsf{Str}(\Sigma), \sqsubseteq \rangle$.

The opposite, however, need not be the case. If Σ contains no ground terms, the minimal Σ-structures, if they exist, will not be reachable.

Example 10.20

Let Σ contain only one binary function symbol $\oplus$. The Σ-structure M with $\underline{M} = \{\bullet\}$ and $[\![\oplus]\!]^M(\bullet, \bullet) = \bullet$ has no proper Σ-substructure. If such a structure N existed, it would require $\underline{N} = \varnothing$, but this is forbidden by Definition 9.1 of Σ-structure, which requires the underlying set to be non-empty. However, M is not Σ-reachable, since $\mathcal{GT}_\Sigma = \varnothing$. □

Fact 10.21 If Σ has at least 1 constant symbol, then any Σ-structure M has a reachable substructure.

Proof. Since $\mathcal{GT}_\Sigma \neq \varnothing$, we can take only the part of $\underline{M}$ consisting of the interpretations of ground terms, keeping the interpretation of relation and function symbols for these elements intact.   QED (10.21)

By Corollary 10.19, such a reachable substructure (of any M) will be minimal element of the partial ordering $\langle \mathsf{Str}(\Sigma), \sqsubseteq \rangle$. One could feel tempted to conclude that this shows that this ordering is well-founded but this is not the case as the following example shows.

Example 10.22

Let Σ contain one constant symbol $\odot$ and let N be a Σ-structure with $\underline{N} = \mathbb{N} = \{0, 1, 2, \ldots\}$ and $[\![\odot]\!]^N = 0$. The structure N_0 with $\underline{N_0} = \{0\}$ is the reachable (and hence minimal) substructure of N. However, we may also form the following chain of substructures: let N_i be given by the underlying set $\underline{N_i} = \mathbb{N} \setminus \{1, 2, \ldots, i\}$ for $i > 0$, and $[\![\odot]\!]^{N_i} = 0 = [\![\odot]\!]^N$. We thus have that $N \sqsupseteq N_1 \sqsupseteq N_2 \sqsupseteq N_3 \sqsupseteq \ldots$, i.e., we obtain an infinite descending chain of substructures. Hence the relation $\sqsubseteq$ is not well-founded (even if we restrict it to the set of substructures of N). □

It follows from Fact 10.21 that for any Σ with at least one constant symbol there is a Σ-reachable structure. A special type of reachable structure is of particular interest, namely the *term structure*. In a term structure over Σ, the domain of interpretation is the set of ground terms over Σ, and moreover every term is interpreted as itself:

Definition 10.23 Let Σ be an alphabet with at least one constant. A *term structure* T_Σ over Σ has the following properties:

- The domain of interpretation is the set of all ground Σ-terms: $\underline{T_\Sigma} = \mathcal{GT}_\Sigma$.
- For each constant symbol $a \in \mathcal{I}$, we let a be its own interpretation:

$$[\![a]\!]^{T_\Sigma} = a.$$

- For each function symbol $f \in \mathcal{F}$ of arity n, and terms $t_1, \ldots, t_n$, we let

$$[\![f]\!]^{T_\Sigma}(t_1, \ldots, t_n) = f(t_1, \ldots, t_n).$$

This may look a bit strange at first but is a perfectly legal specification of (part of) a structure according to Definition 9.1 which admits an arbitrary non-empty interpretation domain. The only special thing here is that, in a sense, we look at terms from two different perspectives. On the one hand, as terms – syntactic objects – to be interpreted, i.e., to be assigned meaning by the operation $[\![\_]\!]^M$. Taking M to be T_Σ, we now find the same terms also on the right-hand sides of the equations above, as the elements – semantic objects – interpreting the syntactic terms, i.e., $[\![t]\!]^{T_\Sigma} = t$.

Such structures are interesting because they provide mechanical means of *constructing* a semantic interpretation from the mere syntax defined by the alphabet.

Example 10.24
Let Σ contain only two constant symbols p, q. The term structure T_Σ will be given by: $T_\Sigma = \{p, q\}$, $[\![p]\!]^{T_\Sigma} = p$, $[\![q]\!]^{T_\Sigma} = q$. (If this still looks confusing, you may think of $\underline{T_\Sigma}$ with all symbols underlined, i.e. $\underline{T_\Sigma} = \{\underline{p}, \underline{q}\}$, $[\![p]\!]^{T_\Sigma} = \underline{p}$, $[\![q]\!]^{T_\Sigma} = \underline{q}$.)

Now extend Σ with one unary function symbol f. The corresponding term structure T_Σ will be now:

$$\underline{T_\Sigma} = \{\ p,\ f(p),\ f(f(p)),\ f(f(f(p))),\ \cdots$$
$$q,\ f(q),\ f(f(q)),\ f(f(f(q))),\ \cdots\ \}$$
$$[\![p]\!]^{T_\Sigma} = p$$
$$[\![q]\!]^{T_\Sigma} = q$$
$$[\![f]\!]^{T_\Sigma}(x) = f(x)\ \text{ for all } x \in \underline{T_\Sigma}.$$

$\square$

Thus every term structure is reachable. Note that nothing is said in Definition 10.23 about the interpretation of relation symbols, so a term structure is not a full FOL-structure. However, for every Σ with at least one constant there is at least one term structure over Σ, for instance, the one where each relation symbol $R \in \mathcal{R}$ is interpreted as the empty set:

$$[\![R]\!]^{T_\Sigma} = \varnothing.$$

Such a structure is, most probably, of little interest. Typically, one is interested in obtaining a *term model*, i.e., in endowing the term structure T_Σ with the interpretation of predicate symbols in such a way that one obtains a model for a given theory.

Remark 10.25 [Term models, reachable models and other models]
Let Σ be given by $\mathcal{I} = \{p, q\}$, $\mathcal{F}_1 = \{f\}$ and $\mathcal{R}_1 = \{P, Q\}$. The term structure T_Σ is then the one from Example 10.24.

Now, consider the following theory:

$$\Gamma = \{P(p),\; Q(q),\; \forall x(P(x) \to Q(f(x))),\; \forall x(Q(x) \to P(f(x)))\}.$$

We can turn T_Σ into a (reachable) model of Γ, the so-called *term model* T_Γ with $\underline{T_\Gamma} = \underline{T_\Sigma} = \mathcal{GT}_\Sigma$, by letting

- $[\![P]\!]^{T_\Gamma} = \{f^{2n}(p) : n \geq 0\} \cup \{f^{2n+1}(q) : n \geq 0\}$ and
- $[\![Q]\!]^{T_\Gamma} = \{f^{2n+1}(p) : n \geq 0\} \cup \{f^{2n}(q) : n \geq 0\}$.

It is easy to verify that, indeed, $T_\Gamma \models \Gamma$. We show that

$$T_\Gamma \models \forall x(P(x) \vee Q(x)). \tag{10.26}$$

We have to show that $[\![P(x) \vee Q(x)]\!]_v^{T_\Gamma} = \mathbf{1}$ for all assignments $v : \{x\} \to \underline{T_\Gamma}$. But all such assignments assign a ground term to x, that is, we have to show that $T_\Gamma \models P(t) \vee Q(t)$ for all ground terms $t \in \mathcal{GT}_\Sigma$. We show this by induction on the complexity of ground terms.

$t \in \mathcal{I} ::$ Since $T_\Gamma \models \Gamma$ we have $T_\Gamma \models P(p)$ and hence $T_\Gamma \models P(p) \vee Q(p)$. In the same way, $T_\Gamma \models Q(q)$ gives that $T_\Gamma \models P(q) \vee Q(q)$.

$f(t) ::$ By IH, we have $T_\Gamma \models P(t) \vee Q(t)$, i.e., either $T_\Gamma \models P(t)$ or $T_\Gamma \models Q(t)$. But also $T_\Gamma \models \Gamma$, so, in the first case, we obtain that $T_\Gamma \models Q(f(t))$, while in the second $T_\Gamma \models P(f(t))$. Hence $T_\Gamma \models P(f(t)) \vee Q(f(t))$.

Thus the claim (10.26) is proved. As a matter of fact, we have proved more than that. Inspecting the proof, we can see that the only assumption we have used was that $T_\Gamma \models \Gamma$, while the inductive proof on $\mathcal{GT}_\Sigma$ was sufficient because any assignment $v : \{x\} \to \underline{T_\Gamma}$ assigned to x an interpretation of some ground term, i.e., $v(x) \in \mathcal{GT}_\Sigma$. In other words, the only assumption was that T_Γ was a *reachable* model of Γ, and what we have proved is:

$$\text{for any reachable } T : T \models \Gamma \Rightarrow T \models \forall x(P(x) \vee Q(x)). \tag{10.27}$$

It is typical that proofs by induction on ground terms like the one above, show us such more general statements like (10.27) and not merely (10.26).

The point now is that the qualification "reachable" is essential and cannot be dropped – it is not the case that $\Gamma \models \forall x(P(x) \vee Q(x))$! Consider a structure M which is exactly like T_Γ but has one additional element, i.e., $\underline{M} = \mathcal{GT}_\Sigma \cup \{*\}$, such that $* \notin [\![P]\!]^M$ and $* \notin [\![Q]\!]^M$ and $[\![f]\!]^M(*) = *$. M is still a model of Γ but not a reachable one due to the presence of $*$. We also see that $M \not\models \forall x (P(x) \lor Q(x))$ since $[\![P(x)]\!]^M_{x \mapsto *} = \mathbf{0}$ and $[\![Q(x)]\!]^M_{x \mapsto *} = \mathbf{0}$. In short, the proof that something holds for all ground terms, shows that the statement holds for all reachable models but, typically, not that it holds for arbitrary models. $\qquad\qquad\qquad\qquad\qquad\qquad\qquad\qquad\qquad\qquad\qquad\square$

Example 10.28

Notice that although a reachable structure always exists (when $\mathcal{GT}_\Sigma \neq \varnothing$), there may be no reachable *model* for some Σ-theory Γ. Let Σ contain only two constants a, b and one predicate R. T_Σ is given by $\{a, b\}$ and $[\![R]\!]^{T_\Sigma} = \varnothing$. Let Γ be $\{\neg R(a), \neg R(b), \exists x R(x)\}$. It is impossible to construct a model for Γ – i.e. interpret $[\![R]\!]$ so that all formulae in Γ are satisfied – which is reachable over Σ.

But let us extend the alphabet to Σ' with a new constant c. Let T' be $T_{\Sigma'}$ but interpret R as $[\![R]\!]^{T'} = \{c\}$. Then, obviously, $T' \models \Gamma$. T' is not a Σ-structure since it contains the interpretation of c which is not in Σ. To turn T' into a Σ-structure satisfying Γ we only have to "forget" the interpretation of c. The resulting structure T is identical to T', except that T is a Σ-structure and the element c of $\underline{T} = \underline{T'}$ has no corresponding term in the alphabet. Thus:

- T' is a reachable Σ'-structure but it is not a Σ-structure
- T' is a Σ'-reachable model of Γ
- T is a Σ-structure but not a Σ'-structure
- T is a Σ-structure but not a Σ-reachable structure
- T is a model of Γ but it is not a Σ-reachable model. $\qquad\qquad\square$

As an important corollary of Theorem 10.13, we obtain the following sufficient conditions for the existence of reachable models.

Corollary 10.29 Let Γ be a collection of Π_1 formulae, over an alphabet Σ with $\mathcal{GT}_\Sigma \neq \varnothing$. If $Mod(\Gamma) \neq \varnothing$ then Γ has a reachable model.

Simply, the existence of some model of Γ allows us, by Fact 10.21, to restrict it to its reachable substructure. By Theorem 10.13, this substructure is also a model of Γ.

Term structures are used primarily as the basis for constructing the domain of interpretation – namely, the ground terms – for some reachable structure, which can, sometimes, be obtained from T_Σ by imposing appropriate interpretation of the relation symbols. Such use is very common in theoretical computer science for constructing canonical models for specifications (theories). This will also be the use we will make of it in the proof of completeness of $\mathcal{N}$ for FOL in the next chapter.

"Syntactic" models have typically an important property of being canonical representatives of the whole model class. When model class comes equipped with the mappings between models (some form of homomorphisms) forming a category, the syntactically constructed models happen typically to be initial ones. We won't describe this concept here and take it simply as a vague synonym for being syntactically generated. This syntactic character makes them useful in mechanizing the proofs and, generally, carrying out mechanical computations. The rest of this section describes a specific computational strategy arising from such models.

10.3.2 HERBRAND'S THEOREM

Herbrand's theorem is intimately related to the so-called Herbrand models. These models are, in fact, term models which we will encounter also in the proof of completeness for FOL. Herbrand's theorem allows, under some general conditions, to reduce provability in FOL to provability in PL. We discuss only a special version of the theorem, making some additional assumptions, which are not necessary in its most general form.

For any set Γ of first order formulae over a given alphabet Σ, let $GI(\Gamma)$ be the set of ground instances of formulae in Γ, i.e., the ground formulae obtained from formulae of Γ by substitution of members of $\mathcal{GT}_\Sigma$ for the free variables.

Theorem 10.30 Suppose Σ contains at least one constant, and let Γ be a set of quantifier-free Σ-formulae. Then: $\Gamma \vdash_{\mathcal{N}}^{\mathsf{FOL}} \bot \iff GI(\Gamma) \vdash_{\mathcal{N}}^{\mathsf{PL}} \bot$.

Requirement on the formulae in Γ to be quantifier-free amounts to their universal closure and is crucial for the reduction of FOL-inconsistency to PL-inconsistency in the theorem. $GI(\Gamma)$ is the set of all ground instances of the formulae in Γ which, by the requirement on the alphabet Σ, is guaranteed to be non-empty. The implication $GI(\Gamma) \vdash_{\mathcal{N}}^{\mathsf{PL}} \bot \Rightarrow \Gamma \vdash_{\mathcal{N}}^{\mathsf{FOL}} \bot$ holds, of course, always, since $GI(\Gamma)$ is a weaker theory than Γ – it axiomatizes *only* ground instances which are also consequences of Γ. Consider, for instance, a language with one constant $\mathbf{a}$ and $\Gamma = \{P(x)\}$. Then $GI(\Gamma) = \{P(\mathbf{a})\}$. Obviously, $GI(\Gamma) = P(\mathbf{a}) \not\vdash_{\mathcal{N}}^{\mathsf{FOL}} \forall x.P(x)$ which, however, is trivially a provable consequence of Γ itself.

The importance of Herbrand's theorem lies in the identification of Γ's allowing also the opposite implication, namely, $\Gamma \vdash_{\mathcal{N}}^{\mathsf{FOL}} \bot \Rightarrow GI(\Gamma) \vdash_{\mathcal{N}}^{\mathsf{PL}} \bot$. It amounts to a kind of reduction of the FOL-theory Γ to its PL-version with all "syntactically generated" instances. Using this reduction, one can attempt to check whether $\Gamma \vdash_{\mathcal{N}}^{\text{FOL}} \phi$ by reducing *ad absurdum* – in PL– the assumption $GI(\Gamma, \neg\phi)$. Such a proof strategy is referred to as "refutational proof" or "proof by contradiction" and proceeds as follows: Assume that ϕ is closed and quantifier-free (i.e., ground). Then, according to the theorem, $\Gamma, \neg\phi \vdash_{\mathcal{N}}^{\text{FOL}} \bot \Leftrightarrow GI(\Gamma), \neg\phi \vdash_{\mathcal{N}}^{\text{PL}} \bot$. Thus, we have to check if some ground instances of (some formulae from) Γ, together with $\neg\phi$ lead to a contradiction. Having succeeded, i.e., showing $GI(\Gamma), \neg\phi \vdash_{\mathcal{N}}^{\text{PL}} \bot$, we obtain $\Gamma, \neg\phi \vdash_{\mathcal{N}}^{\text{FOL}} \bot$ which implies $\Gamma \vdash_{\mathcal{N}}^{\text{FOL}} \neg\phi \to \bot$ and this, again, $\Gamma \vdash_{\mathcal{N}}^{\text{FOL}} \phi$.

This is not yet any effective algorithm but we can imagine that such a checking – if some PL formulae yield contradictions – can be, at least in principle and at least in some cases, performed. The procedure is slightly generalized when ϕ contains variables (which are then interpreted also in a specific way). Various computational mechanisms utilizing this principle will thus restrict their theories to quantifier-free (i.e., universally quantified, according to Exercise 9.11) formulae and alphabets with non-empty set of ground terms. In the following, we will see a particular – and quite central – example of such restrictions.

10.3.3 HORN CLAUSES

An important issue with the utilization of Herbrand's theorem concerns the actual strategy to determine if $GI(\Gamma) \vdash \bot$. Various choices are possible and we show one restriction on the format of the formulae, leading to one such strategy (which does not even follow the proof by contradiction, but derives consequences of a theory directly).

A clause is a formula of the form $L_1 \vee \ldots \vee L_n$ where each L_i is a literal – a positive or negated atom (i.e., a formula $P(\bar{t})$ or $\neg P(\bar{t})$ for a sequence of some (not necessarily ground) terms $\bar{t}$.) By the general fact that $M \models A \Leftrightarrow M \models \forall(A)$, one does not write the universal quantifiers which are present implicitly. A Horn clause is a clause having exactly one positive literal, i.e., $\neg A_1 \vee \ldots \vee \neg A_n \vee A$, which is equivalent to

$$A_1 \wedge A_2 \wedge \ldots \wedge A_n \to A, \tag{10.31}$$

where all A_i's are positive atoms. A theory, in which all formulae are Horn clauses, is a *Horn theory*.

In the context of logic programming, the particular case of a Horn clause with $n = 0$ is called a "fact", the conjunction of the assumptions (negative literals) the "body" and the conclusion the "head" of the clause.

MP and the chaining rule 8.23.(1) can be generalized to the following rule operating on and yielding only Horn clauses:

$$\frac{\Gamma \vdash_{\mathcal{N}} \mathbf{A_1} \wedge \ldots \wedge \mathbf{A_k} \to \mathbf{B_i} \; ; \; \Gamma \vdash_{\mathcal{N}} B_1 \wedge \ldots \wedge \mathbf{B_i} \wedge \ldots \wedge B_n \to C}{\Gamma \vdash_{\mathcal{N}} B_1 \wedge \ldots \wedge (\mathbf{A_1} \wedge \ldots \wedge \mathbf{A_k}) \wedge \ldots \wedge B_n \to C} \quad (10.32)$$

(This is a special case of the general resolution rule which "joins" two clauses removing from each a single literal – an atom occurring positively in one clause and negatively in the other.) In particular, when B_i is a fact, it can be simply removed from the body. When all atoms from the body of a clause get thus removed, its conclusion becomes a new fact.

Suppose, we have the following Horn theory, Γ:

1.	$Parent(Ben, Ada)$	
2.	$Parent(Cyril, Ben)$	
3.	$Parent(Cecilie, Ben)$	
4.	$Parent(David, Cyril)$	(10.33)
5.	$Ancestor(Eve, x)$	
6.	$Parent(y, x) \to Ancestor(y, x)$	
7. $Ancestor(z, y) \wedge Ancestor(y, x) \to Ancestor(z, x)$		

The questions on the left below, have then the answers on the right, and you should have no problems with convincing yourself about that. (Questions with free variables, like 5 or 6, ask for true ground instances, i.e., substitution instances which are derivable from Γ.)

Does $\Gamma \vdash \ldots$

? $Ancestor(Eve, Ada)$	: Yes		1.
? $Parent(David, Ada)$	: No		2.
? $Ancestor(David, Ada)$	: Yes		3. (10.34)
? $Ancestor(Herod, Ben)$	: No		4.
? $Ancestor(Cyril, x)$	: Ben, Ada		5.
? $Ancestor(x, Ben)$	: $Cecilie, Cyril, David, Eve$	6.

The language of Horn clauses has some important properties. The first is that it is the most general sublanguage of FOL which guarantees the existence of *initial* models. These are just Herbrand (term) models obtained by collecting all positive ground facts.

10.3.4 HERBRAND MODELS OF HORN THEORIES

In more detail, every consistent and universal (Π_1) theory over an alphabet with ground terms, has a reachable model by Corollary 10.29. Herbrand model is a term model, i.e., a reachable model which does not identify any distinct terms but interprets a ground term t as itself. In the above example, a possible model could identify all the persons, making both *Parent* and *Ancestor* reflexive (and consequently also transitive and symmetric). This would be a reachable model but not the intended one. Herbrand model will be the one we actually intended writing the program (10.33).

To construct it, we start with the ground instances of all facts and iterate the process of resolving the assumptions of conditional clauses to add more facts, essentially, by applying rule (10.32). Given a Horn clause theory Γ, over an alphabet Σ with $\mathcal{GT}_\Sigma \neq \varnothing$, we define:

$\mathcal{GT}_\Sigma$ the Herbrand universe = the set of all ground terms. In the example (abbreviating the names by their first initials): $\{\mathbf{a}, \mathbf{b}, \mathbf{c}_1, \mathbf{c}_2, \mathbf{d}, \mathbf{e}\}$

HB_Γ the Herbrand base = the set of all ground atoms. The model is a subset of this set, containing all ground atoms which must be true.

H_Γ The construction of the Herbrand model proceeds inductively as follows. View $\Gamma = \mathcal{F} \uplus \mathcal{C}$ as the disjoint union of facts and clauses:

(1) $H_0 = GI(\mathcal{F})$ – all ground instances of all facts;

(2) $H_{i+1} = H_i \cup \{\sigma(C) : A_1 \wedge \ldots \wedge A_n \to C \in \mathcal{C} \ \& \ \sigma(A_1), \ldots, \sigma(A_n) \in H_i\}$ – σ ranges over all ground substitutions (to the variables in $A_1 \ldots A_n, C$);

(3) $H_\Gamma = \bigcup_{i<\omega} H_i$.

For the theory from example (10.33), we would obtain the model:

$\mathcal{GT}_\Sigma = \{\mathbf{a}, \mathbf{b}, \mathbf{c}_1, \mathbf{c}_2, \mathbf{d}, \mathbf{e}\}$

$H_\Gamma :$ $Parent = \{\langle \mathbf{b}, \mathbf{a} \rangle, \langle \mathbf{c}_1, \mathbf{b} \rangle, \langle \mathbf{c}_2, \mathbf{b} \rangle, \langle \mathbf{d}, \mathbf{c}_1 \rangle\}$ and

$Ancestor = \{\langle \mathbf{e}, x \rangle : x \in \mathcal{GT}_\Sigma\} \cup Parent^+$

One sees easily that $H_\Gamma \models \Gamma$. Assignments to $\mathcal{GT}_\Sigma$ amount to ground substitutions so, for all facts, this holds by (1). For a clause $A_1 \ldots A_n \to C$ and an assignment σ, assume that $H_\Gamma \models_\sigma A_1 \wedge \ldots \wedge A_n$. By construction and point (3) this means that, at some step i, we obtained $\sigma(A_k) \in H_i$ for all $1 \leq k \leq n$. But then, by point (2), $\sigma(C)$ is also included in H_Γ at the step H_{i+1}. In fact, we have that

$$H_\Gamma = \{A \in HB_\Gamma : \Gamma \models A\}. \tag{10.35}$$

This is the property of *minimality* – H_Γ does not satisfy more atoms than those which are satisfied in *every* model of the theory. (In general, any model of Γ over the term structure $\mathcal{GT}_\Sigma$ is called Herbrand model – H_Γ is the *minimal* Herbrand model.)

10.3.5 Computing with Horn clauses

The other crucial property of Horn clauses is the possibility of operational interpretation of $\rightarrow$, according to which $A_1 \wedge \ldots \wedge A_n \rightarrow A$ means that, in order to establish A, one has to establish $A_1, \ldots, A_n$. This trivial chaining, expressed by rule (10.32), must be coupled with the treatment of variables. For instance, to establish $Ancestor(e, a)$ one uses the fact $Ancestor(e, x)$ which, however, requires unification of terms x and a, i.e., making them syntactically identical $x \equiv a$. Here, it amounts simply to a substitution of a for x but, in general, may be more involved. For instance, to unify $f(x, g(h, x))$ and $f(d(z), y)$ requires finding the substitutions $x \mapsto d(z)$ and $y \mapsto g(h, d(z))$, after which the two terms become equal to $f(d(z), g(h, d(z)))$.

The query $Ancestor(David, Ada)$ is processed as follows:

(10.33)	goal	justification	unification
	$?Ancestor(\mathbf{d}, \mathbf{a}) \sim Ancestor(e, x)$		fails : $e \not\equiv \mathbf{d}$
6.		$\leftarrow Parent(\mathbf{d}, \mathbf{a})$	no such fact
7.		$\leftarrow Ancestor(\mathbf{d}, y) \wedge Ancestor(y, \mathbf{a})$	?
	the search starts for y satisfying both literals in the body		
	$?Ancestor(\mathbf{d}, y) \sim Ancestor(e, x)$		fails : $e \not\equiv \mathbf{d}$
6.		$\leftarrow Parent(\mathbf{d}, \mathbf{c}_1)$	$y \equiv \mathbf{c}_1$
	$?Ancestor(\mathbf{c}_1, \mathbf{a}) \sim Ancestor(e, x)$		fails : $e \not\equiv \mathbf{c}_1$
6.		$\leftarrow Parent(\mathbf{c}_1, \mathbf{a})$	no such fact
7.		$\leftarrow Ancestor(\mathbf{c}_1, z) \wedge Ancestor(z, \mathbf{a})$	?
	$\vdots$	$\vdots$	$z \equiv \mathbf{b}$
YES			

Thus, we can actually *compute* the facts provable in Horn theories by means of the above mechanism based on the resolution rule (10.32) and unification algorithm: Horn theories can be seen as Horn *programs*. Observe the kind of "backward" process of computing: one starts with the query and performs a "backward chaining" along the available clauses until one manages to resolve all the assumptions by matching them against available facts.

Unification . . . . [optional]
In more detail, the following algorithm implements the rule (10.32). Given Horn clauses: $A_1 \wedge \ldots \wedge A_n \rightarrow A$ and $B_1 \wedge \ldots \wedge B_m \rightarrow B$, when trying to establish A,

we will have to resolve all A_i. We attempt to replace them by facts, and if this fails, by B_i's (until we arrive at facts):

(1) select an atom A_i and

(2) try to unify it with B (see further down for unification)

(3) if unification succeeded, replace A_i by $B_1 \wedge \ldots \wedge B_m$

(4) apply to the resulting clause the unifying substitution from (2).

If unification in (2) fails, try the next A_{i+1}. If none can be unified with B, try other clauses.

Unification of atoms $P(t_1 \ldots t_n)$ and $R(s_1 \ldots s_n)$ requires that P and R are the same symbol, $P \equiv R$, and then amounts to finding a *unifier*, namely, a substitution σ such that for each $i : \overline{\sigma}(t_i) = \overline{\sigma}(s_i)$, Definition 8.18. If two terms have a unifier they also have a *most general unifier*, mgu. For instance, $\alpha = \{x \mapsto d(d(a)), y \mapsto g(a, d(d(a))), z \mapsto d(a)\}$ is a unifier of $t_1 = f(x, g(a, x))$ and $t_2 = f(d(z), y)$, yielding $\overline{\alpha}(t_1) = \overline{\alpha}(t_2) = f(d(d(a)), g(a, d(d(a))))$. However, the unifier $\sigma = \{x \mapsto d(z), y \mapsto g(a, d(z))\}$ is more general – it yields the term $\overline{\sigma}(t_1) = \overline{\sigma}(t_2) = f(d(z), g(a, d(z)))$, from which $\overline{\alpha}(t_1)$ can be obtained by further substitution $\beta = \{z \mapsto d(a)\}$. The mgu of terms t_i is a substitution σ such that for any other unifier α, there is a substitution β such that $\overline{\alpha}(t_i) = \overline{\beta}(\overline{\sigma}(t_i))$.

The following algorithm finds the most general unifier solving a set of equations $\{s_1 = t_1 \ldots s_n = t_n\}$ or reports the nonexistence of any unifier. It chooses repeatedly and nondeterministically one of the equations in the set and, performing the associated action, transforms the set (or halts).

if $s_i = t_i$ has the form		: then
$f(s'_1 \ldots s'_k) = f(t'_1 \ldots t'_k)$		: replace it by k equations $s'_i = t'_i$
$f(s'_1 \ldots s'_k) = g(t'_1 \ldots t'_l)$	and $f \not\equiv g$	: terminate with **failure**
$x = x$		: delete it
$t = x$	and $t \notin \mathcal{V}$	: replace it with $x = t$
$x = t$	and $t \not\equiv x$ and	: if $x \in \mathcal{V}(t)$ – terminate with **failure**
	x occurs elsewhere	: – otherwise, substitute $x \mapsto t$
		: in all other equations

On successful termination, the set has the form $\{x_1 = r_1 \ldots x_m = r_m\}$, with distinct variables x_i's and r_i's determining the needed substitution. (Variable x not occurring in this solution set is substituted by itself.) . . . . [end optional]

Let $\Gamma \rightsquigarrow A$ denote that the ground atom A can be obtained from a set of Horn clauses Γ using the above strategy. The following equivalences express then the soundness and completeness of the strategy with respect to the minimal Herbrand model as well as the whole model class (since the last two are equivalent for ground atoms by (10.35)). For every Horn theory Γ and ground atom A:

$$\Gamma \rightsquigarrow A \iff H_\Gamma \models A \iff \Gamma \models A. \tag{10.36}$$

Thus, our computational strategy, checking if some ground fact holds in the minimal Herbrand model checks, equivalently, if it is a logical consequence of the theory. Notice, however, that these equivalences hold only for the ground atoms. The implication $(H_\Gamma \models A) \Rightarrow (\Gamma \rightsquigarrow A)$ says that every ground atom true in H_Γ can be generated by the strategy. Conversely, we have also that $(H_\Gamma \not\models A) \Rightarrow (\Gamma \not\rightsquigarrow A)$. This, however, says only that if an atom is not satisfied, the strategy may not derive it. But there is then no algorithmic way to ensure derivability of $\neg A$, since such literals are not part of the Horn clause language for which the above results hold.

10.3.6 Computational completeness

The operational interpretation of Horn clauses sketched above, together with the equivalence (10.36), establish Horn theories as a logical counterpart of mechanical computability. Indeed, this mechanism allows to compute everything which can be computed on a Turing machine:

Theorem 10.37 The mechanism of unification with resolution of Horn clauses is Turing complete.

Probably the simplest proof of this fact shows that register machine programs, RMP's, can be simulated by Horn programs.[2] We have to take here for granted the result stating that the RMP's as described below are computationally equivalent to Turing machines.

A register machine operates with a memory consisting of a finite set of registers which can store natural numbers. An RMP, a program for a register machine over m registers $x_1 \ldots x_m$, is a sequence $I_1 \ldots I_n$ of n numbered instructions, each in one of the two forms – a simple increment (inc), or a conditional decrement and jump (cnd):

(inc) $x_i := x_i + 1$

(cnd) **if** $x_i \neq 0$ **then** $x_i := x_i - 1$ **and goto** j.

If, on reaching the instruction of the second form, $x_i = 0$, the program simply proceeds to the next instruction. The program terminates on reaching the **halt** instruction, always implicitly present as the I_{n+1}-th instruction. An RMP is said to compute a (partial) function $f : \mathbb{N}^l \to \mathbb{N}$, $l \leq m$ if $\forall n_1 \ldots n_l \in \mathbb{N}$ the execution starting with the register values $n_1 \ldots n_l, 0 \ldots 0_m$ (the additional registers $x_{l+1} \ldots x_m$ which are not input are initialized to 0), terminates with $x_1 = f(n_1 \ldots n_l)$, whenever $f(n_1 \ldots n_l)$ is defined, and does not terminate otherwise.

An RMP is simulated by a Horn program P as follows. For each instruction I_k, $1 \leq k \leq n + 1$, we have a predicate symbol $P_k(x_1 \ldots x_m, y)$ – the

---

[2] Thanks to Marc Bezem for this argument using RMP's.

x_i's corresponding to the registers and y to the result of the computation. Each I_k is either (inc) or (cnd) and these are simulated, respectively, by:

(inc) $P_{k+1}(x_1 \ldots x_i + 1 \ldots x_m, y) \rightarrow P_k(x_1 \ldots x_i \ldots x_m, y)$

(cnd) $P_{k+1}(x_1 \ldots 0 \ldots x_m, y) \rightarrow P_k(x_1 \ldots 0 \ldots x_m, y)$ (I'_k)

$P_j(x_1 \ldots x_i \ldots x_m, y) \rightarrow P_k(x_1 \ldots x_i + 1 \ldots x_m, y)$

In addition, the **halt** instruction transfers x_1 to the result position:

(hlt) $P_{n+1}(x_1 \ldots x_m, x_1)$ (I'_{n+1})

The query $P_1(n_1 \ldots n_l, 0 \ldots 0_m, y)$ will result in the computation simulating step by step the execution of the corresponding RMP.

As a simple corollary of Theorem 10.37, we obtain:

Theorem 10.38 Validity in first order logic is undecidable.

How does it follow? Having accepted the computational equivalence of Turing machines and RMP's, we have accepted also the equivalence of their halting problems. But halting of an RMP is equivalent to the existence of a result for the initial query, i.e., to the truth of the formula $\exists y P_1(n_1 \ldots n_l, 0 \ldots 0_m, y)$ under the assumptions gathering all the clauses of the program P, in short, to the entailment

$$\forall(I'_1), \ldots, \forall(I'_n), \forall(I'_{n+1}) \models \exists y P_1(n_1 \ldots n_l, 0 \ldots 0_m, y). \qquad (10.39)$$

Since all formulae in (10.39) are closed, it is equivalent to the validity:

$$\models \left(\forall(I'_1) \land \ldots \land \forall(I'_n) \land \forall(I'_{n+1}) \right) \rightarrow \exists y P_1(n_1 \ldots n_l, 0 \ldots 0_m, y).$$

If validity of this sentence could be decided, we could decide if our RMP's halt or not. But such a procedure does not exist by the undecidability of the halting problem for Turing machines and, hence, also for register machines. Note that written in PNF, this last formula is Σ_1 – undecidability does not require more complex formulae than existential sentences.

Remark [Prolog]

Theorem 10.37, and the preceding computation strategy, underlie Prolog – the programming language where programs are Horn theories. It allows one to ask queries about single facts or possible instances making up facts, like the queries 6 or 5 in (10.34), about all x such that $Ancestor(Cyril, x)$. There are, of course, various operational details of not quite logical character which, in some cases, yield unexpected results. They have mainly to do with the treatment of negation.

Wondering about the computational power of such a simple mechanism, one should ask the question: where does the undecidability enter the stage here? The answer, suggested earlier, is: with the treatment of negation. Any positive atom, entailed by a given set of clauses (and facts), can be computed in a finite time according to (10.36). In the above example of family relations the universe of terms was finite so, at least in principle, one can terminate the search for matching pairs of ancestors with the answer 'no' to the query? $Ancestor(Ada, Cyril)$. But in general, in the presence of function symbols, this universe is potentially infinite. Negation turns thus out to be tricky issue: there is no general rule for terminating a prolonged and unsuccessful search for matching substitutions – which would guarantee that the answer 'no' is always correct.   □

EXERCISES 10.

EXERCISE 10.1 Find PNFs for the following formulae
  (1) $\forall x(f(g(x)) = x) \rightarrow \forall x \exists y(f(y) = x)$
  (2) $\exists z \forall x A(x, z) \rightarrow \forall x \exists z A(x, z)$
  (3) $\forall x \exists y(x + y = 0) \land \forall x \forall y(x + y = 0 \rightarrow y + x = 0)$

Since the matrix of a formula in PNF contains no quantifiers, it can be transformed into CNF/DNF by the same manipulations as in PL. Transform the matrix of the result of point (3) into DNF.

EXERCISE 10.2 Two PNFs of a given formula are considered *different* when they have different PNF indices. List all different PNFs of formula (1) from the previous exercise.

EXERCISE 10.3 Show that the remaining, unverified equivalences from Lemma 10.3 do hold.

EXERCISE 10.4 Show reachability of the structures (1)-(3) in Example 9.3.

EXERCISE 10.5 Let Σ contain one constant $\odot$ and one unary function s. Let T_Σ denote its term structure. Show that
  (1) T_Σ is bijective to the set $\mathbb{N}$ of natural numbers,
  (2) T_Σ with the ordering of terms induced by their inductive definition is order-isomorphic (Definition 1.18) to $\mathbb{N}$ with $<$.

EXERCISE 10.6 Show that the class of dense partial orderings, defined in Example 10.15, cannot be axiomatized by Σ_1 sentences.

EXERCISE 10.7 Explain why in point (1) of Theorem 10.13, it is necessary to assume that A is closed. Let, for instance, $A = \exists x R(x, y)$. Is it a Σ_1 formula? Could the statement be proved for this A? (Recall Fact 9.27.)

EXERCISE 10.8 Show that the following sentence has only infinite models:

$$\forall x \neg P(x, x) \wedge (\forall x \forall y \forall z : P(x, y) \wedge P(y, z) \rightarrow P(x, z)) \wedge \forall x \exists y P(x, y).$$

EXERCISE 10.9 A more precise formulation of Lemma 10.2 can be given along the following lines: We may assume that the language of FOL contains propositional variables – viewed as nullary relation symbols. Moreover, it is always possible to substitute a formula, say A, for such a propositional variable a in a formula F, obtaining a new formula, denoted F_A^a. A reformulation of Lemma 10.2 says then that $A \Leftrightarrow B$ implies $F_A^a \Leftrightarrow F_B^a$. Compare this to Lemma 9.9. Would it be possible to formulate also a version saying that $[\![A]\!]_v^M = [\![B]\!]_v^M$ implies $[\![F_A^a]\!]_v^M = [\![F_B^a]\!]_v^M$?

EXERCISE 10.10 [Skolem Normal Form]
Some intricacies in this exercise remind of those from Example 10.28. Let A be the sentence $\forall x \exists y \forall z B(x, y, z)$, where B has no quantifiers.

(1) Let f be a fresh unary function symbol and A_S be $\forall x \forall z B(x, f(x), z)$.
(2) Show that A and A_S are equisatisfiable (either each has a model or none has). (Hint: Show that a model for any one of the two formulae can be transformed into a model for the other.)
(3) Show that A and A_S are not logically equivalent.
(4) For every sentence $A = \forall \overline{x} \exists y B(\overline{x}, y)$, let $A_L = \forall \overline{x} B(\overline{x}, f(\overline{x}))$, where f is a fresh function symbol of appropriate arity (equal length of $\overline{x}$; hence f is a constant if there is no universal quantifier in front of $\exists y$). Show that A and A_L are equisatisfiable.
(5) Use Theorem 10.6, the previous point and induction on the number of existential quantifiers in the prefix of PNF A_P of A to show that for each FOL formula A, there is a formula A_S in PNF with only universal quantifiers (but usually new function symbols) such that A and A_S are equisatisfiable. [Such an A_S is A's *Skolem normal form.*]

EXERCISE 10.11 A *ground instance* of a formula A is a sentence obtained from A by substitution of ground terms for the free variables. Suppose M is a reachable Σ-structure and A a Σ-formula. Show that

$$M \models A \text{ iff } (M \models B \text{ for all ground instances } B \text{ of } A).$$

(Hint: If $x_1, \dots, x_n$ are the free variables of A, then B is a ground instance of A iff there exist ground terms $t_1, \dots, t_n$ such that B equals $A_{t_1, \dots, t_n}^{x_1, \dots, x_n}$, i.e., the result of a simultaneous substitution of each t_i for the corresponding x_i.)

Chapter 11

SOUNDNESS AND COMPLETENESS

11.1 SOUNDNESS OF $\mathcal{N}$

We show the soundness and completeness theorems for the proof system $\mathcal{N}$ for FOL. As in the case of PL, soundness is an easy task.

Theorem 11.1 For every $\Gamma \subseteq \mathsf{WFF_{FOL}}, A \in \mathsf{WFF_{FOL}} : \Gamma \vdash_{\mathcal{N}} A \Rightarrow \Gamma \models A$.

Proof. Axioms A0-A3 and MP are the same as for PL and their validity follows from the proof of the soundness Theorem 6.15 for PL. Validity of A4 was shown in Exercise 9.10.

It suffices to show that $\exists$I preserves truth, i.e. that $M \models B \to C$ implies $M \models \exists x B \to C$ for arbitrary structure M (in particular, one for which $M \models \Gamma$), provided x is not free in C. In fact, it is easier to show the contrapositive implication from $M \not\models \exists x B \to C$ to $M \not\models B \to C$. So suppose $M \not\models \exists x B \to C$, i.e., $M \not\models_v \exists x B \to C$ for some v. Then $M \models_v \exists x B$ and $M \not\models_v C$. Hence $M \models_{v[x \mapsto \underline{a}]} B$ for some $\underline{a}$. Since $M \not\models_v C$ and $x \notin \mathcal{V}(C)$, it follows from Lemma 9.7 that also $M \not\models_{v[x \mapsto \underline{a}]} C$, hence $M \not\models_{v[x \mapsto \underline{a}]} B \to C$, i.e., $M \not\models B \to C$. QED (11.1)

By the same argument as in Corollary 6.16, every satisfiable FOL theory is consistent or, equivalently, inconsistent FOL theory is unsatisfiable:

Corollary 11.2 $\Gamma \vdash_{\mathcal{N}} \bot \Rightarrow Mod(\Gamma) = \varnothing$.

Proof. If $\Gamma \vdash_{\mathcal{N}} \bot$ then, by the theorem, $\Gamma \models \bot$, i.e., for any $M : M \models \Gamma \Rightarrow M \models \bot$. But there is no M such that $M \models \bot$, so $Mod(\Gamma) = \varnothing$.
QED (11.2)

Remark.
Remark 6.17 showed the equivalence of the two soundness notions for PL.
The equivalence holds also for FOL but now the proof of the opposite implication, namely,

$$\Gamma \vdash_{\mathcal{N}} \bot \Rightarrow Mod(\Gamma) = \varnothing \ \text{ implies } \ \Gamma \vdash_{\mathcal{N}} A \Rightarrow \Gamma \models A,$$

involves an additional subtlety concerning possible presence of free variables
in A. Assuming $\Gamma \vdash_{\mathcal{N}} A$ we first observe that then we also have $\Gamma \vdash_{\mathcal{N}} \forall(A)$
by Lemma 8.24.(4). Hence $\Gamma, \neg\forall(A) \vdash_{\mathcal{N}} \bot$ and, by the assumed implication,
it has no models: if $M \models \Gamma$ then $M \not\models \neg\forall(A)$. Since $\forall(A)$ is closed, this
means that $M \models \forall(A)$ (Remark 9.17). By Fact 9.25, we can now conclude
that $M \models A$ and, since M was an arbitrary model of Γ, that $\Gamma \models A$.   □

11.2   COMPLETENESS OF $\mathcal{N}$

As in the case of PL, we prove the opposite of Corollary 11.2, namely, that
every consistent FOL-theory is satisfiable. Starting with a consistent theory
Γ, we have to show that there is a model satisfying Γ. The procedure is
thus very similar to the one applied for PL (which you might repeat before
reading this section). Its main point was expressed in Lemma 6.18, which
has the following counterpart:

Lemma 11.3   The following formulations of completeness are equivalent:

(1) For any $\Gamma \subseteq \mathsf{WFF_{FOL}}$: $\Gamma \not\vdash_{\mathcal{N}} \bot \ \Rightarrow \ Mod(\Gamma) \neq \varnothing$
(2) For any $\Gamma \subseteq \mathsf{WFF_{FOL}}$: $\Gamma \models A \ \Rightarrow \ \Gamma \vdash_{\mathcal{N}} A.$

> **Proof.** (1) $\Rightarrow$ (2). Assuming (1) and $\Gamma \models A$, we consider first the
> special case of a closed A. Then $Mod(\Gamma, \neg A) = \varnothing$ and so $\Gamma, \neg A \vdash_{\mathcal{N}} \bot$
> by (1). By Deduction Theorem $\Gamma \vdash_{\mathcal{N}} \neg A \rightarrow \bot$, so $\Gamma \vdash_{\mathcal{N}} A$ by PL.
> This result for closed A yields the general version: If $\Gamma \models A$ then
> $\Gamma \models \forall(A)$ by Fact 9.27. By the argument above $\Gamma \vdash_{\mathcal{N}} \forall(A)$, and so
> $\Gamma \vdash_{\mathcal{N}} A$ by Lemma 8.24.(1) and MP.
> (2) $\Rightarrow$ (1). This is shown by exactly the same argument as for PL in
> the proof of Lemma 6.18.   QED (11.3)

Although the general procedure, based on the above lemma, is the same,
the details are now more involved as both the language and its semantics are
more complex. The model we will eventually construct will be a term model

for an appropriate extension of Γ. The following definition characterizes the extension we will be looking for.

Definition 11.4 A theory Γ is said to be

- *maximal consistent* iff it is consistent and, for every *closed* formula A, $\Gamma \vdash_{\mathcal{N}} A$ or $\Gamma \vdash_{\mathcal{N}} \neg A$ (cf. Definition 6.19);
- a *Henkin-theory* if for each closed formula of the type $\exists x A$ there is an individual constant c such that $\Gamma \vdash_{\mathcal{N}} \exists x A \to A_c^x$;
- a *max Henkin-theory* if it is both a Henkin-theory and maximal consistent.

In particular, every max Henkin-theory is consistent. The constant c in the definition above is called a *witness* – it witnesses to the truth of the formula $\exists x A$ by providing a ground term which validates the existential quantifier. The precise definition of a Henkin-theory may vary somewhat in the literature. The condition in the lemma below looks slightly weaker than the one in the definition, but turns out to be equivalent. The proof is an easy exercise.

Lemma 11.5 Let Γ be a theory, and suppose that for every formula A with exactly one variable x free, there is a constant c_A such that $\Gamma \vdash_{\mathcal{N}} \exists x A \to A_{c_A}^x$. Then Γ is a Henkin-theory.

The properties of a max Henkin-theory make it easier to construct a model for it. We prove first this special case of the completeness theorem:

Lemma 11.6 Every max Henkin-theory is satisfiable.

Proof. The alphabet of any Henkin-theory will always contain an individual constant. Now consider the term structure T_Γ (we index it with Γ and not merely the alphabet, as in Definition 10.23, since we will make it into a model of Γ) where:

- for each relation symbol $R \in \mathcal{R}$ of arity n, and ground terms $t_1, \ldots, t_n \in \underline{T_\Gamma}$:

$$\langle t_1, \ldots, t_n \rangle \in [\![R]\!]^{T_\Gamma} \iff \Gamma \vdash_{\mathcal{N}} R(t_1, \ldots, t_n).$$

We show, by induction on the number of connectives and quantifiers in a formula, that for any *closed* formula A we have $T_\Gamma \models A$ iff $\Gamma \vdash_{\mathcal{N}} A$. (From this it follows that T_Γ is a model of Γ: if $A \in \Gamma$ then $\Gamma \vdash_{\mathcal{N}} \forall(A)$, so $T_\Gamma \models \forall(A)$, i.e., by Fact 9.27, $T_\Gamma \models A$.)

ATOMIC :: Follows directly from the construction of T_Γ.

$$T_\Gamma \models R(t_1, \ldots, t_n) \iff \langle t_1, \ldots, t_n \rangle \in [\![R]\!]^{T_\Gamma} \iff \Gamma \vdash_{\mathcal{N}} R(t_1, \ldots, t_n).$$

$\neg B$ :: We have the following equivalences:

$$T_\Gamma \models \neg B \;\Leftrightarrow\; T_\Gamma \not\models B \quad \text{definition of } \models,\ A \text{ closed}$$
$$\Leftrightarrow\; \Gamma \not\vdash_\mathcal{N} B \quad \text{IH}$$
$$\Leftrightarrow\; \Gamma \vdash_\mathcal{N} \neg B \quad \text{maximality of } \Gamma$$

$B \to C$ :: Since $T_\Gamma \models A \Leftrightarrow T_\Gamma \not\models B$ or $T_\Gamma \models C$, there are two cases. Each one follows easily by IH and maximality of Γ (Exercise 11.1).

$\exists x B$ :: As A is closed and Γ is Henkin, we have $\Gamma \vdash_\mathcal{N} \exists x B \to B_c^x$ for some c. Now if $\Gamma \vdash_\mathcal{N} A$ then also $\Gamma \vdash_\mathcal{N} B_c^x$ and, by IH, $T_\Gamma \models B_c^x$, hence $T_\Gamma \models \exists x B$ by soundness of A4.

For the converse, assume that $T_\Gamma \models \exists x B$, i.e., there is a $t \in T_\Gamma$ such that $T_\Gamma \models_{x \mapsto t} B$. But then $T_\Gamma \models B_t^x$ by Lemmata 9.9 and 9.7, so by IH, $\Gamma \vdash_\mathcal{N} B_t^x$, and by A4 and MP, $\Gamma \vdash_\mathcal{N} A$.

QED (11.6)

The construction in the above proof is only guaranteed to work for max Henkin-theories. The following examples illustrate why.

Example 11.7

Constructing T_Γ as above for an arbitrary theory Γ, may give a structure which does not satisfy some formulae from Γ (or $Th(\Gamma)$) because:

(1) Some (atomic) formulae are not provable from Γ:
Let Σ contain two constant symbols a and b and one binary relation R. Let Γ be a theory over Σ with one axiom $R(a, b) \lor R(b, a)$. Each model of Γ must satisfy at least one disjunct but, since $\Gamma \not\vdash_\mathcal{N} R(a, b)$ and $\Gamma \not\vdash_\mathcal{N} R(b, a)$, none of these relations will hold in T_Γ.

(2) The interpretation domain $\underline{T_\Gamma}$ has too few elements:
It may happen that $\Gamma \vdash_\mathcal{N} \exists x R(x)$ but $\Gamma \not\vdash_\mathcal{N} R(t)$ for any ground term t. Since only ground terms are in $\underline{T_\Gamma}$, this would again mean that $T_\Gamma \not\models \exists x R(x)$.   □

A general assumption to be made in the following is that the alphabet Σ under consideration is countable (i.e., has at most countably infinitely many symbols). Although not necessary, it is seldom violated and makes the arguments clearer.

We now strengthen Lemma 11.6 by successively removing the extra assumptions about the theory. First we show that the assumption about *maximal* consistency is superfluous; every consistent theory can be extended to a maximal consistent one. (Γ' is an *extension* of Γ if $\Gamma \subseteq \Gamma'$.)

Lemma 11.8 Every consistent theory Γ over countable Σ has a maximal consistent extension $\widehat{\Gamma}$ over the same Σ.

Proof. Since $|\Sigma| \leq \aleph_0$, there are at most $\aleph_0$ Σ-formulae. Choose an enumeration $A_0, A_1, A_2, \ldots$ of all *closed* Σ-formulae and construct an increasing sequence of theories as follows:

BASIS :: $\Gamma_0 = \Gamma$

IND. :: $\Gamma_{n+1} = \begin{cases} \Gamma_n, A_n & \text{if it is consistent} \\ \Gamma_n, \neg A_n & \text{otherwise} \end{cases}$

CLSR. :: $\widehat{\Gamma} = \bigcup_{n \in \mathbb{N}} \Gamma_n$

We show by induction on n that for any n, Γ_n is consistent.

BASIS :: $\Gamma_0 = \Gamma$ is consistent by assumption.

IND. :: Suppose Γ_n is consistent. If Γ_{n+1} is inconsistent, then from the definition of Γ_{n+1} we know that both Γ_n, A_n and $\Gamma_n, \neg A_n$ are inconsistent, hence by Deduction Theorem both $A_n \to \bot$ and $\neg A_n \to \bot$ are provable from Γ_n. By Exercise 4.1.(5), Γ_n proves then both A_n and $\neg A_n$, which contradicts its consistency by Exercise 4.7.

By Theorem 4.28 (holding for FOL by the same argument as in PL), $\widehat{\Gamma}$ is consistent iff each of its finite subtheories is. Any finite subtheory of $\widehat{\Gamma}$ is included in some Γ_n, so $\widehat{\Gamma}$ is consistent. From the definition of $\widehat{\Gamma}$ it now follows that $\widehat{\Gamma}$ is also *maximal* consistent.   QED (11.8)

Corollary 11.9 A consistent Henkin-theory over countable Σ is satisfiable.

Proof. If Γ is a consistent Henkin-theory, it has a maximal consistent extension $\widehat{\Gamma}$. Since $\Gamma \subseteq \widehat{\Gamma}$ and both are theories over the same alphabet Σ, so $\widehat{\Gamma}$ is a Henkin-theory when Γ is. Hence $\widehat{\Gamma}$ is a max Henkin-theory and so has a model, which is also a model of Γ.   QED (11.9)

To bridge the gap between this result and full completeness, we need to show that every consistent theory has a consistent Henkin extension – we shall make use of the following auxiliary notion.

Definition 11.10 Let Σ and Σ' be two alphabets, and assume $\Sigma \subseteq \Sigma'$. Moreover, let Γ be a Σ-theory and let Γ' be a Σ'-theory. Then Γ' is said to be a *conservative extension* of Γ, written $\Gamma \preceq \Gamma'$, if $\Gamma \subseteq \Gamma'$ and for all Σ-formulae A: if $\Gamma' \vdash_{\mathcal{N}} A$ then $\Gamma \vdash_{\mathcal{N}} A$.

A conservative extension Γ' of Γ may prove more formulae over the extended alphabet but any formula over the alphabet of Γ provable from Γ', must be provable already from Γ itself. The next lemma records a few useful facts about conservative extensions. The proof is left as an exercise.

Lemma 11.11 The conservative extension relation $\preceq$:

(1) preserves consistency: if $\Gamma_1 \preceq \Gamma_2$ and Γ_1 is consistent, then so is Γ_2.
(2) is transitive: if $\Gamma_1 \preceq \Gamma_2$ and $\Gamma_2 \preceq \Gamma_3$ then $\Gamma_1 \preceq \Gamma_3$.
(3) is preserved in limits: if $\Gamma_1 \preceq \Gamma_2 \preceq \Gamma_3 \preceq \ldots$ is an infinite sequence with each theory being a conservative extension of the previous one, then $\bigcup_{n \in \mathbb{N}} \Gamma_n$ is a conservative extension of each Γ_i.

We shall make use of these facts in a moment, but first we record the following important lemma, stating that adding a single Henkin witness and formula to a theory yields its conservative extension.

Lemma 11.12 Let Γ be a Σ-theory, A a Σ-formula with at most x free, c an individual constant that does not occur in Σ and let $\Sigma' = \Sigma \cup \{c\}$. Then the Σ'-theory $\Gamma \cup \{\exists x A \to A_c^x\}$ is a conservative extension of Γ.

Proof. Let B be an arbitrary Σ-formula, which is provable from the extended theory, i.e.,

$1 : \Gamma, \exists x A \to A_c^x \vdash_{\mathcal{N}} B$
$2 : \Gamma \vdash_{\mathcal{N}} (\exists x A \to A_c^x) \to B \quad DT + \text{ at most } x \text{ free in } A$

This means that Γ, with axioms *not involving any* occurrences of c, proves the indicated formula which has such occurrences. Thus we may choose a fresh variable y (not occurring in this proof) and replace all occurrences of c in this proof by y. We then obtain

$3 : \Gamma \vdash_{\mathcal{N}} (\exists x A \to A_y^x) \to B$
$4 : \Gamma \vdash_{\mathcal{N}} \exists y (\exists x A \to A_y^x) \to B \quad \exists \mathrm{I} + y \text{ not free in } B$
$5 : \Gamma \vdash_{\mathcal{N}} \exists y (\exists x A \to A_y^x) \qquad \text{Exc. 8.3.(2)} + \text{Exc.8.7}$
$6 : \Gamma \vdash_{\mathcal{N}} B \qquad\qquad MP : 5, 4 \qquad\qquad\qquad \text{QED (11.12)}$

Lemma 11.13 Let Σ be a countable alphabet, and let Γ be a Σ-theory. Then there exists a countable alphabet Σ_H and a Σ_H-theory Γ_H such that $\Sigma \subseteq \Sigma_H$, $\Gamma \preceq \Gamma_H$, and Γ_H is a Henkin-theory over Σ_H.

Proof. Let Γ be a Σ-theory. Extend the alphabet Σ to $H(\Sigma)$ by adding, for each Σ-formula A with exactly one variable free, a new constant c_A. Let $H(\Gamma)$ be the $H(\Sigma)$-theory obtained by adding to Γ, for each such A, the new Henkin axiom $\exists x A \to A_{c_A}^x$, where x is the free variable of A.

In particular, $H(\Gamma)$ can be obtained by the following inductive construction. Enumerate all formulae A with exactly one variable free, getting $A_0, A_1, A_2, \ldots$ and let, for any n, x_n be the free variable of A_n. Define inductively:

$$\begin{aligned}
\text{BASIS} &:: \Gamma_0 = \Gamma \text{ and } \Sigma_0 = \Sigma \\
\text{IND.} &:: \Gamma_{n+1} = \Gamma_n, \exists x_n A_n \to (A_n)_{c_{A_n}}^{x_n}, \text{ and} \\
&\quad\ \Sigma_{n+1} = \Sigma_n \cup \{c_{A_n}\} \\
\text{CLSR.} &:: H(\Gamma) = \bigcup_{n \in \mathbb{N}} \Gamma_n \text{ and } H(\Sigma) = \bigcup_{n \in \mathbb{N}} \Sigma_n.
\end{aligned} \qquad (11.14)$$

By Lemma 11.12, each theory Γ_{n+1} is a conservative extension of Γ_n, and hence by Lemma 11.11, $H(\Gamma)$ is a conservative extension of Γ. It is also clear that $H(\Sigma)$ is countable, since only countably many new constants are added.

However, $H(\Gamma)$ need not be a Henkin-theory, since we have not ensured the provability of Henkin formulae $\exists x A \to A_c^x$ for $H(\Sigma)$-formulae A that are not Σ-formulae. For instance, for $R \in \Sigma$

$$H(\Gamma) \vdash_{\mathcal{N}} \exists x \exists y R(x, y) \to \exists y R(c_{\exists y R(x,y)}, y),$$

but there may be no c such that

$$H(\Gamma) \vdash_{\mathcal{N}} \exists y R(c_{\exists y R(x,y)}, y) \to R(c_{\exists y R(x,y)}, c).$$

To obtain a Henkin-theory, the construction (11.14) has to be iterated, i.e, the sequence of theories $\Gamma, H(\Gamma), H^2(\Gamma), H^3(\Gamma), \ldots$ is constructed (where $H^{n+1}(\Gamma)$ is obtained by starting (11.14) with $\Gamma_0 = H^n(\Gamma)$), and Γ_H is defined as the union of them all $\bigcup_{n \in \mathbb{N}} H^n(\Gamma)$. The sequence of alphabets $\Sigma, H(\Sigma), H^2(\Sigma), H^3(\Sigma), \ldots$ is collected into the corresponding union Σ_H. Γ_H is a Σ_H-theory and Σ_H, being the union of countably many countable sets, is itself countable.

Since each theory $H^{n+1}(\Gamma)$ is a conservative extension of $H^n(\Gamma)$, it follows by Lemma 11.11 that Γ_H is a conservative extension of Γ.

Finally we check that Γ_H is a Henkin-theory over Σ_H: let A be any Σ_H-formula with exactly x free. A contains only finitely many symbols, so A is also a $H^n(\Sigma)$-formula for some n. But then $\exists x A \to A_{c_A}^x$ is contained in $H^{n+1}(\Gamma)$, and hence in Γ_H. By Lemma 11.5, this proves that Γ_H is a Henkin-theory. QED (11.13)

Gathering all the pieces we thus obtain the main result.

Theorem 11.15 A consistent theory over a countable alphabet is satisfiable.

Proof. Let Σ be a countable alphabet. Suppose Γ is a consistent Σ-theory. Then there exist Γ_H and Σ_H with the properties described in Lemma 11.13. Since Γ is consistent, Γ_H, being a conservative extension, must be consistent as well. By Corollary 11.9, Γ_H (and hence Γ) has a Σ_H-model. This can be converted to a Σ-model by "forgetting" the interpretation of symbols in $\Sigma_H \setminus \Sigma$.   QED (11.15)

This is the strongest version that we prove here. The assumption about countability is however unnecessary, and the following version is also true.

Theorem 11.16 Every consistent theory is satisfiable.

Lemma 11.3 and soundness yield then the final result:

Corollary 11.17 For any $\Gamma \subseteq \mathsf{WFF_{FOL}}$, $A \in \mathsf{WFF_{FOL}}$ the following hold:
(1) $\Gamma \models A$ iff $\Gamma \vdash_{\mathcal{N}} A$, and
(2) $Mod(\Gamma) \neq \varnothing$ iff $\Gamma \nvdash_{\mathcal{N}} \bot$, i.e. Γ is satisfiable iff it is consistent.

11.2.1: COMPLETENESS OF GENTZEN'S SYSTEM ....[optional]
Recall Exercise 6.14 and the discussion just after the introduction of Gentzen's system for **FOL** in Section 8.4. Also, recall from Section 9.3.1 that the semantics of $\vdash_{\mathcal{G}}$ does not correspond to $\models$ but to $\Rightarrow$. We thus want to show that whenever $\bigwedge \Gamma \Rightarrow \bigvee \Delta$ then $\Gamma \vdash_{\mathcal{G}} \Delta$, where Γ, Δ are finite sets of formulae.

(A) By Exercise 6.14, the propositional rules of Gentzen's system are sound and invertible, and so are the quantifier rules by Exercise 9.17. Invertibility plays a crucial role in the argument below.

(B) We proceed as we did in Exercise 6.14, constructing a counter-model for any unprovable sequent. But now we have to handle the additional complications of possibly non-terminating derivations. To do this, we specify the following strategy for an exhaustive bottom-up proof search. (It refines the strategy suggested in the proof of Theorem 4.31 showing decidability of $\mathcal{G}$ for **PL**.)

(B.1) First, we have to ensure that even if a branch of a (bottom-up) proof does not terminate, all formulae in the sequent are processed. Let us therefore view a sequent as a pair of (finite) sequences $\Gamma = G_1, ..., G_a$ and $\Delta = D_1, ..., D_c$. Such a sequent is processed by applying bottom-up the appropriate rule first to G_1, then to G_2, to G_3, etc. until G_a, and followed by D_1 through D_c. If no rule is applicable to a formula, it is skipped and one continues with the next formula. New formulae arising from rule applications are always placed at the start of the appropriate sequence (on the left or on the right of $\vdash_{\mathcal{G}}$). These restrictions are particularly important for the quantifier rules which introduce new formulae, i.e.:

$$(\vdash\exists)\ \frac{\Gamma\vdash_\mathcal{G} A_t^x\ldots\exists xA\ldots}{\Gamma\vdash_\mathcal{G}\ldots\exists xA\ldots}\ A_t^x\ \text{legal} \qquad (\forall\vdash)\ \frac{A_t^x\ldots\forall xA\ldots\vdash_\mathcal{G}\Delta}{\ldots\forall xA\ldots\vdash_\mathcal{G}\Delta}\ A_t^x\ \text{legal}$$

These two rules might start a non-terminating, repetitive process introducing new substitution instances of A without ever considering the remaining formulae. Processing formulae from left to right, and placing new formulae to the left of the actually processed ones, ensures that all formulae in the sequent will be processed, before starting the processing of the newly introduced formulae.

(B.2) We must also ensure that all possible substitution instances of quantified formulae are attempted in search for axioms. To do this, we assume an enumeration of all terms and require that the formula A_t^x, introduced in the premise of rule $(\vdash\exists)$ or $(\forall\vdash)$, is the smallest which (uses the smallest term t, so that A_t^x) does not already occur in the sequence to which it is introduced.

(C) Let now $\Gamma\vdash_\mathcal{G}\Delta$ be an arbitrary sequent for which the above strategy does not yield a proof. There are two cases.

(C.1) If every branch in the obtained proof tree is finite, then all its leaves contain irreducible sequents, some of which are non-axiomatic. Select such a non-axiomatic leaf, say, with $\Phi\vdash_\mathcal{G}\Psi$. Irreducibility means that no rule can be applied to this sequent, i.e., all its formulae are atomic. That it is non-axiomatic means that $\Phi\cap\Psi=\varnothing$. Construct a counter-model M by taking all terms occurring in the atoms of Φ,Ψ as the interpretation domain $\underline{M}$. (In particular, if there are open atomic formulae, their variables are treated as elements on line with ground terms.) Interpret the predicates over these elements by making all atoms in Φ true and all atoms in Ψ false. This is possible since $\Phi\cap\Psi=\varnothing$. (E.g., a leaf $P(x)\vdash_\mathcal{G}P(t)$ gives the structure with $\underline{M}=\{x,t\}$ where $t\notin[\![P]\!]^M=\{x\}$.)

Invertibility of the rules implies now that this is also a counter-model to the initial sequent, i.e., $\bigwedge\Gamma\not\models\bigvee\Delta$.

(C.2) In the other case, the resulting tree has an infinite branch. We then select an arbitrary infinite branch B and construct a counter-model M from all the terms occurring in the atomic formulae on B. (Again, variables occurring in such formulae are taken as elements on line with ground terms.) The predicates are defined by

$$\bar{t}\in[\![P]\!]^M\ \text{iff there is a node on}\ B\ \text{with}\ P(\bar{t})\ \text{on the left of}\vdash_\mathcal{G}. \qquad (*)$$

Hence, atomic statements not occurring on the left of $\vdash_\mathcal{G}$ are made false. This gives a well-defined structure, since atomic formulae remain unchanged (moving bottom-up) once they appear in any branch of the proof – no node is axiomatic on B, so no formula occurring there on the left of $\vdash_\mathcal{G}$, occurs also on the right.

The claim is now that M is a counter-model to every sequent on the whole B. We show, by induction on the complexity of the formulae, that all those occurring on B on the left of $\vdash_\mathcal{G}$ are true and all those on the right false. (Complexity reflects essentially the number of logical symbols. In particular, quantified formula QxA is more complex than any substitution instance A_t^x.) The claim is obvious for the atomic formulae by the definition $(*)$. The claim is easily verified for the propositional connectives by the invertibility of the propositional rules. Consider now any quantified formula occurring on the left. Due to fairness strategy **(B.1)**, it has been processed. If it is universal, it has been processed by the rule:

$$(\forall \vdash) \quad \frac{G_t^x \dots \forall x G \dots \vdash_{\mathcal{G}} \Delta}{\dots \forall x G \dots \vdash_{\mathcal{G}} \Delta} \quad G_t^x \text{ legal.}$$

Then $[\![G_t^x]\!]^M = \mathbf{1}$ by IH and, moreover, since by **(B.2)** all such substitution instances are tried on every infinite branch, so $[\![G_{t_i}^x]\!]^M = \mathbf{1}$ for all terms t_i. But this means that $M \models \forall x G$. If the formula is existential, it is processed by the rule:

$$(\exists \vdash) \quad \frac{\Gamma, G_{x'}^x \vdash_{\mathcal{G}} \Delta}{\Gamma, \exists x G \vdash_{\mathcal{G}} \Delta} \quad x' \text{ fresh.}$$

Then x' occurs in the atomic subformula(e) of G and is an element of $\underline{M}$. By IH, $[\![G_{x'}^x]\!]^M = \mathbf{1}$ and hence also $[\![\exists x G]\!]^M = \mathbf{1}$.

Similarly, by **(B.1)** every quantified formula on the right of $\vdash_{\mathcal{G}}$ has been processed. Universal one was processed by the rule

$$(\vdash \forall) \quad \frac{\Gamma \vdash_{\mathcal{G}} D_{x'}^x, \Delta}{\Gamma, \vdash_{\mathcal{G}} \forall x D, \Delta} \quad x' \text{ fresh.}$$

Then $x' \in \underline{M}$ and, by IH, $[\![D_{x'}^x]\!]^M = \mathbf{0}$. Hence also $[\![\forall x D]\!]^M = \mathbf{0}$. An existential formula on the right is processed by the rule

$$(\vdash \exists) \quad \frac{\Gamma \vdash_{\mathcal{G}} D_t^x \dots \exists x D \dots}{\Gamma \vdash_{\mathcal{G}} \dots \exists x D \dots} \quad D_t^x \text{ legal}$$

and, by the exhaustive fairness strategy **(B.2)**, all such substitution instances $D_{t_i}^x$ appear on the right of $\vdash_{\mathcal{G}}$ in B. By IH, $[\![D_{t_i}^x]\!]^M = \mathbf{0}$ for all t_i, which means that $[\![\exists x D]\!]^M = \mathbf{0}$.

(D) Since the sequent $\Gamma \vdash_{\mathcal{G}} \Delta$ is itself on the branch B, being the root of the whole tree, **(C.1)** and **(C.2)** together show that its unprovability, $\Gamma \nvdash_{\mathcal{G}} \Delta$, implies the existence of a counter-model, $M \nvDash \bigwedge \Gamma \to \bigvee \Delta.$, i.e., $\bigwedge \Gamma \nRightarrow \bigvee \Delta$. Formulated contrapositively, if a sequent is valid (has no counter-model), $\bigwedge \Gamma \Rightarrow \bigvee \Delta$, then it is provable, $\Gamma \vdash_{\mathcal{G}} \Delta$..[end optional]

11.3 Some applications

We list some typical questions, the answers to which may be significantly simplified by using soundness and completeness theorem. These are the same questions which were listed earlier in Subsection 6.5 after the respective theorems for propositional logic. The schemata of the arguments are also the same as before, because they are based exclusively on the soundness and completeness of the respective axiomatic system. The differences concern, of course, the semantic definitions which are more complicated for FOL, than they were for PL.

1. Is a formula provable?
If it is, it may be worth trying to construct a syntactic proof of it. Gentzen's

system is easiest to use, so it can be most naturally used for this purpose. However, one should first try to make a "justified guess". To make a guess, we first try to see if we can easily construct a counter-example, i.e., a structure which falsifies the formula. For instance, trying to determine if

$$\vdash_{\mathcal{N}} (\exists x P(x) \to \exists x Q(x)) \to \forall x (P(x) \to Q(x)), \qquad (11.18)$$

we can just start looking for a syntactic proof. But we better think first. Perhaps, we can falsify this formula, i.e., find a structure M such that

$$(i) \ M \models \exists x P(x) \to \exists x Q(x) \quad \text{and} \quad (ii) \ M \not\models \forall x (P(x) \to Q(x)). \quad (11.19)$$

More explicitly, (i) requires that

$$\begin{aligned} \text{either}: \quad &\text{for all } m_1 \in \underline{M} : [\![P(x)]\!]^M_{x \mapsto m_1} = \mathbf{0} \\ \text{or}: \ &\text{for some } m_2 \in \underline{M} : [\![Q(x)]\!]^M_{x \mapsto m_2} = \mathbf{1} \end{aligned} \qquad (11.20)$$

while (ii) that

$$\text{for some } m \in \underline{M} : [\![P(x)]\!]^M_{x \mapsto m} = \mathbf{1} \text{ and } [\![Q(x)]\!]^M_{x \mapsto m} = \mathbf{0}. \qquad (11.21)$$

But this should be easy to do. Let $\underline{M} = \{m_1, m_2\}$ with $[\![P]\!]^M = \{m_1\}$ and $[\![Q]\!]^M = \{m_2\}$. This makes (11.20) true since $[\![Q(x)]\!]^M_{x \mapsto m_2} = \mathbf{1}$. On the other hand (11.21) holds for $m_1 : [\![P(x)]\!]^M_{x \mapsto m_1} = \mathbf{1}$ and $[\![Q(x)]\!]^M_{x \mapsto m_1} = \mathbf{0}$. Thus, the formula in (11.18) is not valid and, **by soundness** of $\mathcal{N}$, is not provable.

This is the main way of showing that a formula is *not* provable in a sound system which is not decidable (and since FOL is undecidable by Theorem 10.38, so sound and complete $\mathcal{N}$ cannot be decidable) – to find a structure providing a counter-example to the validity of the formula.

If such an analysis fails, i.e., if we are unable to find a counter-example, it may indicate that we should rather try to construct a proof of the formula in our system. **By completeness** of this system, such a proof will exist, if the formula is valid.

2. Is a formula valid?

For instance, is it the case that

$$\models \forall x (P(x) \to Q(x)) \to (\exists x P(x) \to \exists x Q(x))? \qquad (11.22)$$

We may first try to see if we can find a counter-example. In this case, we need a structure M such that $M \models \forall x (P(x) \to Q(x))$ and $M \not\models \exists x P(x) \to \exists x Q(x)$ – since both (sub)formulae are closed we need not consider particular assignments. Thus, M should be such that

$$\text{for all } m \in \underline{M} : [\![P(x) \to Q(x)]\!]^M_{x \mapsto m} = \mathbf{1}. \qquad (11.23)$$

To falsify the other formula we have to find an

$$m_1 \in \underline{M} \text{ such that } [\![P(x)]\!]_{x \mapsto m_1}^{M} = \mathbf{1} \qquad (11.24)$$

and such that

$$\text{for all } m_2 \in \underline{M} : [\![Q(x)]\!]_{x \mapsto m_2}^{M} = \mathbf{0}. \qquad (11.25)$$

Assume that m_1 is as required by (11.24). Then (11.23) implies that we also have $[\![Q(x)]\!]_{x \mapsto m_1}^{M} = \mathbf{1}$. But this means that (11.25) cannot be forced, m_1 being a witness contradicting this statement. Thus, the formula from (11.22) cannot be falsified in any structure, i.e., it is valid. This is sufficient argument – direct, semantic proof of validity of the formula.

However, such semantic arguments involve complicating subtleties which we may try avoiding. Namely, if we have a strong conviction that the formula indeed is valid, we may instead attempt a syntactic proof. Below, we are doing it in Gentzen's system – as we have seen, it is also sound and complete. (Notice that we first eliminate the quantifier from $\exists x P(x)$ since this requires a fresh variable y; the subsequent substitutions must be legal but need not introduce fresh variables.)

$$
\begin{array}{ll}
(\rightarrow\vdash) & \dfrac{P(y) \vdash_{\mathsf{G}} Q(y), P(y) \qquad Q(y), P(y) \vdash_{\mathsf{G}} Q(y)}{} \\
(\vdash\exists) & \dfrac{P(y) \rightarrow Q(y), P(y) \vdash_{\mathsf{G}} Q(y)}{} \\
(\forall\vdash) & \dfrac{P(y) \rightarrow Q(y), P(y) \vdash_{\mathsf{G}} \exists x Q(x)}{} \text{ legal subst.} \\
(\exists\vdash) & \dfrac{\forall x(P(x) \rightarrow Q(x)), P(y) \vdash_{\mathsf{G}} \exists x Q(x)}{} \text{ legal subst.} \\
(\vdash\rightarrow) & \dfrac{\forall x(P(x) \rightarrow Q(x)), \exists x P(x) \vdash_{\mathsf{G}} \exists x Q(x)}{} \; y \text{ fresh} \\
(\vdash\rightarrow) & \dfrac{\forall x(P(x) \rightarrow Q(x)) \vdash_{\mathsf{G}} \exists x P(x) \rightarrow \exists x Q(x)}{\vdash_{\mathsf{G}} \forall x(P(x) \rightarrow Q(x)) \rightarrow (\exists x P(x) \rightarrow \exists x Q(x))}
\end{array}
$$

Having this proof we conclude, **by soundness** of $\vdash_{\mathsf{G}}$, that the formula is indeed valid.

Summarising these two points.
In most axiomatic systems the relation $X \vdash Y$ is semi-decidable: to establish that it holds, it is enough to generate all the proofs until we encounter one which proves Y from X. Therefore, if this actually holds, it may be natural to try to construct a syntactic proof (provided that the axiomatic system is easy to use, like $\vdash_{\mathsf{G}}$) – completeness of the system guarantees that there exists a proof of a valid formula. If, however, the relation does not hold, it is always easier to find a semantic counter-example. If it is found, and the system is sound, it allows us to conclude that the relation $X \vdash Y$ does *not* hold. That is, in order to know what is easier to do, we have to know what

the answer is! This is, indeed, a vicious circle, and the best one can do is to "guess" the right answer before proving it. The quality of such "guesses" increases only with exercise and work with the system itself and cannot be given in the form of a ready-made recipe.

3. Is a rule admissible?

Suppose that we have an axiomatic system and a rule $R : \dfrac{\Gamma \vdash A_1 \ldots \Gamma \vdash A_n}{\Gamma \vdash C}$

The question whether R is admissible can be answered by trying to verify by purely proof theoretic means that any given proofs for the premises entitle the existence of a proof for the conclusion C. This, however, is typically a cumbersome task.

If the system is sound and complete, there is a much better way to do that. The schema of the proof is as follows. For the first, we verify if the rule is sound. If it isn't, we can immediately conclude, **by soundness** of our system, that it is not admissible. If, on the other hand, the rule is sound, the following schematic argument allows us to conclude that it is admissible:

$$\frac{\Gamma \vdash A_1 \ \ldots \ \Gamma \vdash A_n}{\Gamma \vdash C} \quad \overset{\text{soundness}}{\underset{\text{completeness}}{\underset{\Longleftarrow}{\Longrightarrow}}} \quad \frac{\Gamma \models A_1 \ldots \Gamma \models A_n}{\Gamma \models C} \ \Downarrow \text{ soundness of } R$$

For instance, are the following rules admissible in $\vdash_{\mathcal{N}}$:

$$i) \ \frac{\vdash_{\mathcal{N}} \exists x (P(x) \to Q(x)) \quad ; \quad \vdash_{\mathcal{N}} \forall x P(x)}{\vdash_{\mathcal{N}} \exists x Q(x)}$$

$$ii) \ \frac{\vdash_{\mathcal{N}} \exists x (P(x) \to Q(x)) \quad ; \quad \vdash_{\mathcal{N}} \exists x P(x)}{\vdash_{\mathcal{N}} \exists x Q(x)}?$$

The first thing to check is whether the rules are sound. So assume that M is an arbitrary structure which satisfies the premises. For the rule i), this means

$$M \models \exists x (P(x) \to Q(x)) \quad \text{and} \quad M \models \forall x P(x)$$

i.e. for some $m \in \underline{M}$:   and   for all $n \in \underline{M}$:   (11.26)

$$[\![P(x) \to Q(x)]\!]^M_{x \mapsto m} = 1 \qquad [\![P(x)]\!]^M_{x \mapsto n} = 1$$

Will M satisfy the conclusion? Let m be a witness making the first assumption true, i.e., either $[\![P(x)]\!]^M_{x \mapsto m} = 0$ or $[\![Q(x)]\!]^M_{x \mapsto m} = 1$. But from the second premise, we know that for all n, in particular for the chosen $m : [\![P(x)]\!]^M_{x \mapsto m} = 1$. Thus, it must be the case that $[\![Q(x)]\!]^M_{x \mapsto m} = 1$. But then m is also a witness to the fact that $[\![\exists x Q(x)]\!]^M = 1$, i.e., the rule is sound. By the above argument, i.e., **by soundness and completeness** of $\vdash_{\mathcal{N}}$, the rule i) is admissible.

For the second rule ii), we check first its soundness. Let M be an arbitrary structure satisfying the premises, i.e.:

$$M \models \exists x (P(x) \to Q(x)) \quad \text{and} \quad M \models \exists x P(x)$$

i.e. for some $m \in \underline{M}$: $\qquad$ and $\qquad$ for some $n \in \underline{M}$: $\qquad$ (11.27)

$$[\![P(x) \to Q(x)]\!]^M_{x \mapsto m} = 1 \qquad\qquad [\![P(x)]\!]^M_{x \mapsto n} = 1$$

Here it is possible that $m \neq n$ and we can utilize this fact to construct an M which does not satisfy the conclusion. Let $\underline{M} = \{m, n\}$ with $[\![P]\!]^M = \{n\}$ and $[\![Q]\!]^M = \varnothing$. Both assumptions from (11.27) are now satisfied. However, $[\![Q]\!]^M = \varnothing$, and so $M \not\models \exists x Q(x)$. Thus the rule is not sound and, **by soundness** of $\vdash_{\!\mathcal{N}}$, cannot be admissible there.

EXERCISES 11.

EXERCISE 11.1 Show the inductive step for the case $B \to C$, which was omitted in the proof of Lemma 11.6.

EXERCISE 11.2 (1) Give two *different* PNFs (cf. Exercise 10.2) for the formula $F : (\exists x \forall y : R(x, y)) \to (\forall y \exists x : R(x, y))$.

(2) Let M be the structure with $\underline{M} = \{a, b\}$ and $R^M = \{\langle a, b\rangle, \langle b, b\rangle\}$. Does $M \models F$?

(3) Is F valid? Either give a syntactic proof of F and explain (in one sentence) why this entails F's validity, or provide a counter-model for F.

(4) Consider now the formula $G : (\exists x \forall y : R(x, y)) \to (\exists x : R(x, y))$. Repeat the points above for this formulae, i.e.:

(a) Give two different PNFs for G.

(b) With the same M as above in (2), does $M \models G$?

(c) Is G valid? Do not repeat earlier arguments but explain briefly how/why it can be obtained from the answer to (3) above.

EXERCISE 11.3 Show that the formula (1) from Lemma 8.24 is provable, i.e., that $\vdash_{\!\mathcal{N}} \forall x A \to A$, without constructing the actual syntactic proof.

EXERCISE 11.4 Without constructing any syntactic proofs, show that the rules (2), (3) and (4) from Lemma 8.24 are admissible in $\mathcal{N}$.

EXERCISE 11.5 Show admissibility in $\mathcal{N}$ of the following form of Deduction Theorem (1) and its inversion (2), for arbitrary A, B and Γ:

$$(1) \; \frac{\Gamma, A \vdash_{\!\mathcal{N}} B}{\Gamma \vdash_{\!\mathcal{N}} \forall(A) \to B} \qquad\qquad (2) \; \frac{\Gamma \vdash_{\!\mathcal{N}} \forall(A) \to B}{\Gamma, A \vdash_{\!\mathcal{N}} B} \;.$$

EXERCISE 11.6 Prove the three statements of Lemma 11.11.

(Hint: In the proof of the last point, you will need (a form of) compactness, i.e., if $\Gamma \vdash_{\mathcal{N}} A$, then there is a finite subtheory $\Delta \subseteq \Gamma$ such that $\Delta \vdash_{\mathcal{N}} A$.)

EXERCISE 11.7 Suppose Definition 11.4 of maximal consistency is strengthened to the requirement that $\Gamma \vdash_{\mathcal{N}} A$ or $\Gamma \vdash_{\mathcal{N}} \neg A$ for *all* (not only closed) formulae. Then Lemma 11.8 would no longer be true. Explain why.

(Hint: Let P be a unary predicate, a, b two constants and $\Gamma = \{P(a), \neg P(b)\}$. Γ is consistent, but what if you add to it open formula $P(x)$, resp. $\neg P(x)$? Recall discussion from Remark 9.17, in particular, fact (9.20).)

EXERCISE 11.8 Prove the following theorem of Robinson: If Γ, Γ' are two satisfiable sets of sentences (over the same alphabet), but $\Gamma \cup \Gamma'$ has no model, then there is a sentence F such that $\Gamma \models F$ and $\Gamma' \models \neg F$.

(Hint: Formulate first semantic version of compactness theorem for FOL, analogous to Exercise 6.12, and then show that if no sentence like F exists, then every finite subset of $\Gamma \cup \Gamma'$ has a model.)

EXERCISE 11.9 Monadic FOL, mFOL, is the sublanguage of FOL admitting only unary relation symbols (e.g., $P(x)$ are allowed, but not $R(x, y)$.) It has the so-called *finite model property*, namely, if a sentence F of mFOL is not valid, then there is a finite structure M (with a finite interpretation domain $\underline{M}$) such that $M \not\models F$.

Give an argument that validity in mFOL is decidable.

EXERCISE 11.10 Prove Herbrand's theorem, 10.30, i.e., for any Γ consisting of quantifier-free formulae over Σ with $\mathcal{GT}_\Sigma \neq \varnothing$: $\Gamma \vdash_{\mathcal{N}}^{\mathrm{FOL}} \bot$ iff $GI(\Gamma) \vdash_{\mathcal{N}}^{\mathrm{PL}} \bot$.

(Hint: The implication to the left is easiest: use the fact that everything provable in $\vdash_{\mathcal{N}}^{\mathrm{PL}}$ is also provable in $\vdash_{\mathcal{N}}^{\mathrm{FOL}}$, and Deduction Theorem to prove that $A \vdash_{\mathcal{N}}^{\mathrm{FOL}} B$ for every ground instance B of A. For the other direction use the completeness theorem for PL to deduce the existence of a valuation model for $GI(\Gamma)$ whenever $GI(\Gamma) \not\vdash_{\mathcal{N}}^{\mathrm{PL}} \bot$ and show how to convert this into a term structure. Finally apply the result from Exercise 10.11 and soundness theorem for FOL.)

EXERCISE 11.11 [(downward) Löwenheim-Skolem theorem]

Prove that every consistent theory over a countable alphabet has a countable model.

(Hint: You need only find the relevant lemmata. Essentially, you repeat the proof of completeness verifying that each step preserves countability and, finally, that this leads to a countable model (in the proof of Lemma 11.6). Specify only which adjustments are needed at which places.)

WHY IS FIRST ORDER LOGIC
"FIRST ORDER"?

FOL is called "predicate logic", since its atomic formulae consist of applications of predicate/relation symbols to terms. Why is it also called "first order"? Because its variables range only over individual elements from the interpretation domain. In other words, if $A \in \mathsf{WFF_{FOL}}$ then $\exists x A \in \mathsf{WFF_{FOL}}$ only when x *is an individual variable* $x \in \mathcal{V}$.

Consider the formula $\forall x\ P(x) \rightarrow P(x)$. Obviously, no matter which unary predicate $P \in \mathcal{R}$ we choose, this formula will be valid. Thus, we might be tempted to write $\forall P\ \forall x\ P(x) \rightarrow P(x)$, i.e., try to quantify a variable P which ranges over predicates. This is exactly what is impossible in first order language. The formula would not be first order – but second order, since it quantifes not only over individuals but also over sets thereof. The schema continues, so third order language allows also quantification over sets of sets of individuals, etc., but in practice one seldom steps beyond second order. It suffices to do most of mathematics and to axiomatize most structures of interest, which cannot be axiomatized in first order logic. The best known example of such a structure are the natural numbers $\mathbb{N}$.

Example.
The following definitions and axioms were introduced by Italian mathematician Giuseppe Peano in the years 1894-1908. The language of natural numbers is given by the terms $\mathcal{T}_{\mathbb{N}}$ over the alphabet $\Sigma_{\mathbb{N}} = \{0, s, \equiv\}$. We are trying here to define the property "n *is a natural number*", which we denote $\mathbb{N}(n)$:

(1) $\mathbb{N}(0)$
    0 is a natural number.

(2) $\mathbb{N}(n) \rightarrow \mathbb{N}(s(n))$
    s(uccessor) of a natural number is a natural number.

We want all these terms to denote different numbers, for instance, we should not consider the structures satisfying $s(0) \equiv 0$. Thus we have to add the identity axioms (as in Example 8.14) and the axioms:

(3) $s(n) \not\equiv 0$
   0 is not a successor of any natural number.

(4) $s(n) \equiv s(m) \rightarrow n \equiv m$
   If two numbers have the same successor then they are equal – the s function is injective.

This is fine but it does not guarantee that the models we obtain will contain *only* elements interpreting the ground terms. Assuming that $\equiv$ is taken as a logical symbol, with the fixed interpretation as identity, any model will have to contain distinct elements interpreting $\{0, s(0), s(s(0)), \ldots\}$. But we also obtain unreachable models with additional elements! We would like to pick only the reachable models, but reachability itself is a semantic notion which is not expressible in the syntax of FOL. In order to ensure that no such "junk" elements appear in any model it is sufficient to require that

(5) For any property P, if
   - $P(0)$, and
   - $\forall n \ (P(n) \rightarrow P(s(n)))$
   then $\forall n \ \mathbb{N}(n) \rightarrow P(n)$.

This last axiom is actually the induction property, as we saw it in (2.29) – if 0 satisfies P and whenever an n satisfies P then its successor satisfies P, then all natural numbers satisfy P. In other words, what is true about all terms generated by (1) and (2) is true about all natural numbers, because there are no more natural numbers than what can be obtained by applying s an arbitrary number of times to 0.

   These five axioms actually describe the natural numbers up to isomorphism. (Any model of the five axioms is isomorphic to the $\{0, s, \mathbb{N}\}$-model with the natural numbers as the interpretation domain, and with 0 and s interpreted as zero and the successor function.) Unfortunately, the fifth axiom is not first order. Written as a single formula it reads:

$$\forall P \ \left(\left(P(0) \wedge \forall n \ (P(n) \rightarrow P(s(n))) \right) \rightarrow \forall n \ (\mathbb{N}(n) \rightarrow P(n)) \right),$$

i.e., we have to introduce the quantification over predicates, thus moving to second order logic.   □

It can also be shown that the fifth axiom is necessary to obtain the natural numbers. In Exercise 2.21 we have shown by induction that the binary $+$ operation defined in Example 2.26, is commutative. This is certainly something we would expect for the natural numbers. However, if we take only the first four axioms above, and define $+$ as in 2.26, we won't be able to prove commutativity. (The inductive proof in Exercise 2.21 established commutativity only in reachable models. But we can take, for instance, square matrices with matrix multiplication interpreting $+$ as a model of these four axioms, where the diagonal matrix of 1's interprets 0, and multiplication by the diagonal matrix of, say, 2's represents the successor operation. In this model, $+$ is not commutative, though it is commutative on the reachable elements.)

One of the most dramatic consequences of moving to second order logic was discovered by Kurt Gödel in 1931, to a big surprise of everybody, and in particular, of the so-called formalists and logical positivists. We only quote a reformulated version of the famous Gödel's (first) incompleteness theorem:

Theorem. There is no decidable axiomatic system which is both sound and complete for second order logic.

A decidable system is one where the set of axioms is recursive, i.e., there exists a Turing machine which, given any formula A, eventually halts with a correct verdict YES or NO regarding A being an axiom.

We can thus set up the following comparison between the three levels of logical systems (SOL stands for second order logic) with respect to their provability relations:

$\vdash$	sound	complete	decidable
SOL	$+$	$-$	$-$
FOL	$+$	$+$	$-$
PL	$+$	$+$	$+$

Moving upwards, i.e., increasing the expressive power of the logical language, we lose some desirable properties. Although SOL allows us to axiomatize natural numbers, the price we pay for this is that we no longer can prove all valid formulae. The choice of adequate logical system for a given application involves often the need to strike a balance between the expressive power, on the one hand, and other desirable properties like completeness or decidability, on the other.

Index

Index of symbols

The Greek alphabet

upper	lower		upper	lower	
A	α	alpha	N	ν	nu
B	β	beta	Ξ	ξ	xi
Γ	γ	gamma	O	o	omicron
Δ	δ	delta	Π	π	pi
E	ϵ	epsilon	R	ρ	rho
Z	ζ	zeta	Σ	σ	sigma
H	η	eta	T	τ	tau
Θ	θ	theta	Y	υ	upsilon
I	ι	iota	Φ	ϕ	phi
K	κ	kappa	X	χ	chi
Λ	λ	lambda	Ψ	ψ	psi
M	μ	mu	Ω	ω	omega

Printed in the United States
By Bookmasters